T0281516

Lecture Notes in Computer Science   13933

Founding Editors

Gerhard Goos
Juris Hartmanis

The series Lecture Notes in Computer Science (LNCS), including its subseries Lecture Notes in Artificial Intelligence (LNAI) and Lecture Notes in Bioinformatics (LNBI), has established itself as a medium for the publication of new developments in computer science and information technology research, teaching, and education.

LNCS enjoys close cooperation with the computer science R & D community, the series counts many renowned academics among its volume editors and paper authors, and collaborates with prestigious societies. Its mission is to serve this international community by providing an invaluable service, mainly focused on the publication of conference and workshop proceedings and postproceedings. LNCS commenced publication in 1973.

Minming Li · Xiaoming Sun · Xiaowei Wu
Editors

Frontiers of Algorithmics

17th International Joint Conference, IJTCS-FAW 2023
Macau, China, August 14–18, 2023
Proceedings

 Springer

Editors
Minming Li
City University of Hong Kong
Hong Kong SAR, China

Xiaoming Sun 🔟
Chinese Academy of Sciences
Beijing, China

Xiaowei Wu 🔟
University of Macau
Macau, China

ISSN 0302-9743   ISSN 1611-3349 (electronic)
Lecture Notes in Computer Science
ISBN 978-3-031-39343-3   ISBN 978-3-031-39344-0 (eBook)
https://doi.org/10.1007/978-3-031-39344-0

This Springer imprint is published by the registered company Springer Nature Switzerland AG
The registered company address is: Gewerbestrasse 11, 6330 Cham, Switzerland

Preface

This volume contains the contributed, accepted papers presented at International Joint Conference on Theoretical Computer Science - Frontiers of Algorithmic Wisdom (IJTCS-FAW 2023), which combined the 17th International Conference on Frontiers of Algorithmic Wisdom (FAW) and the 4th International Joint Conference on Theoretical Computer Science (IJTCS), held in Macau, China, during August 14–18, 2023. For the first time in the past four years, the conference was run in a fully in-person mode. FAW started as the Frontiers of Algorithmics Workshop in 2007 at Lanzhou, China, and was held annually from 2007 to 2022 and published archival proceedings. IJTCS, the International Joint Theoretical Computer Science Conference, started in 2020, aiming to attract presentations covering active topics in selected tracks in theoretical computer science.

To accommodate the diversified new research directions in theoretical computer science, FAW and IJTCS joined their forces together to organize an event for information exchange of new findings and work of enduring value in the field. The conference had both contributed talks submitted to the three tracks in IJTCS-FAW 2023, namely,

- Track A: The 17th Conference on Frontiers of Algorithmic Wisdom,
- Track B: Blockchain Theory and Technology,
- Track C: Computational Economics and Algorithmic Game Theory,

and invited talks in focused tracks on Blockchain Theory; Multi-agent Learning, Multi-agent Systems, Multi-agent Games; Learning Theory; Quantum Computing; and Conscious AI. Furthermore, the CSIAM forum, Female Forum, Young PhD Forum, Undergraduate Forum, Young Faculty in TCS and the CCF Annual Conference on Computational Economics 2023 (CCF CE 2023) were also organized during the five-day event.

For the three tracks that accepted submissions, the Program Committee, consisting of 45 top researchers from the field, reviewed 34 submissions and decided to accept 21 of them as full papers. These are presented in this proceedings volume. Each paper had three reviews. The review process was double blind and conducted entirely electronically via the EasyChair system. The conference papers included in the volume have taken the Springer Nature policies into account. After a review process and a thorough discussion among the Program Committee members, we gave out a Best Paper Award and a Best Student Paper Award.

The Best Paper Award goes to

- Max-Min Greedy Matching Problem: Hardness for the Adversary and Fractional Variant

authored by Hubert T-H. Chan, Zhihao Gavin Tang and Quan Xue.

The Best Student Paper Award goes to

- EFX Allocations Exist for Binary Valuations

authored by Xiaolin Bu, Jiaxin Song and Ziqi Yu.

Besides the regular talks, IJTCS-FAW 2023 had keynote talks from Kazuhisa Makino (Kyoto University), Xiaotie Deng (Peking University) and Jianwei Huang (Chinese University of Hong Kong, Shenzhen). We are very grateful to all the people who made this conference possible: the authors for submitting their papers, the Program Committee members and the Track Chairs for their excellent work in coordinating the review process and inviting the speakers, and all the keynote speakers and invited speakers. We also thank the Advisory Committee and Steering Committee for providing the timely advice about running the conference. In particular, we would like to thank the Local Organization Committee members from the University of Macau and Peking University for providing organizational support. Finally, we would like to thank Springer for their encouragement and cooperation throughout the preparation of this conference.

August 2023

Minming Li
Xiaoming Sun
Xiaowei Wu

Organization

General Chair

Chengzhong Xu   University of Macau, Macau, China

Program Committee Chairs

Minming Li   City University of Hong Kong, Hong Kong, China
Xiaoming Sun   Chinese Academy of Sciences, China
Xiaowei Wu   University of Macau, Macau, China

Track Chairs

The 17th Conference on Frontiers of Algorithmic Wisdom

Zhiyi Huang   University of Hong Kong, Hong Kong, China
Chihao Zhang   Shanghai Jiao Tong University, China

Blockchain Theory and Technology

Jing Chen   Stony Brook University, USA
Xiaotie Deng   Peking University, China

Computational Economics and Algorithmic Game Theory

Yukun Cheng   Suzhou University of Science and Technology, China
Zhengyang Liu   Beijing Institute of Technology, China
Biaoshuai Tao   Shanghai Jiao Tong University, China

Steering Committee

Xiaotie Deng   Peking University, China
Jian Li   Tsinghua University, China

Pinyan Lu	Shanghai University of Finance and Economics, China
Jianwei Huang	Chinese University of Hong Kong Shenzhen, China
Lijun Zhang	Chinese Academy of Sciences, China

Program Committee

Xiaohui Bei	Nanyang Technological University, Singapore
Zhigang Cao	Beijing Jiaotong University, China
Xue Chen	University of Science and Technology of China, China
Xujin Chen	Chinese Academy of Sciences, China
Donglei Du	University of New Brunswick, Canada
Muhammed F. Esgin	Monash University, Australia
Mingyu Guo	University of Adelaide, Australia
Chao Huang	University of California Davis, USA
Moran Koren	Harvard University, USA
Shi Li	Nanjing University, China
Bo Li	Hong Kong Polytechnic University, Hong Kong, China
Bingkai Lin	Nanjing University, China
Jingcheng Liu	Nanjing University, China
Shengxin Liu	Harbin Institute of Technology Shenzhen, China
Jinyan Liu	Beijing Institute of Technology, China
Luchuan Liu	BNU & HKBU United International College, China
Xinhang Lu	UNSW Sydney, Australia
Qi Qi	Renmin University of China, China
Shuai Shao	University of Science and Technology of China, China
Xiaorui Sun	Columbia University, USA
Yi Sun	Chinese Academy of Science, China
Changjun Wang	Chinese Academy of Sciences, China
Ye Wang	University of Macau, Macau, China
Zihe Wang	Renmin University of China, China
Jiayu Xu	Oregon State University, USA
Yingjie Xue	Brown University, USA
Kuan Yang	Shanghai Jiao Tong University, China
Haoran Yu	Beijing Institute of Technology, China
Yang Yuan	Tsinghua University, China

Zhijie Zhang	Fuzhou University, China
Yuhao Zhang	Shanghai Jiao Tong University, China
Jinshan Zhang	Zhejiang University, China
Jie Zhang	University of Bath, UK
Yong Zhang	Chinese Academy of Science Shenzhen Institutes of Advanced Technology, China
Zhenzhe Zheng	Shanghai Jiao Tong University, China

Keynote Speeches

Majority Game in Blockchain

Xiaotie Deng

Peking University

Abstract. Majority Equilibrium has made its way in Economic Systems in its implementation of Bitcoin. Its economic stability or security has met a challenge in the Selfish mining attack by Ittay Eyal and Emin Gün Sirer. This talk is a presentation on the cognitive level view of the majority game, based on a recent joint work "Insightful Mining Equilibria" on WINE 2022, with Mengqian Zhang, Yuhao Li, Jichen Li, Chaozhe Kong.

Mechanism Design with Data Correlation

Jianwei Huang

Chinese University of Hong Kong (Shenzhen)

Abstract. High-quality data collection is essential for various data-driven analysis scenarios. However, this process can compromise user privacy, especially when data across individuals are correlated. In such cases, one may suffer privacy loss even without reporting his own data directly. This talk considers the design of privacy-preserving mechanisms in two application scenarios: non-verifiable data with analysis-based incentives, and verifiable data with payment-based incentives. In both cases, we discuss how data correlation affects users' data contribution behavior and how the data collector should optimize the mechanism accordingly.

Optimal Composition Ordering for 1-Variable Functions

Kazuhisa Makino

Kyoto University

Abstract. We outline the composition ordering problem of 1-variable functions, i.e., given n 1-variable functions, we construct a minimum composition ordering for them. We discuss applications and related problems for the problem as well as the current status of the complexity issue.

Contents

Understanding the Relationship Between Core Constraints and Core-Selecting Payment Rules in Combinatorial Auctions

Robin Fritsch[1], Younjoo Lee[2], Adrian Meier[1], Ye Wang[3]([✉]),
and Roger Wattenhofer[1]

[1] ETH Zurich, Zurich, Switzerland
{rfritsch,wattenhofer}@ethz.ch, meiera@student.ethz.ch
[2] Seoul National University, Seoul, Korea
youlee@student.ethz.ch
[3] University of Macau, Macau, China
wangye@um.edu.mo

Abstract. Combinatorial auctions (CAs) allow bidders to express complex preferences for bundles of goods being auctioned, which are widely applied in the web-based business. However, the behavior of bidders under different payment rules is often unclear. In this paper, we aim to understand how core constraints interact with different core-selecting payment rules. In particular, we examine the natural and desirable non-decreasing property of payment rules, which states that bidders cannot decrease their payments by increasing their bids. Previous work showed that, in general, the widely used VCG-nearest payment rule violates the non-decreasing property in single-minded CAs. We prove that under a single effective core constraint, the VCG-nearest payment rule is non-decreasing. In order to determine in which auctions single effective core constraints occur, we introduce a conflict graph representation of single-minded CAs and find sufficient conditions for the single effective core constraint in CAs. We further show that the VCG-nearest payment rule is non-decreasing with no more than five bidders.

Keywords: Combinatorial auctions · Core-selecting payment rules · VGC-nearest payment rule · Non-decreasing payment rules · Overbidding

1 Introduction

Combinatorial Auctions (CAs) [19] are widely used to sell multiple goods with unknown value at competitive market prices. CAs permit bidders to fully express their preferences by allowing them to bid on item *bundles* instead of being limited to bidding on individual items. A CA consists of an allocation algorithm

M. Li et al. (Eds.): IJTCS-FAW 2023, LNCS 13933, pp. 1–14, 2023.
https://doi.org/10.1007/978-3-031-39344-0_1

that chooses the winning bidders, and a payment function that determines the winner's payments. CAs are popular, sometimes with a total turnover in the billions of US dollars [1], especially in the web-based business [16]. Often auction designers want an auction to be truthful, in the sense that all bidders are incentivized to reveal their true value.

Table 1. An example of an auction with 3 bidders and 2 items. The two local bidders win the auction because they bid $6 + 7 = 13$, whereas the global bidder only bids 9.

	Local Bidder 1	Local Bidder 2	Global Bidder
Bundle	$\{A\}$	$\{B\}$	$\{A, B\}$
Bid	6	7	9
Allocation	$\{A\}$	$\{B\}$	$\{\}$
First-price payment	6	7	0
VCG payment	2	3	0
VN payment	4	5	0

Consider the example in Table 1. This is a so-called Local-Local-Global (LLG) auction; two bidders are local in the sense that they are only interested in one good each, while the global bidder wants to buy all goods. If the payment scheme is the first-price payment, the winning local bidders would need to pay $6+7 = 13$. They would have been better off by lying, for instance, by bidding a total amount of 10 only.

The well-known Vickrey-Clarke-Groves (VCG) payment scheme [9,14,21] is the unique payment function to guarantee being truthful under the optimal welfare allocation. In our example, the VCG payments are much lower. Indeed, VCG payments are often not plausible in practice because of too low payments [3]. In our example, the VCG payments of $2 + 3 = 5$ are less valuable than the bid of the global bidder. Therefore, the global bidder and the seller should ignore the VCG mechanism and make a direct deal.

Core-selecting payment rules, in particular the VCG-nearest (VN) payment [11], have been introduced to improve the situation and to guarantee the seller a reasonable revenue [12]. The VN payment rule selects the closest point to the VCG payments in the *core*, where the core is the set of payments, for which no coalition is willing to pay more than the winners [10] (see Fig. 1).

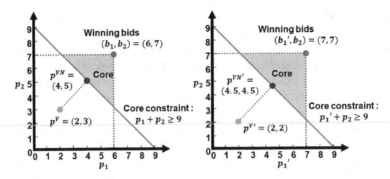

Fig. 1. Left: payment space of winning bidders in Table 1. The green point p^V is the VCG payment point, the red point P^{VN} is the VN payment point, the orange line is the core constraint on payments of local bidders 1 and 2, and the gray triangle is the core given by core constraints. Right: If bidder 1 increases their bid from 6 to 7, their payment increases as well, from 4 to 4.5.

In this paper, we study payment rules for welfare maximizing known single-minded CAs in which each bidder is interested in a single known bundle. The profile of desired bundles together with the profile of bids define a number of linear constraints (the core constraints) which form a polytope (the core).

However, also core-selecting payments such as the VN payment are not perfect. It has been shown that bidders can sometimes decrease their payments by announcing *higher-than-truthful* bids under the VN payment rule. Examples which show such overbidding behavior already need a non-trivial amount of goods and bidders [7]. In this paper, we study the limitations of VN payments. How complicated can CAs get such that bidders cannot profit from higher-than-truthful bids when using VN payments? What is the relation between different core constraints and core-selecting payment rules?

In particular, we study when the non-decreasing property holds, which is a natural and desirable property of payment rules. This property requires that a bidder cannot decrease their payment by increasing their bid. We examine for which kind of core constraints VN payments are non-decreasing. More precisely, we show that the non-decreasing property holds whenever a single effective core constraint exists.

Our second result determines which kinds of auctions are non-decreasing. To do so, we introduce a graph-based representation of CAs. We construct a conflict graph based on the overlap between the desired bundles of the bidders. We find sufficient conditions on the conflict graph to have a single effective core constraint. In particular, we show that this is the case if the conflict graph is a complete multipartite graph or if any maximal independent set in the conflict graph has at most two nodes. Furthermore, we show that for auctions with at most five bidders, the VN payment is non-decreasing, without relying on the existence of a single effective core constraint.

2   Related Work

The incentives of bidders in CAs with core-selecting payment rules are not under-stood well [13]. Day and Milgrom claimed that core-selecting payment rules minimize incentives to misreport [10]. However, it is not known under which cir-cumstances certain incentive properties, like the non-decreasing property, hold. The non-decreasing property has been observed for the VN payment rule in LLG auctions [2], but does not hold in other single-minded CAs [7]. Markakis and Tsikiridis examined two other payment rules, 0-nearest and b-nearest, which select the closest point in the minimum-revenue core to the origin and to the actual bids, respectively [15]. They prove that these two payment rules satisfy the non-decreasing property in single-minded CAs.

To analyze the effectiveness of core-selecting payment rules in Computational Auctions (CAs), Day and Raghavan introduced a constraint generation approach to succinctly formulate the pricing problem [12]. Building upon this work, B"unz et al. presented an enhanced algorithm that expedites the generation of core con-straints by leveraging conflict graphs among auction participants [4]. Niazadeh et al. [17] and Cheng et al. [8] further extended these advancements by develop-ing rapid algorithms tailored for specific use cases, such as advertising auctions and path auctions, respectively.

Payment properties strongly influence incentive behaviors in CAs. Previous research focused on game-theoretic analysis [12] and showed that bidders might deviate from their truthful valuation to *under*-bidding strategies (bid shading) or *over*-bidding strategies, where bidders place a bid lower or higher than their valuation, respectively. Ryuji Sano [20] proved that the truthful strategy is not dominant in proxy and bidder-optimal core-selecting auctions without a trian-gular condition. However, whether over-bidding strategies exist in any NE is still an open question.

Previous work has shown that in both full and incomplete information set-ting, under-bidding strategies always exist in Pure Nash equilibria (PNE) and Bayesian Nash equilibria (BNE) for core-selecting CA. Beck and Ott exam-ined over-bidding strategies in a general full-information setting and proved that every minimum-revenue core-selecting CA has a PNE, which only contain over-bids [18]. Although the existence of over-bidding strategies in PNE has been proven, incentives for over-bidding when values are private are not very well understood. In BNE, bidders choose from their action space to respond to oth-ers' expected strategies with a common belief about the valuation distribution among all bidders. One of the few known facts is that bidders might over-bid on a losing bundle to decrease their payment for a winning bundle [5,6,18].

Compared to previous studies our work fills the following three gaps. First, previous studies have not fully considered how core constraints influence core-selecting payment property. This paper examines how core constraints interact with core-selecting payment rules, which motivates better designs of CA models. Second, since we believe that graph representations are at the heart of under-standing the core constraints and core-selecting payment rules, we represent conflicts as a graph. Finally, the relationship between non-decreasing payment rules and incentive behaviors in CAs has not been studied yet. Our work provides new insight into the existence of over-bidding strategies in Nash equilibria, underlining the importance of the non-decreasing property.

3 Formal Model

We study auctions under the assumption that all bidders as well as the auctioneer act independently, rationally, and selfishly. Each bidder aims to maximize personal utility.

3.1 Combinatorial Auctions

In a combinatorial auction (CA) a set $M = \{1, \ldots, m\}$ of goods is sold to a set $N = \{1, \ldots, n\}$ of bidders. In this paper, we consider single-minded CAs (SMCAs) in which every bidder only bids on a single bundle. Let $k_i \subset M$ be the single bundle that bidder i is bidding for and denote $k = (k_1, \ldots, k_n)$ as the interest profile of the auction. We assume that the interest profile of an auction is known and fixed. Furthermore, let $v_i \in \mathbb{R}_{\geq 0}$ be the true (private) value of k_i to bidder i and $b_i \in \mathbb{R}_{\geq 0}$ the bid bidder i submits for k_i. The bids of all bidders are summarized in the bid profile $b = (b_1, \ldots, b_n)$. We denote the bid profile of all bids except bidder i's b_{-i}, and in general, the bid profile of a set $L \subset N$ of bidders b_L.

A CA mechanism (X, P) consists of a winner determination algorithm X and a payment function P. The winner determination selects the winning bids while the payment function determines how much each winning bidder must pay.

3.2 Winner Determination

The allocation algorithm $X(b)$ returns an efficient allocation x, i.e. a set of winning bidders who receive their desired bundles. All other bidders receive nothing. An allocation is called efficient if it maximizes the reported social welfare which is defined as the sum of all winning bids. We denoted the reported social welfare as $W(b, x) = \sum_{i \in x} b_i$. This optimization problem is subject to the constraint that every item is contained in at most one winning bundle.

Every bidder intends to maximize their utility which is the difference between their valuation of the bundle they acquire and the payment they make. The social optimum would be to choose the allocation that maximizes the sum of valuations of the winning bundles. However, since the valuations are private, the auctioneer can only maximize the reported social welfare.

3.3 Payment Functions

We assume the payment function satisfies voluntary participation, i.e., no bidder pays more than they bid. So the payment p_i of bidder i satisfies $p_i \leq b_i$ for every $i \in N$.

The Vickrey-Clarke-Groves (VCG) payment is the unique payment rule which always guarantees truthful behavior of bidders in CAs. We denote bidder i's VCG payment as p_i^V.

Definition 1 (VCG payment). *For an efficient allocation $x = X(b)$, the VCG payment of bidder i is*

$$p_i^V(b, x) := W(b, X(b_{-i})) - W(b, x_{-i})$$

where $x_{-i} = x \setminus \{i\}$ is the set of all winning bidder except i. Note that $X(b_{-i})$ is an efficient allocation in the auction with all bids except bidder i's bid.

The VCG payment p_i^V is a measurement of bidder i's contribution to the solution. It represents the difference between the maximum social welfare in an auction without i and the welfare of all winners except i in the original auction.

Definition 2 (Core-selecting Payment Rule). *For an efficient allocation $x = X(b)$, the core is the set of all points $p(b, x)$ which satisfies the following core constraint for every subset $L \subseteq N$:*

$$\sum_{i \in N \setminus L} p_i(b, x) \geq W(b, X(b_L)) - W(b, x_L)$$

Here, $x_L = x \cap L$ is the set of winning bidders in L under the allocation x. Note that $X(b_L)$ is an efficient allocation in the auction with only the bids of bidders in L.

A payment rule is called core-selecting *if it selects a point within the core. The* minimum revenue core *forms the set of all points $p(b, x)$ minimizing $\sum_{i \in N} p_i(b, x)$ subject to being in the core.*

The core is described by lower bound constraints (the core constraints) on the payments such that no coalition can form a mutually beneficial renegotiation among themselves. Those core constraints impose that any set of winning bidders must pay at least as much as their opponents would be willing to pay to get their items. The VN payment rule selects a payment point in the core closest to the VCG point.

Definition 3 (VCG-nearest Payment). *The VCG-nearest payment rule (quadratic payment, VN payment) picks the closest point to the VCG payment within the minimum-revenue core with respect to Euclidean distance.*

Definition 4 (Non-decreasing Payment Rule). *For any allocation x, let $\mathcal{B}_x$ be the set of bid profiles for which x is efficient. The payment-rule p is non-decreasing if, for any bidder i, any allocation x, and bid profiles $b, b' \in \mathcal{B}_x$ with $b_i' \geq b_i$ and $b_{-i} = b_{-i}'$, the following holds:*

$$p_i(b', x) \geq p_i(b, x)$$

4 Non-decreasing Payment Rules and Single Effective Core Constraints

We begin by proving a sufficient condition on the core constraints that guarantees that VN is a non-decreasing payment rule.

For core selecting payment rules, the core constraints bound the payments from below to ensure that no coalition has a higher reported price than the winners. However, many of the constraints are redundant since other constraints are more restrictive. For example, consider an LLG auction such as the one shown in Fig. 1 in which the local bidders win. The core constraints on their payment are then

$$p_1 + p_2 \geq b_G \tag{1}$$

$$p_1 \geq b_G - b_2 \tag{2}$$

$$p_2 \geq b_G - b_1 \tag{3}$$

where b_G is the bid of the global bidder. Of these constraints, (2) is immediately satisfied, as soon (1) holds since $p_2 \leq b_2$. The same is true for (3). So (1) is the only effective constraint. We will formalize this idea in the following. Note that the constraints (2) and (3) discussed above are of the form $p_i \geq p_i^V$. Such a constraint arises for every winning bidder from the core constraint for $N \setminus L = \{i\}$. However, when calculating the VN payments, we can disregard core constraints of the form $p_i \geq p_i^V$ which we call VCG-constraints since we are minimizing the distance between p and p^V and no other constraint forces $p_i < p_i^V$.

Definition 5. *Consider an SMCA with a fixed interest profile and a fixed winner allocation. Intuitively, we say a* single effective core constraint *(SECC) exists, if for all bid profiles, the fact that a single core constraint holds implies that all other core constraints are satisfied. More formally, an SECC exists, if for all bid profiles, the polytope defined by this core constraint together with the voluntary participation constraints exactly equals the core (which is defined by all core constraints).*

Theorem 1. *The VN payment rule is non-decreasing for SMCAs with a single effective core constraint.*

Proof. To prove this theorem, we will first compute an explicit formula for the VN payments. The payments of all losing bidders are 0. For all winning bidders whose payment is not part of the SECC, the VN payment simply equals the VCG payment. Let S be the set of winners whose payment is part of the SECC. Then we have the following constraints on the VN payments to S, where (4) is the SECC with some lower bound B.

$$\sum_{i \in S} p_i^{VN} \geq B \tag{4}$$

$$p_i^{VN} \leq b_i \quad \text{for } i \in S \tag{5}$$

The quadratic optimization problem to be solved is minimizing the Euclidean distance between p^{VN} and p^V under the constraints above. For the solution of this optimization, the voluntary participation constraint (5) will be active for some i. Let A be the set of indices for which (5) is active, i.e. $p_i^{VN} = b_i$ for $i \in A$.

For the remaining $i \in S \setminus A$, we write $p_i^{VN} = p_i^V + \delta_i$. The single effective core constraint (4) can now be rewritten as

$$\sum_{i \in S \setminus A} \delta_i \geq B - \sum_{i \in S \setminus A} p_i^V - \sum_{i \in A} b_i.$$

Minimizing the Euclidean distance between p^{VN} and p^V is equivalent to minimizing $\sum_{i \in S \setminus A} \delta_i^2$. Since we have a lower bound on the sum of the δ_i, the minimum possible value of $\sum_{i \in S \setminus A} \delta_i^2$ is achieved when all δ_i are equal, i.e.

$$\delta_i = \delta = \frac{1}{|S \setminus A|} \left(B - \sum_{j \in S \setminus A} p_j^V - \sum_{j \in A} b_j \right) \tag{6}$$

for $i \in S \setminus A$. With that we conclude

$$p_i^{VN} = \begin{cases} b_i & \text{for } i \in A \\ p_i^V + \delta & \text{for } i \in S \setminus A. \end{cases} \tag{7}$$

Finally, we verify that VN is non-decreasing. Assume bidder i increases their bid and this does not change the allocation x. If i is a losing bidder in x, their VN payment is 0 and can obviously not decrease. Furthermore, if i is a winning bidder, but i's payment is not part of the SECC, i's VN payment will equal their VCG payment which does not change as it only depends on the other bids. From now on, we assume that bidder i is a winning bidder whose payment is part of the SECC, i.e. $i \in S$.

Consider how the quadratic optimization problem changes when increasing bidder i's bid. One constraint and the point p^V move continuously with this change. So clearly the solution, i.e. p^{VN}, also moves continuously. During this move, some of the constraints (5) will become active or inactive. We call the moments when this happens *switches* and examine the *steps* between two consecutive switches.

As p^{VN} changes continuously around switches, Eq. (7) will yield the same result at the switch, no matter if we consider the switching constraint to be active or not. So for every single step, we can assume that the set of active constraints is the same at the beginning and the end of the step. If suffices to show that bidder i's payment does not decrease in every step between two switches. (In short, p^{VN} is a continuous function which is piecewise defined and we verify all of its pieces are non-decreasing.)

Assume bidder i's bid increases from b_i to b_i' in a certain step and let $b = (b_i, b_{-i})$ and $b' = (b_i', b_{-i})$ denote the corresponding bid profiles. We distinguish two case based on if i is in the set of active constraints in this step or not. If

i's constraint is active, i.e. $i \in A$, we have $p_i^{VN}(b, x) = b_i$ and $p_i^{VN}(b', x) = b_i'$ in (7). Then the voluntary participation constraint implies

$$p_i^{VN}(b, x) = b_i \leq b_i' = p_i^{VN}(b', x).$$

Otherwise, for $i \notin A$, we have $p_i^{VN}(b, x) = p_i^V + \delta$ and $p_i^{VN}(b', x) = p_i^V + \delta'$ where δ' is the term in (6) for the bidding profile b' with the increased bid. Then it remains to argue that $\delta \leq \delta'$. This is true since neither B nor $|S \setminus A|$ in (6) change. The sum $\sum_{j \in A} b_j$ also stays the same since $i \notin A$. Furthermore, $\sum_{j \in S \setminus A} p_j^V$ decreases or stays the same because the VCG payments of all other bidders decrease or stay the same when a winning bidder increases their bid.

So the existence of an SECC is a sufficient condition for the VN payment to be non-decreasing. It is however, not a necessary condition as can be shown by an appropriate example.

5   Graph Representation of Auction Classes

In the following, we examine for which auction classes there is guaranteed to exist only a single effective core constraint. To this end, we consider a representation of the auction classes as graphs. More precisely, we construct a conflict graph from the interest profile of an auction which represents the overlap between the bundles as follows. For an interest profile $k = (k_1, \ldots, k_n)$ of an SMCA, consider the graph $G = (V, E)$, where $V = \{k_1, \ldots, k_n\}$, i.e. each node represents a bidder. Two nodes are connected by an edge if and only if the corresponding bundles intersect in at least one item. Two simple examples are shown in Fig. 2.

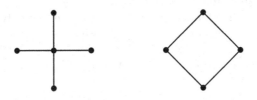

Fig. 2. Two examples of conflict graphs. The left one corresponds to the interest profile $(\{A\}, \{B\}, \{C\}, \{D\}, \{A, B, C, D\})$, the right one to the $(\{A, B\}, \{B, C\}, \{C, D\}, \{D, A\})$.

Every set of winners of the auction corresponds to a maximal independent set (MIS) in the graph.

Note that every graph with n nodes is the conflict graph of an SMCA with n bidders, i.e. the mapping is surjective: Given a graph, we associate a distinct item with every edge. For every node, we then choose the bundle containing all items of adjacent edges. While different interest profiles are mapped to the same conflict graph, auctions with the same conflict graph lead to equivalent core constraints.

Lemma 1. *Interest profiles with the same conflict graph have equivalent core constraints (up to renaming the bidders) for all possible bid profiles.*

Proof. For two interest profiles with isomorphic conflict graphs, let us renumber the bidders in one profile such that the isomorphism maps the i-th bidder in one graph to the i-th bidder in the other graph for all $i \in \{1, \dots, n\}$. Remember Definition 2 of the core constraints:

$$\sum_{i \in N \setminus L} p_i(b, x) \geq W(b, X(b_L)) - W(b, x_L)$$

Note that for every L, the sets x_L and $X(b_L)$ depend only on the conflict graphs and the bid profile. So the same is true for the whole right side on the inequality.

Only by looking at the conflict graph, we can tell by the following sufficient conditions if an SECC exists.

Lemma 2. *Every auction whose conflict graph is a complete multipartite graph has a single effective core constraint.*

Proof. If the conflict graph of an auction is a complete multipartite graph, the bidders can be grouped into k bidder groups $B_1, \dots, B_k$, where no edge between two bidders within the same group exists, but any two bidders in different groups are connected by an edge.

We argue that the winning set must be one of these bidder groups. A winning set clearly can not contain bidders from different groups since their bundles overlap. Moreover, if the winning set is only a subset of a bidder group, the current winner allocation does not maximize reported social welfare, since the rest of the group can simply be added to the winners.

Let B_w be the winning bidder group. We now argue that only a single effective core constraint exists. For any subset $L \subseteq N$, we have the core constraint

$$\sum_{i \in N \setminus L} p_i(b, x) \geq W(b, X(b_L)) - W(b, x_L). \tag{8}$$

First, note that the core constraint is not effective if $N \setminus L$ contains a losing bidder. Adding this losing bidder to L does not change the left-hand side (LHS) of (8) since this losing bidder's payment must be 0. On the other hand, the right-hand side (RHS) will not decrease since $W(b, x_L)$ does not change. Hence, the new constraint covers the previous one.

So we only need to consider the core constraints with $N \setminus L \subset B_w$. Choose L' such that $B_w \setminus L' = N \setminus L$. Furthermore, let B_o be the winning bidder group in the auction with only the bidders $N \setminus B_w$. The term $W(b, X(b_L))$ on the RHS equals either $\sum_{i \in L'} b_i$ or $\sum_{i \in B_o} b_i$. If the former is true, the RHS is 0 and the constraint is clearly not effective. In the latter case, the constraint is of the form

$$\sum_{i \in B_w \setminus L'} p_i(b, x) \geq \sum_{i \in B_o} b_i - \sum_{i \in L'} b_i.$$

Because of $p_i(b, x) \leq b_i$, any such constraints is covered by the constraint

$$\sum_{i \in B_w} p_i(b, x) \geq \sum_{i \in B_o} b_i$$

which is therefore the single effective core constraint.

Note, both graphs in Fig. 2 are complete bipartite meaning an SECC exists for any auctions with such a conflict graph. Another sufficient condition for the existence of an SECC is the following.

Lemma 3. *If every MIS in the conflict graph contains at most 2 nodes, the auction has a single effective core constraint.*

Proof. As seen in the previous proof, we only need to consider core constraints where $N \setminus L$ contains only winning bidders. So we get a constraint of the form $p_i + p_j \geq B$, and one each for p_i and p_j. These are either $p_i \geq 0$ or $p_i \geq B - b_j$ (and similarly for p_j). Because $p_i \leq b_i$ and $p_j \leq b_j$, $p_i + p_j \geq B$ is the only effective core constraint.

The Lemmas 2 and 3 show two sufficient conditions for the existence of a single effective core constraint. However, there do exist examples with a SECC, where the conflict graph has a MIS of size larger than 2 and is not a complete multipartite graph, meaning the conditions are not necessary.

So looking at the conflict graph can tell us when the auction has a SECC and consequently, if certain payment rules are non-decreasing for this auction. On the other hand, by understanding induced subgraphs of the conflict graph, we can also determine that the non-decreasing property of a payments rule is violated for this auction.

Lemma 4. *Consider two interest profiles k and k' with corresponding conflict graphs G and G'. If G' is an induced subgraph of G and a payment rule is not non-decreasing for k', then the payment rule is also not non-decreasing for k.*

Proof. According to Definition 4, a payment rule not being non-decreasing for k' means there exists an allocation x and bid profiles b and b' with $b_i' \geq b_i$ and $b_{-i} = b'_{-i}$ such that $p_i(b', x) < p_i(b, x)$. By simply choosing zero (or arbitrarily small) bids for all bidders in $G \setminus G'$, we also find two bid profiles with the same property for k.

Bosshard et al. proved that the VN payment violates the non-decreasing property by proposing an interest profile and corresponding bids [7]. Hence, VN is also not non-decreasing for any auction that contains the graph of this example as an induced subgraph. This principle motivates the search for minimal examples of overbidding, as well as proving further sufficient conditions for when overbidding does not occur. In the following, we show a sufficient condition for the non-decreasing property, without relying on the existence of a single effective core constraint.

Theorem 2. *The VN-payment rule is non-decreasing for all auctions that have an interest profile for which every winner allocation contains at most three winners.*

Proof. The case that the auction is won by two bidders is already treated in Lemma 3. Assume three bidders win the auction, and without loss of generality, let the winners be bidders 1, 2 and 3. Then the core constraints are

$$p_1^{VN} + p_2^{VN} + p_3^{VN} \geq W(b, X(b_{N\setminus\{1,2,3\}})) \tag{9}$$

$$p_1^{VN} + p_2^{VN} \geq W(b, X(b_{N\setminus\{1,2\}})) - b_3 \tag{10}$$

$$p_2^{VN} + p_3^{VN} \geq W(b, X(b_{N\setminus\{2,3\}})) - b_1 \tag{11}$$

$$p_1^{VN} + p_3^{VN} \geq W(b, X(b_{N\setminus\{1,3\}})) - b_2. \tag{12}$$

Remember, that we can ignore core constraints of the form $p_i^{VN} \geq p_i^{V}$ (VCG-constraints). Furthermore, assume without loss of generality that bidder 3 increases their bid.

Let M be the minimum revenue determined by the core constraints. There are two possibilities for the minimum revenue core: First, if the plane described by (9) is not fully covered by the constraints (10), (11) and (12), the minimum revenue is $M = W(b, X(b_{N\setminus\{1,2,3\}}))$. We further discuss this case in the next paragraph. The second possibility is that the plane described by (9) is fully covered by the other constraints, and $M > (b, X(b_{N\setminus\{1,2,3\}}))$. Then the minimum revenue core is a single point determined by equality holding in (10), (11) and (12). Since the right sides of (10), (11) and (12) are not larger than the right side of (9), all three constraints are needed to fully cover the plain. In particular, bidder 3 must be part of $X(b_{N\setminus\{1,2\}})$, otherwise (9) implies (10), and the plane is not fully covered. But this means, that increasing b_3 does not change the right sides of (10). As the same is true for (11) and (12), increasing b_3 does not move the minimum revenue core and thereby the VN payment point.

In the following, we assume that constraint (9) is active. Moreover, we assume that $M = W(b, X(b_{N\setminus\{1,2,3\}}))$. We argue similarly to the proof of Theorem 1: All changes in the VN payments are continuous in the change of the bid b_3. At any time, a number of constraints are active, and this set of active constraints changes at certain switches. To prove, the payment does not decrease overall, it suffices to prove it does not decrease between two switches, when the set of active constraints does not change. In the following, we distinguish three possible cases.

1st Case. Only constraint (9) is active. Hence, the VN payments are

$$\begin{pmatrix} p_1^{V} + \frac{M - (p_1^{V} + p_2^{V} + p_3^{V})}{3} \\ p_2^{V} + \frac{M - (p_1^{V} + p_2^{V} + p_3^{V})}{3} \\ p_3^{V} + \frac{M - (p_1^{V} + p_2^{V} + p_3^{V})}{3} \end{pmatrix}$$

When b_3 is increased, $M = W(b, X(b_{N\setminus\{1,2,3\}}))$, the minimum revenue does not change. Furthermore, p_1^{V} and p_2^{V} stay the same or decrease. So bidder 3's payment does not decrease according to the formula above.

2nd Case. The constraints (9) and (10) are active. This implies

$$p_3^{VN} = b_3 + M - W(b, X(b_{N \setminus \{1,2\}}))$$
$$p_1^{VN} + p_2^{VN} = W(b, X(b_{N \setminus \{1,2\}})) - b_3.$$

As b_3 increases, $W(b, X(b_{N \setminus \{1,2\}}))$ can increase by at most as much as b_3. Hence, the payment p_3^{VN} will not decrease.

3rd Case. Constraints (9) and (11) are active. (Note that the case when constraints (9) and (12) are active is equivalent due to symmetry.) This implies

$$p_1^{VN} = b_1 + M - W(b, X(b_{N \setminus \{2,3\}}))$$
$$p_2^{VN} + p_3^{VN} = W(b, X(b_{N \setminus \{2,3\}})) - b_1.$$

These equations describe a line on which the VN payment points lies. Increasing b_3 does not change the right side of the equation. Furthermore, it may decrease p_2^V, but does not change p_3^V. Since p^{VN} is the closest point to p^V on the line, this can, if it causes a change, only lead to a decrease of p_2^{VN} and an increase of p_3^{VN}.

Corollary 1. *The VN-payment rule is non-decreasing for all auctions with at most five bidders.*

Proof. If the auction has at most three winners, the previous theorem applies. Furthermore, if there are four bidders and four winners or five bidders and five winners, the conflict graphs are not connected. For non-connected conflict graphs, the problem reduces to multiple smaller, independent auctions. So the only remaining case is five bidders and four winners. Then the only possible connected conflict graph is the complete bipartite graph $K_{1,4}$. And this case is covered by Lemma 2.

6   Conclusion

In this paper, we study the relationship between payment rules and core constraints in CAs. We show how core constraints interact with an incentive property of payment rules in SMCAs, more precisely, that a single effective core constraint results in the non-decreasing property of the VN payment rule. Additionally, we introduce a conflict graph representation of SMCAs and prove sufficient conditions on it for the existence of a single effective core constraint.

References

1. Ausubel, L.M., Baranov, O.: A practical guide to the combinatorial clock auction. Econ. J. **127**(605), F334–F350 (2017). https://doi.org/10.1111/ecoj.12404
2. Ausubel, L.M., Baranov, O.: Core-selecting auctions with incomplete information. Int. J. Game Theory **49**(1), 251–273 (2020)

3. Ausubel, L.M., Milgrom, P., et al.: The lovely but lonely Vickrey auction. Comb. Auctions **17**, 22–26 (2006). https://www.researchgate.net/profile/Paul_Milgrom/publication/247926036_The_Lovely_but_Lonely_Vickrey_Auction/links/54bdcfe10cf27c8f2814ce6e/The-Lovely-but-Lonely-Vickrey-Auction.pdf

4. Bünz, B., Seuken, S., Lubin, B.: A faster core constraint generation algorithm for combinatorial auctions. In: Proceedings of the AAAI Conference on Artificial Intelligence, vol. 29, no. 1 (2015). https://doi.org/10.1609/aaai.v29i1.9289. https://ojs.aaai.org/index.php/AAAI/article/view/9289

5. Bosshard, V., Bünz, B., Lubin, B., Seuken, S.: Computing Bayes-Nash equilibria in combinatorial auctions with verification. J. Artif. Intell. Res. **69**, 531–570 (2020)

6. Bosshard, V., Seuken, S.: The cost of simple bidding in combinatorial auctions. arXiv:2011.12237 (2020)

7. Bosshard, V., Wang, Y., Seuken, S.: Non-decreasing payment rules for combinatorial auctions. In: IJCAI, pp. 105–113 (2018)

8. Cheng, H., Zhang, W., Zhang, Y., Zhang, L., Wu, J., Wang, C.: Fast core pricing algorithms for path auction. Auton. Agent. Multi-Agent Syst. **34**(1), 1–37 (2020). https://doi.org/10.1007/s10458-019-09440-y

9. Clarke, E.H.: Multipart pricing of public goods. Public Choice **11**(1), 17–33 (1971)

10. Day, R., Milgrom, P.: Core-selecting package auctions. Int. J. Game Theory **36**(3–4), 393–407 (2008). https://doi.org/10.1007/s00182-007-0100-7

11. Day, R.W., Cramton, P.: Quadratic core-selecting payment rules for combinatorial auctions. Oper. Res. **60**(3), 588–603 (2012)

12. Day, R.W., Raghavan, S.: Fair payments for efficient allocations in public sector combinatorial auctions. Manag. Sci. **53**(9), 1389–1406 (2007). https://pubsonline.informs.org/doi/pdf/10.1287/mnsc.1060.0662

13. Goeree, J.K., Lien, Y.: On the impossibility of core-selecting auctions. Theor. Econ. **11**(1), 41–52 (2016)

14. Groves, T.: Incentives in teams. Econometrica: J. Econometric Soc. 617–631 (1973)

15. Markakis, E., Tsikiridis, A.: On core-selecting and core-competitive mechanisms for binary single-parameter auctions. In: Caragiannis, I., Mirrokni, V., Nikolova, E. (eds.) WINE 2019. LNCS, vol. 11920, pp. 271–285. Springer, Cham (2019). https://doi.org/10.1007/978-3-030-35389-6_20

16. Narahari, Y., Dayama, P.: Combinatorial auctions for electronic business. Sadhana **30**(2), 179–211 (2005)

17. Niazadeh, R., Hartline, J., Immorlica, N., Khani, M.R., Lucier, B.: Fast core pricing for rich advertising auctions. Oper. Res. **70**(1), 223–240 (2022)

18. Ott, M., Beck, M., et al.: Incentives for overbidding in minimum-revenue core-selecting auctions. In: VfS Annual Conference 2013 (Duesseldorf): Competition Policy and Regulation in a Global Economic Order, no. 79946, Verein für Sozialpolitik/German Economic Association (2013)

19. Rassenti, S.J., Smith, V.L., Bulfin, R.L.: A combinatorial auction mechanism for airport time slot allocation. Bell J. Econ. 402–417 (1982)

20. Sano, R.: Incentives in core-selecting auctions with single-minded bidders. Games Econom. Behav. **72**(2), 602–606 (2011)

21. Vickrey, W.: Counterspeculation, auctions, and competitive sealed tenders. J. Financ. **16**(1), 8–37 (1961)

An Improved Analysis
of the Greedy+Singleton Algorithm
for k-Submodular Knapsack
Maximization

Zhongzheng Tang[1], Jingwen Chen[2], and Chenhao Wang[2,3]([✉])

[1] School of Science, Beijing University of Posts and Telecommunications,
Beijing 100876, China
[2] BNU-HKBU United International College, Zhuhai 519087, China
chenhwang@bnu.edu.cn
[3] Beijing Normal University, Zhuhai 519087, China

Abstract. We focus on maximizing a non-negative k-submodular function under a knapsack constraint. As a generalization of submodular functions, a k-submodular function considers k distinct, non-overlapping subsets instead of a single subset as input. We explore the algorithm of Greedy+Singleton, which returns the better one between the best singleton solution and the fully greedy solution. When the function is monotone, we prove that Greedy+Singleton achieves an approximation ratio of $\frac{1}{4}(1-\frac{1}{e^2}) \approx 0.216$, improving the previous analysis of 0.158 in the literature. Further, we provide the first analysis of Greedy+Singleton for non-monotone functions, and prove an approximation ratio of $\frac{1}{6}(1-\frac{1}{e^3}) \approx 0.158$.

Keywords: k-submodularity · Greedy · Knapsack · Approximation

1 Introduction

A k-submodular function generalizes a submodular function in a natural way that captures interactions among k subsets. While a submodular function takes a single subset of a finite nonempty set V as input, a k-submodular function considers k disjoint subsets of V, and exhibits the property of diminishing marginal returns common to many problems in operations research.

Given a finite nonempty set V of n items, let $(k+1)^V := \{(X_1, \ldots, X_k) \mid X_i \subseteq V \ \forall i \in [k], X_i \cap X_j = \varnothing \ \forall i \neq j\}$ be the family of k disjoint sets, where $[k] := \{1, \ldots, k\}$. A function $f : (k+1)^V \to \mathbb{R}$ is k-submodular if and only if for every k-tuples $\mathbf{x} = (X_1, \ldots, X_k)$ and $\mathbf{y} = (Y_1, \ldots, Y_k)$ in $(k+1)^V$,

$$f(\mathbf{x}) + f(\mathbf{y}) \geq f(\mathbf{x} \sqcup \mathbf{y}) + f(\mathbf{x} \sqcap \mathbf{y}),$$

where

$$\mathbf{x} \sqcup \mathbf{y} := (X_1 \cup Y_1 \setminus (\bigcup_{i \neq 1} X_i \cup Y_i), \ldots, X_k \cup Y_k \setminus (\bigcup_{i \neq k} X_i \cup Y_i)),$$

© Springer Nature Switzerland AG 2023
M. Li et al. (Eds.): IJTCS-FAW 2023, LNCS 13933, pp. 15–28, 2023.
https://doi.org/10.1007/978-3-031-39344-0_2

$$\mathbf{x} \sqcap \mathbf{y} := (X_1 \cap Y_1, \ldots, X_k \cap Y_k).$$

For a k-tuple $\mathbf{x} = (X_1, \ldots, X_k) \in (k+1)^V$, we define its size by $|\mathbf{x}| = |\cup_{i \in [k]} X_i|$. We say that $f : (k+1)^V \to \mathbb{R}$ is *monotone*, if $f(\mathbf{x}) \leq f(\mathbf{y})$ holds for any $\mathbf{x} = (X_1, \ldots, X_k)$ and $\mathbf{y} = (Y_1, \ldots, Y_k)$ with $X_i \subseteq Y_i$ for $i \in [k]$.

Since Huber and Kolmogorov [6] proposed the notion of k-submodularity one decade ago, there have been increased theoretical and algorithmic interests in the study of k-submodular functions, as various combinatorial optimization problems and practical problems can be formulated as k-submodular function maximization. The applications include influence maximization with k topics in social networks [22], sensor placement with k types of sensors [13], multi-document summarization [11] and multi-class feature selection [27]. For example, given k topics or rumors in a social network, each topic has a different spread model, and we want to select several influential people for each topic to start its spread, in order to maximize the population influenced by at least one topic. This objective function can be modeled as a k-submodular function. More detailed discussions can be found in [22,27].

As a generalization of the NP-hard submodular maximization problem, the k-submodular maximization problem is also NP-hard. Compared with the sub-modular maximization where we determine which elements/items are incorporated into the solution, additionally, for k-submodular maximization we need to specify which subsets/dimensions they belongs to. Extensive research has been devoted to developing efficient algorithms and proving their approximation ratios in different settings. In addition to the unconstrained setting [7,17,23], researchers also investigate this problem under cardinality constraints [13,22], matroid constraints [16,18], and certainly, knapsack constraints [15,21], which is the focus of this article.

Our Contributions

In this paper, we study the k-submodular maximization problem under a knapsack constraint, called *k-submodular knapsack maximization* (kSKM). Each item $a \in V$ has a cost $c(a)$, and the total cost of the items selected in the solution cannot exceed a given budget $B \in \mathbb{R}_+$. We consider the combination of two natural heuristics for knapsack problems, **Singleton** and **Greedy**. The former returns the best singleton solution $\arg\max_{\mathbf{x}:|\mathbf{x}|=1} f(\mathbf{x})$, that is, it selects a single item and assigns it to a dimension in a way that maximizes the gain of function value. The latter algorithm adds an item to a dimension that maximizes the marginal density (i.e., the marginal gain divided by its cost), until no item fits. Both heuristics are well known to have unbounded approximations, even for linear objectives.

Chen *et al.* [1] first notice that their combination **Greedy+Singleton** achieves a bounded approximation for kSKM, and prove an approximation ratio $\frac{1}{4}(1 - \frac{1}{e}) \approx 0.158$ when the function is monotone. This algorithm compares the outcomes of **Singleton** and **Greedy**, and returns the one with greater value.

We re-consider the **Greedy+Singleton** algorithm and prove an approximation ratio of $\frac{1}{4}(1 - \frac{1}{e^2}) \approx 0.216$ by a more careful analysis for monotone k-submodular functions, improving the result in [1]. Furthermore, for non-monotone k-submodular functions, we derive an approximation ratio of $\frac{1}{6}(1 - \frac{1}{e^3}) \approx 0.158$.

Though there are several algorithms with proven performance guarantees that are better than ours kSKM in the monotone case, the main advantage of the **Greedy+Singleton** algorithm is the time complexity. Tang *et al.* [21] provide a greedy algorithm with approximation $\frac{1}{2}(1 - \frac{1}{e}) \approx 0.316$ (which combines **Singleton** with a routine that completes all feasible solutions of size 2 greedily), but it takes $O(n^4 k^3)$ queries of the function. Wang and Zhou [22] provide an asymptotically-optimal $(\frac{1}{2} - \epsilon)$-approximation, but it involves designing a continuous extension of the discrete problem and rounding the fractional solution to recover the discrete solution. Compared with them, **Greedy+Singleton** requires only $O(n^2 k)$ queries.

This paper is organized as follows. In Sect. 2 we present the model and preliminaries. In Sect. 3 we consider the maximization problem without any constraint, of which the result will be used in the analysis of kSKM. In Sect. 4 we analyze the approximation of **Greedy+Singleton** for kSKM. In Sect. 5 we compare it with the method in [1].

Related Work
Huber and Kolmogorov [6] proposed k-submodular functions to express submodularity on choosing k disjoint sets of elements instead of a single set. Recently, this has become a popular subject of research [2,4,5,9,12,17].

For the kSKM problem, Tang *et al.* [21] were the first to consider it in the community. When the function is monotone, they provided a $\frac{1}{2}(1 - \frac{1}{e})$-approximation algorithm that combines **Singleton** with a greedy algorithm that completes all feasible solutions of size 2 greedily. Their analysis framework follows from that of Sviridenko [19] for submodular knapsack maximization problems. Xiao *et al.* [24] later improved the ratio of the same algorithm to $\frac{1}{2}(1 - e^{-2})$ and $\frac{1}{3}(1 - e^{-3})$ for the monotone and non-monotone case, respectively. Wang and Zhou [22] presented an algorithm with asymptotically optimal ratio of $\frac{1}{2} - \epsilon$ by multilinear extension techniques (relaxing the optimization to the continuous space and then rounding the fractional solution). Pham *et al.* [15] proposed a streaming algorithm with approximation ratios $\frac{1}{4} - \epsilon$ and $\frac{1}{5} - \epsilon$ for the monotone and non-monotone cases, respectively, which requires $O(\frac{n}{\epsilon} \log n)$ queries. Other works related to kSKM include [20,25,26].

Chen *et al.* first analyzed the performance of **Greedy+Singleton** for kSKM, and proved an approximation ratio $\frac{1}{4}(1 - \frac{1}{e})$. Before them, due to its simplicity and efficiency, **Greedy+Singleton** has received lots of attention for *submodular knapsack maximization*. This algorithm was first suggested in [8] for coverage functions, and adapted to monotone submodular function in [11]. Feldman *et al.* [3] showed that the approximation ratio is within $[0.427, 0.462]$, and Kulik *et al.* [10] presented an improved upper bound of 0.42945. Hence, it limits the approximation ratio of **Greedy+Singleton** for submodular knapsack maximization to the interval $[0.427, 0.42945]$.

The k-submodular maximization problem is also studied in different unconstrained or constrained settings. For unconstrained k-submodular maximization,

Ward and Živný [23] proposed a $\max\{\frac{1}{3}, \frac{1}{1+a}\}$-approximation algorithm with $a = \max\{1, \sqrt{(k-1)/4}\}$. Later, Iwata et $al.$ [7] improved it to $\frac{1}{2}$, which is more recently improved to $\frac{k^2+1}{2k^2+1}$ by Oshima [14]. For monotone k-submodular maximization, Iwata et $al.$ [7] also proposed a randomized $\frac{k}{2k-1}$-approximation algorithm, and showed that the ratio is asymptotically tight. For monotone k-submodular maximization under a total size constraint (i.e., $\sum_{i\in[k]}|X_i| \le B$ for an integer budget B), Ohsaka and Yoshida [13] proposed a $\frac{1}{2}$-approximation algorithm, and a $\frac{1}{3}$-approximation algorithm for that under individual size constraints (i.e., $|X_i| \le B_i$ $\forall i \in [k]$ with budgets B_i). Under a matroid constraint, Sakaue [16] proposed a $\frac{1}{2}$-approximation algorithm for the monotone case, which is asymptotically tight, and Sun et $al.$ [18] gave a $\frac{1}{3}$-approximation algorithm for the non-monotone case.

2   Preliminaries

We introduce more characteristics of k-submodular functions. If two k-tuples $\mathbf{x} = (X_1, \ldots, X_k)$ and $\mathbf{y} = (Y_1, \ldots, Y_k)$ in $(k+1)^V$ with $X_i \subseteq Y_i$ for each $i \in [k]$, we denote $\mathbf{x} \preceq \mathbf{y}$. Define the marginal gain when adding item a to the i-th dimension of $\mathbf{x} = (X_1, \ldots, X_k)$ to be

$$\Delta_{a,i}(\mathbf{x}) := f(X_1, \ldots, X_{i-1}, X_i \cup \{a\}, X_{i+1}, \ldots, X_k) - f(\mathbf{x}),$$

and thus $\frac{\Delta_{a,i}(\mathbf{x})}{c(a)}$ is the marginal density. A k-submodular function f clearly satisfies the $orthant$ $submodularity$

$$\Delta_{a,i}f(\mathbf{x}) \ge \Delta_{a,i}f(\mathbf{y}), \ \forall \mathbf{x}, \mathbf{y} \in (k+1)^V \ \text{ with } \ \mathbf{x} \preceq \mathbf{y}, a \notin \bigcup_{j\in[k]} Y_j, i \in [k],$$

and the $pairwise$ $monotonicity$

$$\Delta_{a,i_1}f(\mathbf{x}) + \Delta_{a,i_2}f(\mathbf{x}) \ge 0, \ \forall \mathbf{x} \in (k+1)^V \ \text{ with } \ a \notin \bigcup_{j\in[k]} X_j, i_1, i_2 \in [k], i_1 \ne i_2.$$

Ward and Živný [23] show that the converse is also true.

Lemma 1 ([23]). *A function $f : (k+1)^V \to \mathbb{R}$ is k-submodular if and only if f is orthant submodular and pairwise monotone.*

It is easy to see that when f is monotone, the k-submodularity degenerates into orthant submodularity.

   Every k-tuple $\mathbf{x} = (X_1, \ldots, X_k) \in (k+1)^V$ uniquely corresponds to a set $S = \{(a, d) \mid \exists d \in [k] \ a \in X_d\}$ that consists of $item$-$index$ $pairs$. That is, an item-index pair (a, d) belongs to S (called a $solution$) if and only if there is an index d so that $a \in X_d$. From now on, with a slight abuse of notations, we write $\mathbf{x}$ and its corresponding solution S interchangeably, for example, $\Delta_{a,d}(S)$ means the marginal gain $f(S \cup \{(a, d)\}) - f(S)$. For any solution S, we define

---

Algorithm 1. Unconstrained Greedy

Input: Set V', k-submodular function f
Output: A solution $S \in (k+1)^V$
1: $S \leftarrow \varnothing$
2: **for** each item $a \in V'$ **do**
3:      $d_a \leftarrow \arg\max_{d \in [k]} \Delta_{a,d}(S)$
4:      $S \leftarrow S \cup \{(a, d_a)\}$
5: **end for**
6: **return** S

$U(S) := \{a \in V \mid \exists d \in [k] \ (a, d) \in S\}$ to be the set of items included, and the *size* of S is $|S| = |U(S)|$. In this paper, let f be a non-negative k-submodular function, and we further assume w.l.o.g. that $f(\varnothing) = 0$.

We point out the following lemma that will repeatedly and implicitly used in our analysis.

Lemma 2 ([21]). *For any solutions S, S' with $S \subseteq S'$, we have*

$$f(S') - f(S) \leq \sum_{(a,d) \in S' \setminus S} \Delta_{a,d}(S).$$

3   A Key Lemma for Unconstrained k-Submodular Maximization

In this section, we consider the problem of maximizing the function value in the unconstrained setting, for an arbitrary subset of items $V' = \{e_1, e_2, \ldots, e_m\} \subseteq V$. Algorithm 1 (**Unconstrained Greedy**) considers items in V' in an arbitrary order, and assigns each item the *best* index that brings the largest marginal gain in each iteration. We will introduce a lemma that is important for the analysis in Sect. 4 for kSKM.

Let $T = \{(e_1, d_1^*), \ldots, (e_m, d_m^*)\}$ be an optimal solution that maximizes the function value over V' (such an optimal solution must exist due to the pairwise monotonicity). We dictate that **Unconstrained Greedy** considers the items in an order of $e_1, e_2, \ldots, e_m$, and denote the returned greedy solution by $S = \{(e_1, d_1), \ldots, (e_m, d_m)\}$.

For $j = 0, 1, \ldots, m$, define

$$S_j = \{(e_1, d_1), \ldots, (e_j, d_j)\} \text{ and} \tag{1}$$

$$T_j = \big(T \setminus \{(e_1, d_1^*), \ldots, (e_j, d_j^*)\}\big) \cup S_j. \tag{2}$$

That is, S_j is the first j item-index pairs in the greedy solution S, and T_j is obtained from the optimal solution T by replacing the first j item-index pairs with S_j. Clearly, $S_0 = \varnothing, S_m = S, T_0 = T$ and $T_m = S$.

The following key lemma bounds the optimal value $f(T)$ in terms of $f(S_t)$ and marginal gains. This conclusion is firstly noticed by Ward and Živný (implicitly in Theorem 5.1 [23]) and formalized by Xiao *et al.* [24]. For completeness, we write down the proof and credit [23,24].

We point out the following lemma

Lemma 3. *For* $t = 0, 1, \ldots, m$,

(a) if f is monotone, then $f(T) \leq 2f(S_t) + \sum_{(a,d) \in T_t \setminus S_t} \Delta_{a,d}(S_t)$;
(b) if f is non-monotone, then $f(T) \leq 3f(S_t) + \sum_{(a,d) \in T_t \setminus S_t} \Delta_{a,d}(S_t)$;

Proof. For $j = 0, \ldots, t-1$, we introduce an intermediate $P_j := T_j \setminus (e_{j+1}, d_{j+1}^*) = T_{j+1} \setminus (e_{j+1}, d_{j+1})$. That is, P_j consists of $m - 1$ items (excluding e_{j+1}), where the indices of items $e_1, \ldots, e_j$ coincide those in S, and the indices of other items coincide those in T. Then

$$f(T_j) = f(P_j) + \Delta_{e_{j+1}, d_{j+1}^*}(P_j),$$

$$f(T_{j+1}) = f(P_j) + \Delta_{e_{j+1}, d_{j+1}}(P_j).$$

When f is monotone, the difference of $f(T_j)$ and $f(T_{j+1})$ is

$$
\begin{aligned}
f(T_j) - f(T_{j+1}) &= \Delta_{e_{j+1}, d_{j+1}^*}(P_j) - \Delta_{e_{j+1}, d_{j+1}}(P_j) \\
&\leq \Delta_{e_{j+1}, d_{j+1}^*}(S_j) \quad &(3) \\
&\leq \Delta_{e_{j+1}, d_{j+1}}(S_j) \quad &(4) \\
&= f(S_{j+1}) - f(S_j).
\end{aligned}
$$

Equation (3) follows from the fact of $S_j \subseteq P_j$ and the monotonicity of f. Equation (4) follows from the fact that the greedy algorithm always assign the index with maximum marginal gain to the item considered, and (e_{j+1}, d_{j+1}) is the $(j + 1)$-th pair added. Summing this inequality from $j = 0$ to $t - 1$, we obtain

$$f(T_0) - f(T_t) \leq f(S_t) - f(S_0) = f(S_t).$$

Since $S_t \subseteq T_t$ and Lemma 2, we have

$$f(T) \leq f(S_t) + f(T_t) \leq 2f(S_t) + \sum_{(a,d) \in T_t \setminus S_t} \Delta_{a,d}(S_t).$$

When f is non-monotone, Eq. (3) no longer holds. Instead, we bound the difference of $f(T_j)$ and $f(T_{j+1})$ by

$$
\begin{aligned}
f(T_j) - f(T_{j+1}) &= \Delta_{e_{j+1}, d_{j+1}^*}(P_j) - \Delta_{e_{j+1}, d_{j+1}}(P_j) \\
&= 2\Delta_{e_{j+1}, d_{j+1}^*}(P_j) - [\Delta_{e_{j+1}, d_{j+1}^*}(P_j) + \Delta_{e_{j+1}, d_{j+1}}(P_j)] \\
&\leq 2\Delta_{e_{j+1}, d_{j+1}^*}(P_j) \quad &(5) \\
&\leq 2\Delta_{e_{j+1}, d_{j+1}^*}(S_j) \leq 2\Delta_{e_{j+1}, d_{j+1}}(S_j) \\
&= 2f(S_{j+1}) - 2f(S_j),
\end{aligned}
$$

where Eq. (5) follows from the pairwise monotonicity. Summing it from $j = 0$ to $t - 1$, we obtain

$$f(T_0) - f(T_t) \leq 2f(S_t) - 2f(S_0) = 2f(S_t).$$

Since $S_t \subseteq T_t$ and Lemma 2, we have

$$f(T) \leq 2f(S_t) + f(T_t) \leq 3f(S_t) + \sum_{(a,d) \in T_t \setminus S_t} \Delta_{a,d}(S_t).$$

$\square$

Letting $t = m$ in the above lemma, it is easy to see that the greedy solution S is 2-approximation of $f(T)$ when f is monotone, and 3-approximation of $f(T)$ when f is non-monotone.

4   Greedy+Singleton for k-Submodular Knapsack

We consider the kSKM problem. Each item $a \in V$ has a cost $c(a)$, and the total cost of selected items must not exceed a given budget $B \in \mathbb{R}_+$. For any solution $S \in (k+1)^V$, define $c(S) = \sum_{a \in U(S)} c(a)$ to be the total cost of all items in S.

We consider **Greedy+Singleton** (Algorithm 2). It returns the better solution between **Greedy** and **Singleton**, where the former greedily chooses the item-index pair of maximum marginal density in every iteration until no item fits (Line 2–11), and the latter chooses the single item-index pair of maximum marginal gain (Line 1).

Next, we prove approximation ratios $\frac{1}{4}(1 - \frac{1}{e^2})$ and $\frac{1}{6}(1 - \frac{1}{e^3})$ for the monotone and non-monotone cases, respectively. The general framework follows from Khuller *et al.* [8] for the budgeted maximum coverage problem, which gives a $\frac{1}{2}(1 - \frac{1}{e})$ approximation for the submodular knapsack maximization. We adapt it to kSKM, and utilize the characteristics of k-submodularity.

Algorithm 2. Greedy+Singleton

1: Let $S^* \in \arg\max_{S:\ |S|=1, c(S) \leq B} f(S)$ be a singleton solution giving the largest value.
2: $G_0 \leftarrow \varnothing$, $V^0 \leftarrow V$
3: **for** t from 1 to n **do**
4:    Let $(a_t, d_t) = \arg\max_{a \in V^{t-1}, d \in [k]} \frac{\Delta_{a,d}(G_{t-1})}{c(a)}$ be the pair maximizing the marginal density
5:    **if** $c(G_{t-1}) + c(a_t) \leq B$ **then**
6:       $G_t = G_{t-1} \cup \{(a_t, d_t)\}$
7:    **else**
8:       $G_t = G_{t-1}$
9:    **end if**
10:    $V^t = V^{t-1} \setminus \{a_t\}$
11: **end for**
12: $S^* \leftarrow G_n$ if $f(G_n) > f(S^*)$
13: **return** S^*

Let OPT be the optimal solution, and $f(OPT)$ be the optimal value. In each iteration $t = 1, \ldots, n$, a pair (a_t, d_t) is considered, and G_t is called the partial greedy solution. Let $l + 1$ be the first time when Algorithm 2 does not add an item in $U(OPT)$ to the current solution because its addition would violate the budget (i.e., $a_{l+1} \in U(OPT)$ and $c(a_{l+1}) + c(G_l) > B$). We can further assume that $l + 1$ is the first time t for which $G_t = G_{t-1}$. This assumption is without loss of generality, because if it happens earlier for some $t' < l + 1$, then $a_{t'}$ does not belong to the optimal solution T, nor the approximate solution we are interested in; thus, we can remove $a_{t'}$ from the ground set V, without affecting the analysis, the optimal solution T, and the approximate solution. Thus, we have $G_t = G_{t-1} \cup \{(a_t, d_t)\}$ for $t = 1, \ldots, l$.

For each $t = 1, \ldots, l$, we define $\bar{G}_t = G_t$ to be the partial greedy solution after the t-th iteration, and define $\bar{G}_{l+1} = G_l \cup \{(a_{l+1}, d_{l+1})\}$ to be the solution obtained by adding (a_{l+1}, d_{l+1}) to G_l. Note that $\bar{G}_{l+1}$ violates the budget, and $G_l = G_{l+1} \neq \bar{G}_{l+1}$ by our assumption.

Next, we prove the approximation ratio of **Greedy+Singleton** by a series of lemmas and Theorem 1. In Lemma 4, we show that a selected item which occupies a large proportion of the budget gives a good approximation. In Lemma 5 we bound the marginal gain in every iteration, and then Lemma 6 gives a lower bound on every $f(\bar{G}_t)$.

Lemma 4. For $t = 1, \ldots, l$, if $c(a_t) \geq \alpha \cdot B$, then the partial greedy solution $\bar{G}_t$ is $\min\{\frac{1}{2}, \alpha\}$-approximation if f is monotone, and $\min\{\frac{1}{3}, \alpha\}$-approximation if f is non-monotone.

Proof. For each $t = 1, \ldots, l$, we consider the unconstrained maximization over the items in $V' := U(\bar{G}_{t-1}) \cup U(OPT) = \{e_1 \ldots, e_m\}$. Assume w.l.o.g. that $e_1 = a_1, e_2 = a_2, \ldots, e_{t-1} = a_{t-1}$. Let Algorithm 1 consider the items in the order of $e_1, \ldots, e_m$. Recall the notations in Eq. (1) and (2), and note that $\bar{G}_j = S_j$ for $j = 1, \ldots, t-1$, that is, the partial greedy solutions in Algorithm 2 coincide those in Algorithm 1. Denote by OPT' the optimal solution of the unconstrained maximization over $U(\bar{G}_{t-1}) \cup U(OPT)$, and we apply Lemma 3 to bound $f(OPT')$.

When f is monotone, by Lemma 3 we have

$$f(OPT) \leq f(OPT') \leq 2f(\bar{G}_{t-1}) + \sum_{(a,d) \in T_{t-1} \setminus \bar{G}_{t-1}} \Delta_{a,d}(\bar{G}_{t-1})$$

$$\leq 2f(\bar{G}_{t-1}) + \sum_{(a,d) \in T_{t-1} \setminus \bar{G}_{t-1}} c(a) \cdot \frac{\Delta_{a_t,d_t}(\bar{G}_{t-1})}{c(a_t)} \quad (6)$$

$$\leq 2f(\bar{G}_{t-1}) + \frac{\Delta_{a_t,d_t}(\bar{G}_{t-1})}{c(a_t)} \cdot B, \quad (7)$$

where Eq. (6) is because (a_t, d_t) is the pair of maximum marginal density by the greedy algorithm, and Eq. (7) is because the items in $T_{t-1} \setminus \bar{G}_{t-1}$ must belong to OPT and their total cost is at most B. Combining with the value of the partial greedy solution $f(\bar{G}_t) = f(\bar{G}_{t-1}) + \Delta_{a_t,d_t}(\bar{G}_{t-1})$, it is easy to see that

$$\frac{f(\bar{G}_t)}{f(OPT)} \geq \frac{1}{2} \cdot \frac{f(\bar{G}_{t-1}) + \Delta_{a_t,d_t}(\bar{G}_{t-1})}{f(\bar{G}_{t-1}) + \frac{\Delta_{a_t,d_t}(\bar{G}_{t-1})B}{2c(a_t)}}.$$

If $\frac{B}{2c(a_t)} \leq 1$, then clearly $\frac{f(\bar{G}_t)}{f(OPT)} \geq \frac{1}{2}$. If $\frac{B}{2c(a_t)} > 1$, then

$$\frac{f(\bar{G}_t)}{f(OPT)} \geq \frac{1}{2} \cdot \frac{\Delta_{a_t,d_t}(\bar{G}_{t-1})}{\frac{\Delta_{a_t,d_t}(\bar{G}_{t-1})B}{2c(a_t)}} = \frac{c(a_t)}{B} \geq \alpha.$$

Therefore, $\bar{G}_t$ is $\min\{\frac{1}{2}, \alpha\}$-approximation.

When f is non-monotone, using Lemma 3 for $t = 1, \ldots, l$, similarly we have

$$f(OPT) \leq f(OPT') \leq 3f(\bar{G}_{t-1}) + \sum_{(a,d) \in T_{t-1} \setminus \bar{G}_{t-1}} \Delta_{a,d}(\bar{G}_{t-1})$$

$$\leq 3f(\bar{G}_{t-1}) + \sum_{(a,d) \in T_{t-1} \setminus \bar{G}_{t-1}} c(a) \cdot \frac{\Delta_{a_t,d_t}(\bar{G}_{t-1})}{c(a_t)}$$

$$\leq 3f(\bar{G}_{t-1}) + \frac{\Delta_{a_t,d_t}(\bar{G}_{t-1})}{c(a_t)} \cdot B,$$

Combining with the value of the partial greedy solution $f(\bar{G}_t) = f(\bar{G}_{t-1}) + \Delta_{a_t,d_t}(\bar{G}_{t-1})$, it is easy to see that

$$\frac{f(\bar{G}_t)}{f(OPT)} \geq \frac{1}{3} \cdot \frac{f(\bar{G}_{t-1}) + \Delta_{a_t,d_t}(\bar{G}_{t-1})}{f(\bar{G}_{t-1}) + \frac{\Delta_{a_t,d_t}(\bar{G}_{t-1})B}{3c(a_t)}}.$$

If $\frac{B}{3c(a_t)} \leq 1$, then clearly $\frac{f(\bar{G}_t)}{f(OPT)} \geq \frac{1}{3}$. If $\frac{B}{3c(a_t)} > 1$, then

$$\frac{f(\bar{G}_t)}{f(OPT)} \geq \frac{1}{3} \cdot \frac{\Delta_{a_t,d_t}(\bar{G}_{t-1})}{\frac{\Delta_{a_t,d_t}(\bar{G}_{t-1})B}{3c(a_t)}} = \frac{c(a_t)}{B} \geq \alpha.$$

Therefore, $\bar{G}_t$ is $\min\{\frac{1}{3}, \alpha\}$-approximation. $\square$

The following lemma bounds the marginal gain in every iteration.

Lemma 5. *For each $t = 1, \ldots, l + 1$,*

(a) if f is monotone, then

$$f(\bar{G}_t) - f(\bar{G}_{t-1}) \geq \frac{c(a_t)}{B} (f(OPT) - 2f(\bar{G}_{t-1}))$$

(b) if f is non-monotone, then

$$f(\bar{G}_t) - f(\bar{G}_{t-1}) \geq \frac{c(a_t)}{B} (f(OPT) - 3f(\bar{G}_{t-1}))$$

Proof. As in the proof of Lemma 4, we again consider the unconstrained maximization over $U(\bar{G}_{t-1}) \cup U(OPT)$ for each $t = 1, \ldots, l+1$, and assume that the partial greedy solutions in Algorithm 2 coincide those in Algorithm 1. Denote by OPT' the optimal solution of this unconstrained maximization problem.

When f is monotone, by Lemma 3 (a), for $t = 1, \ldots, l+1$ we have

$$f(OPT) \le f(OPT') \le 2 \cdot f(\bar{G}_{t-1}) + \sum_{(a,d) \in T_{t-1} \setminus \bar{G}_{t-1}} \Delta_{a,d}(\bar{G}_{t-1})$$

$$\le 2 \cdot f(\bar{G}_{t-1}) + B \cdot \frac{\Delta_{a_t,d_t}(\bar{G}_{t-1})}{c(a_t)}$$

$$= 2 \cdot f(\bar{G}_{t-1}) + B \cdot \frac{f(\bar{G}_t) - f(\bar{G}_{t-1})}{c(a_t)},$$

where the last inequality follows from the facts that the marginal density is maximized in each iteration and the capacity remained is at most B. Then immediately we have $f(\bar{G}_t) - f(\bar{G}_{t-1}) \ge \frac{c(a_t)}{B}\big(f(OPT) - 2f(\bar{G}_{t-1})\big)$.

When f is non-monotone, by Lemma 3 (b), a similar analysis gives

$$f(OPT) \le 3 \cdot f(\bar{G}_{t-1}) + B \cdot \frac{f(\bar{G}_t) - f(\bar{G}_{t-1})}{c(a_t)}.$$

$\square$

Lemma 6. *For each* $t = 1, \ldots, l+1$, *we have*

$$f(\bar{G}_t) \ge (1 - x_t) \cdot f(OPT),$$

where $x_1 = 1 - \frac{c(a_1)}{B}$, $x_t = (1 - \frac{2c(a_t)}{B})x_{t-1} + \frac{c(a_t)}{B}$ *if* f *is monotone, and* $x_t = (1 - \frac{3c(a_t)}{B})x_{t-1} + \frac{2c(a_t)}{B}$ *if* f *is non-monotone.*

Proof. We prove it by induction. Firstly, when $t = 1$, clearly we have $f(\bar{G}_1) \ge \frac{c(a_1)}{B} f(OPT)$. Assume that the statement holds for iterations $1, 2, \ldots, t-1$. We show that it also holds for iteration t. When f is monotone, by Lemma 5 (a),

$$f(\bar{G}_t) = f(\bar{G}_{t-1}) + f(\bar{G}_t) - f(\bar{G}_{t-1})$$

$$\ge f(\bar{G}_{t-1}) + \frac{c(a_t)}{B}\big(f(OPT) - 2f(\bar{G}_{t-1})\big)$$

$$= (1 - \frac{2c(a_t)}{B})f(\bar{G}_{t-1}) + \frac{c(a_t)}{B}f(OPT)$$

$$\ge (1 - \frac{2c(a_t)}{B})(1 - x_{t-1}) \cdot f(OPT) + \frac{c(a_t)}{B}f(OPT)$$

$$= \big[1 - ((1 - \frac{2c(a_t)}{B})x_{t-1} + \frac{c(a_t)}{B})\big]f(OPT).$$

When f is non-monotone, a similar analysis follows from Lemma 5 (b). $\square$

It is not hard to see that the recurrence relation $x_t = (1 - \frac{2c(a_t)}{B})x_{t-1} + \frac{c(a_t)}{B}$ with initial state $x_1 = 1 - \frac{c(a_1)}{B}$ can be written as

$$x_t - \frac{1}{2} = (1 - \frac{2c(a_t)}{B})x_{t-1} - \frac{1}{2}(1 - \frac{2c(a_t)}{B}) = (1 - \frac{2c(a_t)}{B})(x_{t-1} - \frac{1}{2}).$$

Hence, for the monotone case we can easily get a general formula

$$x_t = (\frac{1}{2} - \frac{c(a_1)}{B})\prod_{j=2}^{t}(1 - \frac{2c(a_j)}{B}) + \frac{1}{2}. \tag{8}$$

For the non-monotone case, similarly we can write the recurrence relation as

$$x_t - \frac{2}{3} = (1 - \frac{3c(a_t)}{B})x_{t-1} - \frac{2}{3}(1 - \frac{3c(a_t)}{B}) = (1 - \frac{3c(a_t)}{B})(x_{t-1} - \frac{2}{3}),$$

and get a general formula

$$x_t = (\frac{1}{3} - \frac{c(a_1)}{B})\prod_{j=2}^{t}(1 - \frac{3c(a_j)}{B}) + \frac{2}{3}. \tag{9}$$

Now we are ready to prove our main theorem.

Theorem 1. *For the kSKM, **Greedy+Singleton** achieves an approximation ratio of $\frac{1}{4}(1 - \frac{1}{e^2}) \approx 0.216$ and $\frac{1}{6}(1 - \frac{1}{e^3}) \approx 0.158$ when the function is monotone and non-monotone, respectively, within $O(n^2 k)$ queries.*

Proof. When f is monotone, by Lemma 6 and Eq. (8), we have

$$f(\bar{G}_{l+1}) \geq (1 - x_{l+1}) \cdot f(OPT)$$

$$= \left(\frac{1}{2} - (\frac{1}{2} - \frac{c(a_1)}{B})\prod_{j=2}^{l+1}(1 - \frac{2c(a_j)}{B})\right) \cdot f(OPT)$$

$$= \left(\frac{1}{2} - \frac{1}{2}\prod_{j=1}^{l+1}(1 - \frac{2c(a_j)}{B})\right) \cdot f(OPT). \tag{10}$$

If $1 - \frac{2c(a_j)}{B} \geq 0$ for all $j \in [l+1]$, since $c(\bar{G}_{l+1}) = c(\bar{G}_l) + c(a_{l+1}) > B$, we obtain

$$f(\bar{G}_{l+1}) \geq \left(\frac{1}{2} - \frac{1}{2}\prod_{j=1}^{l+1}(1 - \frac{2c(a_j)}{c(\bar{G}_{l+1})})\right) \cdot f(OPT)$$

$$\geq \left(\frac{1}{2} - \frac{1}{2} \cdot (1 - \frac{2}{l+1})^{l+1}\right) \cdot f(OPT)$$

$$\geq \left(\frac{1}{2} - \frac{1}{2e^2}\right) \cdot f(OPT). \tag{11}$$

If $1 - \frac{2c(a_j)}{B} < 0$ for exactly one $j \in [l+1]$ and $1 - \frac{2c(a_j)}{B} \geq 0$ for all $i \neq j$, it immediately follows from Eq. (10) that $f(\bar{G}_{l+1}) \geq \frac{1}{2}f(OPT)$. It remains to consider the case when $1 - \frac{2c(a_j)}{B} < 0$ for exactly one $j \in [l]$ and $1 - \frac{2c(a_{l+1})}{B} < 0$. By Lemma 4, the large cost of item a_j implies that $\bar{G}_j$ has an approximation at least $\min\{\frac{1}{2}, \frac{c(a_j)}{B}\} \geq \frac{1}{2}$. By the monotonicity we have $f(\bar{G}_{l+1}) \geq f(\bar{G}_j) \geq \frac{1}{2}f(OPT)$.

Hence, we always have

$$f(\bar{G}_{l+1}) = f(\bar{G}_l) + \Delta_{a_{l+1},d_{l+1}}(\bar{G}_l) \geq \left(\frac{1}{2} - \frac{1}{2e^2}\right) \cdot f(OPT).$$

Note that $\Delta_{a_{l+1},d_{l+1}}(\bar{G}_l)$ is no more than the maximum profit of a single item, i.e., the outcome of **Singleton**, say (a^*, d^*). Therefore, the better solution between $\bar{G}_l$ and $\{(a^*, d^*)\}$ has a value

$$\max\{f(\bar{G}_l), f(\{(a^*, d^*)\})\} \geq \frac{1}{2}\left(\frac{1}{2} - \frac{1}{2e^2}\right) \cdot f(OPT).$$

Since $\bar{G}_l$ is a part of the solution returned by **Greedy+Singleton** when **Greedy** performs better than **Singleton**, it establishes an approximation ratio $\frac{1}{4}(1 - \frac{1}{e^2})$.

When f is non-monotone, by Lemma 6 and Eq. (9), we have

$$f(\bar{G}_{l+1}) \geq (1 - x_{l+1}) \cdot f(OPT)$$

$$= \left(\frac{1}{3} - (\frac{1}{3} - \frac{c(a_1)}{B})\prod_{j=2}^{l+1}(1 - \frac{3c(a_j)}{B})\right) \cdot f(OPT)$$

$$= \left(\frac{1}{3} - \frac{1}{3}\prod_{j=1}^{l+1}(1 - \frac{3c(a_j)}{B})\right) \cdot f(OPT). \tag{12}$$

If $1 - \frac{3c(a_j)}{B} \geq 0$ for all $j \in [l+1]$, since $c(\bar{G}_{l+1}) > B$, we obtain

$$f(\bar{G}_{l+1}) \geq \left(\frac{1}{3} - \frac{1}{3}\prod_{j=1}^{l+1}(1 - \frac{3c(a_j)}{c(\bar{G}_{l+1})})\right) \cdot f(OPT)$$

$$\geq \left(\frac{1}{3} - \frac{1}{3} \cdot (1 - \frac{3}{l+1})^{l+1}\right) \cdot f(OPT)$$

$$\geq \left(\frac{1}{3} - \frac{1}{3e^3}\right) \cdot f(OPT).$$

If $1 - \frac{3c(a_j)}{B} < 0$ holds for one j or three j's in $[l+1]$, then it immediately follows from Eq. (12) that $f(\bar{G}_{l+1}) \geq \frac{1}{3}f(OPT)$. It remains to consider the case when $1 - \frac{3c(a_{j_1})}{B} < 0$, $1 - \frac{3c(a_{j_2})}{B} < 0$, and $1 - \frac{3c(a_i)}{B} \geq 0$ for all $i \notin \{j_1, j_2\}$. Assume $j_1 \in [l]$. By Lemma 4, the large cost of item a_{j_1} implies that $\bar{G}_{j_1}$ has an approximation at least $\min\{\frac{1}{3}, \frac{c(a_{j_1})}{B}\} \geq \frac{1}{3}$. By the pairwise monotonicity and the greedy procedure we know that the function value of partial greedy solutions is non-decreasing, and thus $f(\bar{G}_{l+1}) \geq f(\bar{G}_{j_1}) \geq \frac{1}{3}f(OPT)$.

Therefore, **Greedy+Singleton** has a value at least

$$\max\{f(\bar{G}_l), f(\{(a^*, i^*)\})\} \geq \frac{1}{2}\left(\frac{1}{3} - \frac{1}{3e^3}\right) \cdot f(OPT).$$

$\square$

5   Conclusion

We provided a novel analysis of **Greedy+Singleton** for the kSKM, and proved approximation ratios $\frac{1}{4}(1 - \frac{1}{e^2})$ and $\frac{1}{6}(1 - \frac{1}{e^3})$ for monotone and non-monotone functions, respectively. Compared with the $\frac{1}{4}(1 - \frac{1}{e})$-approximation in [1], our improvement heavily replies on the key proposition of Lemma 3, which gives upper bounds on the optimum $f(T)$ (for unconstrained maximization) in terms of every partial greedy solution S_t, instead of the simple 2-approximation achieved by the final greedy solution in [1]. Moreover, our Lemma 4 shows that a selected item with a large cost gives a good approximation, which is also useful for proving the improved approximation ratios. Future directions include further improving the approximation ratio of **Greedy+Singleton** and looking for other efficient algorithms.

Acknowledgements. This work is partially supported by Artificial Intelligence and Data Science Research Hub, BNU-HKBU United International College (UIC), No. 2020KSYS007, and by a grant from UIC (No. UICR0400025-21). Zhongzheng Tang is supported by National Natural Science Foundation of China under Grant No. 12101069. Chenhao Wang is supported by NSFC under Grant No. 12201049, and is also supported by UIC grants of UICR0400014-22, UICR0200008-23 and UICR0700036-22.

References

1. Chen, J., Tang, Z., Wang, C.: Monotone k-submodular knapsack maximization: an analysis of the Greedy+Singleton algorithm. In: Ni, Q., Wu, W. (eds.) AAIM 2022. LNCS, vol. 13513, pp. 144–155. Springer, Cham (2022). https://doi.org/10.1007/978-3-031-16081-3_13
2. Ene, A., Nguyen, H.: Streaming algorithm for monotone k-submodular maximization with cardinality constraints. In: Proceedings of the 39th International Conference on Machine Learning (ICML), pp. 5944–5967. PMLR (2022)
3. Feldman, M., Nutov, Z., Shoham, E.: Practical budgeted submodular maximization. Algorithmica 1–40 (2022)
4. Gridchyn, I., Kolmogorov, V.: Potts model, parametric maxflow and k-submodular functions. In: Proceedings of the IEEE International Conference on Computer Vision (ICCV), pp. 2320–2327 (2013)
5. Hirai, H., Iwamasa, Y.: On k-submodular relaxation. SIAM J. Discret. Math. **30**(3), 1726–1736 (2016)
6. Huber, A., Kolmogorov, V.: Towards minimizing k-submodular functions. In: Mahjoub, A.R., Markakis, V., Milis, I., Paschos, V.T. (eds.) ISCO 2012. LNCS, vol. 7422, pp. 451–462. Springer, Heidelberg (2012). https://doi.org/10.1007/978-3-642-32147-4_40
7. Iwata, S., Tanigawa, S., Yoshida, Y.: Improved approximation algorithms for k-submodular function maximization. In: Proceedings of the 27th Annual ACM-SIAM Symposium on Discrete Algorithms (SODA), pp. 404–413 (2016)
8. Khuller, S., Moss, A., Naor, J.S.: The budgeted maximum coverage problem. Inf. Process. Lett. **70**(1), 39–45 (1999)
9. Kudla, J., Živný, S.: Sparsification of monotone k-submodular functions of low curvature. arXiv preprint arXiv:2303.03143 (2023)

10. Kulik, A., Schwartz, R., Shachnai, H.: A refined analysis of submodular greedy. Oper. Res. Lett. **49**(4), 507–514 (2021)
11. Lin, H., Bilmes, J.: Multi-document summarization via budgeted maximization of submodular functions. In: Human Language Technologies: The 2010 Annual Conference of the North American Chapter of the Association for Computational Linguistics, pp. 912–920 (2010)
12. Nguyen, L., Thai, M.T.: Streaming k-submodular maximization under noise subject to size constraint. In: Proceedings of the 37th International Conference on Machine Learning (ICML), pp. 7338–7347. PMLR (2020)
13. Ohsaka, N., Yoshida, Y.: Monotone k-submodular function maximization with size constraints. In: Proceedings of the 28th International Conference on Neural Information Processing Systems (NeurIPS), vol. 1, pp. 694–702 (2015)
14. Oshima, H.: Improved randomized algorithm for k-submodular function maximization. SIAM J. Discret. Math. **35**(1), 1–22 (2021)
15. Pham, C.V., Vu, Q.C., Ha, D.K., Nguyen, T.T., Le, N.D.: Maximizing k-submodular functions under budget constraint: applications and streaming algorithms. J. Comb. Optim. **44**(1), 723–751 (2022)
16. Sakaue, S.: On maximizing a monotone k-submodular function subject to a matroid constraint. Discret. Optim. **23**, 105–113 (2017)
17. Soma, T.: No-regret algorithms for online k-submodular maximization. In: Proceedings of the 22nd International Conference on Artificial Intelligence and Statistics (AISTATS), pp. 1205–1214. PMLR (2019)
18. Sun, Y., Liu, Y., Li, M.: Maximization of k-submodular function with a matroid constraint. In: Du, D.Z., Du, D., Wu, C., Xu, D. (eds.) TAMC 2022. LNCS, vol. 13571, pp. 1–10. Springer, Cham (2022). https://doi.org/10.1007/978-3-031-20350-3_1
19. Sviridenko, M.: A note on maximizing a submodular set function subject to a knapsack constraint. Oper. Res. Lett. **32**(1), 41–43 (2004)
20. Tang, Z., Wang, C., Chan, H.: Monotone k-submodular secretary problems: cardinality and knapsack constraints. Theor. Comput. Sci. **921**, 86–99 (2022)
21. Tang, Z., Wang, C., Chan, H.: On maximizing a monotone k-submodular function under a knapsack constraint. Oper. Res. Lett. **50**(1), 28–31 (2022)
22. Wang, B., Zhou, H.: Multilinear extension of k-submodular functions. arXiv preprint arXiv:2107.07103 (2021)
23. Ward, J., Živný, S.: Maximizing k-submodular functions and beyond. ACM Trans. Algorithms **12**(4), 1–26 (2016)
24. Xiao, H., Liu, Q., Zhou, Y., Li, M.: Small notes on k-submodular maximization with a knapsack constraint. Technical report (2023)
25. Yu, K., Li, M., Zhou, Y., Liu, Q.: Guarantees for maximization of k-submodular functions with a knapsack and a matroid constraint. In: Ni, Q., Wu, W. (eds.) AAIM 2022. LNCS, vol. 13513, pp. 156–167. Springer, Cham (2022). https://doi.org/10.1007/978-3-031-16081-3_14
26. Yu, K., Li, M., Zhou, Y., Liu, Q.: On maximizing monotone or non-monotone k-submodular functions with the intersection of knapsack and matroid constraints. J. Comb. Optim. **45**(3), 1–21 (2023)
27. Yu, Q., Küçükyavuz, S.: An exact cutting plane method for k-submodular function maximization. Discret. Optim. **42**, 100670 (2021)

Generalized Sorting with Predictions Revisited

T.-H. Hubert Chan⬤, Enze Sun⬤, and Bo Wang[(✉)]⬤

The University of Hong Kong, Pok Fu Lam, Hong Kong, China
{hubert,ezsun,bwang}@cs.hku.hk

Abstract. This paper presents a novel algorithm for the generalized sorting problem with predictions, which involves determining a total ordering of an underlying directed graph using as few probes as possible. Specifically, we consider the problem of sorting an undirected graph with predicted edge directions. Our proposed algorithm is a Monte Carlo approach that has a polynomial-time complexity, which uses $O(n \log w + w)$ probes with probability at least $1 - e^{-\Theta(n)}$, where n is the number of vertices in the graph and w is the number of mispredicted edges. Our approach involves partitioning the vertices of the graph into $O(w)$ disjoint verified directed paths, which can reduce the number of probes required. Lu et al. [11] introduced a bound of $O(n \log n + w)$ for the number of probes, which was the only known result in this setting. Our bound reduces the factor $O(\log n)$ to $O(\log w)$.

Keywords: Forbidden Comparison · Predictions · Generalized Sorting

1 Introduction

Sorting is a fundamental task in computer science, in which a collection of data is required to be arranged according to some specified total ordering. *Comparison-based* sorting algorithms are most widely used since they are generally regarded as more flexible and efficient for a broader range of data types and application scenarios, particularly when the input data is not structured in any particular way. In the comparison-based sorting framework, an algorithm can get information about this total ordering only via pair-wise elements comparison. The standard model of comparison-based sorting assumes that all pairs of elements can be compared, and the cost of each comparison is the same. However, this might not always be the case. For instance, some pairs may be prohibited in certain scenarios [1,9], or the cost of comparisons may vary [4–6,8].

The generalized sorting problem (with *forbidden comparisons*) was proposed by Huang et al. [7], which involves sorting a set of elements with the restriction that certain pairs of elements cannot be compared (while it is guaranteed that

This research was partially supported by the Hong Kong RGC grants 17201220, 17202121 and 17203122.

M. Li et al. (Eds.): IJTCS-FAW 2023, LNCS 13933, pp. 29–41, 2023.
https://doi.org/10.1007/978-3-031-39344-0_3

allowed pairs will induce a total ordering on the items) and comparisons are of unit cost. The goal is to produce a sorted list or ordering of the elements that satisfies the restriction on forbidden comparisons. The first non-trivial bound on the number of comparisons for the generalized sorting problem of n items has been given in [7]. Specifically, a randomized algorithm with $\widetilde{O}(n^{1.5})$ comparisons was provided to solve the worst-case version generalized sorting problem with high probability; while in the stochastic version where each pair is included independently and randomly with probability p, an $\widetilde{O}(n^{1.4})$ bound was achievable with high probability, for the worst-case choice of p.

Lu et al. [11] considered a variant of the problem called generalized sorting with *predictions* in which the algorithm is given an undirected graph $G = (V, E)$ together with predictions of the directions of all the edges in E, where $|V| = n$. Each edge has some hidden *true* direction, and it is guaranteed that there exists a Hamiltonian path in the underlying directed graph. By conducting a *probe* on an edge, the predicted direction of an edge can be identified as either correct (*consistent*) or incorrect (*inconsistent*). The goal of the problem is to compute a total ordering on the underlying directed graph using as few number of probes as possible. We use w to denote the number of mispredicted edges, which is unknown in advance. Lu et al. [11] provided a randomized algorithm with $O(n \log n + w)$ probes and a deterministic algorithm with $O(nw)$ probes, which outperform the bound $\widetilde{O}(n^{1.5})$ when w is small.

In summary, we give an improved algorithm for the generalized sorting with predictions problem in the paper. The main contribution of the paper is formally stated as follows:

Theorem 1 (Main Theorem). *There is a polynomial-time Monte Carlo algorithm which solves the generalized sorting with prediction problem using $O(n \log w + w)$ probes with probability at least $1 - e^{-\Theta(n)}$.*

Note that w is at most n^2 (the number of edges in E), hence the performance of our proposed algorithm is not worse than neither $O(n \log n + w)$ nor $O(nw)$ for any w. Notably, $O(n \log w + w)$ is better than both $O(n \log n + w)$ and $O(nw)$ when $w = \log n$.

High Level Ideas. The overall framework is to for each node, to check directly or indirectly the true direction of each of its incoming edges in the predicted graph. The main novel idea is that, it is possible to partition vertices in G into $O(w)$ disjoint *verified* directed paths using $O(n)$ probes if there are w mispredicted edges, where edges in a verified path all have true directions. Roughly speaking, for each node, it is sufficient to only consider an in-neighbor of the node from each path as representatives instead of considering all in-neighbors by taking advantage of the verified paths so as to reducing the number of probes. This idea allows us to reduce one factor $O(\log n)$ in [7] to $O(\log w)$.

Other Related Work. Banerjee and Richards [2] considered the worst-case version generalized sorting problem under the setting where $\binom{n}{2} - q$ pairs are assumed to be comparable. They gave an algorithm with $O((q + n) \log n)$ comparisons. The paper also studied stochastic version generalized sorting problem and gave a lower bound $\widetilde{O}(\min\{n^{\frac{3}{2}}, pn^2\})$ to the number of comparisons. Biswas et al. [3] further improved the lower bound for the worst-case version generalized sorting problem to $O(q + n \log n)$ (note that it is assumed that most pairs are comparable). Kuszmaul and Narayanan [10] improved the lower bounds for the worst-case and stochastic version generalized sorting problem to $O(n \log(pn))$ and $\widetilde{O}(\sqrt{nm})$, separately, where m is the number of all comparable pairs.

Roadmap. In Sect. 2, we will revisit generalized sorting with predictions problem. In Sect. 3, we will introduce an algorithm to partition n vertices into $O(w)$ disjoint verified directed paths, which will then be utilized to achieve the $O(n \log w + w)$ randomized algorithm in Sect. 4.

2   Preliminaries

We revisit the generalized sorting problem with predictions that has been introduced in [11], whose notation will also be used in this paper.

Definition 1 (Generalized Sorting with Prediction [11]). *An adversary picks a directed acyclic graph $\vec{G} = (V, \vec{E})$ (where $n = |V|$ and $m = |\vec{E}|$) that contains a (directed) Hamiltonian path. A directed edge $(u, v) \in \vec{E}$ represents the true direction and means that $u < v$, and a Hamiltonian paths means that there is a total ordering on V.*

The adversary may reverse the directions of an arbitrary subset of edges in $\vec{E}$ to produce $\vec{G}_P = (V, \vec{P})$ that is given to the algorithm. Note that $|\vec{P}| = |\vec{E}|$ and $w = |\vec{P} \setminus \vec{E}|$ is the number of edges whose directions have been reversed, but the algorithm does not know w in advance.

An algorithm may only probe an edge in $\vec{P}$ and it will be told whether the direction of that edge has been reversed. The algorithm may adaptively decide the next edge to be probed, based on the answers of all previous probes. If the algorithm uses randomness, the randomness is generated after the adversary has picked $\vec{G}$ and $\vec{G}_P$.

The goal of the algorithm is to find the (unique) Hamiltonian path in $\vec{E}$ using a minimum number of probes.

Probe Charging. A probe for an edge $(u, v) \in \vec{P}$ is consistent if the direction $(u, v) \in \vec{E}$ is correct; otherwise, the probe is inconsistent. Since we allow a term $O(w)$ in the number of probes, we focus on bounding the number of consistent probes. Moreover, for each inconsistent probe, we may charge $O(1)$ number of consistent probes to it. Therefore, if there are X number of remaining uncharged consistent probes, we can bound the total number of probes by $X + O(w)$.

High Level Approach. The idea in [11] is to, for each $u \in V$, check directly or indirectly the true direction of each of its incoming edges in $\vec{P}$. Since the approach focuses on in-neighbors, we will use succinct terminology as follows.

Definition 2 (Succinct Neighbor Terminology). *For any $u \in V$, we define the following terminology.*

- *Denote $N_u := \{v : (v, u) \in \vec{P}\}$ as the p-neighbors of u.*
- *Denote $S_u := \{v : (v, u) \in \vec{P} \cap \vec{E}\}$ as the true p-neighbors of u. A vertex $v \in N_u \setminus S_u$ is a false p-neighbor of u.*
- *Observe that after the algorithm has made some probes, it may be able to directly or indirectly infer whether a p-neighbor is true or false. In this case, we say that the p-neighbor is revealed.*
- *A p-neighbor v of u is plausible at some moment if it is either a true p-neighbor or unrevealed at that particular moment. We use $\widetilde{T}_u$ to denote the collection of plausible p-neighbors of u. Observe that $\widetilde{T}_u$ changes as the algorithm gains more information.*
- *For simplicity, the reader may consider the more conservative definition that $v \in \widetilde{T}_u$ if $v \in S_u$ or the edge $(v, u) \in \vec{P}$ has not been probed.*

Checked Vertices. The algorithm maintains a collection A of *checked* vertices. Initially, $A := \emptyset$. A vertex u is checked if all its p-neighbors are revealed. The algorithm adds checked vertices to A one by one, and eventually terminates when $A = V$. Observe that if an edge $(u, v) \in \vec{E}$ has both endpoints in A, then its direction is known to the algorithm. Therefore, when $A = V$, the standard topological sort can be performed to return the desired Hamiltonian path.

How to Check a Vertex with a Small Number of Consistent Probes? To check a vertex u, we can avoid probing all its remaining plausible p-neighbors if we can use information derived from edges totally contained in A.

Definition 3 (Partial Order $<_A$). *A partial order $<_A$ is defined on the vertices in A such that $a <_A b$ iff there is a directed path from a to b consisting only of vertices in A and edges in $\vec{E}$.*

Ideal Vertices. A vertex u is *ideal* if (i) all its remaining plausible p-neighbors in $\widetilde{T}_u$ are in A, and (ii) if $\widetilde{T}_u$ is non-empty, the elements in $\widetilde{T}_u$ are totally ordered with respect to $<_A$.

Lemma 1 (Checking an Ideal Vertex). *An ideal vertex can be checked with at most 1 consistent probe (i.e., there is no need to charge it to any inconsistent probe).*

Proof. Starting from the largest element in $\widetilde{T}_u$, we can probe (v, u) in decreasing order of $v \in \widetilde{T}_u$ with respect to $<_A$. Hence, we can identify the largest true p-neighbor (if any) using at most one consistent probe, while removing all false p-neighbors from $\widetilde{T}_u$.

Lemma 1 says that an ideal vertex u can be checked with at most one (uncharged) consistent probe and can be added to A. The following notion of certificate shows that a vertex u cannot be ideal.

Definition 4 (Certificate). *The following is a valid certificate for u with respect to A.*

- ***Type 1.*** *There is a true p-neighbor $v \in S_u$ such that $v \notin A$.*
- ***Type 2.*** *If all plausible p-neighbors $\widetilde{T}_u$ are contained in A, there are two incomparable elements in $\widetilde{T}_u$ with respect to $<_A$ that are both true p-neighbors of u.*

Lemma 2 (Finding Certificate). *For any vertex $u \notin A$, either a certificate for u can be found or the vertex u becomes ideal. If x inconsistent probes are made in the process, then at most $x + 2$ consistent probes are made. Hence, x of these consistent probes can be charged to the inconsistent probes, which means that there are at most 2 uncharged consistent probes.*

Moreover, the outcome of whether u becomes ideal does not depend on the choices made in the process.

Proof. If a vertex u is not ideal because $\widetilde{T}_u \setminus A$ is non-empty, we sequentially consider each $v \in \widetilde{T}_u \setminus A$ and probe (v, u), until we find a true p-neighbor $v \in S_u$ (which is a valid Type-1 certificate for u), or $\widetilde{T}_u \setminus A$ becomes empty. Observe that if all p-neighbors in $\widetilde{T}_u \setminus A$ turns out to be false, then we have not made any consistent probe.

Next, we try to find a Type-2 certificate. If there are two different v_1 and v_2 in $\widetilde{T}_u$ that are incomparable with respect to $<_A$, we probe (v_1, u) and (v_2, u) (if necessary) to reveal the true direction of the edges. If both v_1 and v_2 are true p-neighbors of u, then we have found a Type-2 certificate. Otherwise, we remove false p-neighbors from $\widetilde{T}_u$ and repeat until either a Type-2 certificate is found, or we conclude that u is ideal.

Finally, observe that the order in which we probe the edges in the above process cannot affect the outcome of whether u becomes ideal.

Remark 1. As demonstrated in [11], if we try to find a certificate for the smallest vertex in $V \setminus A$ with respect to the true global ordering $<_V$, the process described in Lemma 2 will conclude that it is ideal using zero number of consistent probes. This implies that the algorithm will always terminate.

Algorithm 1. An algorithm based on certificates [11]

1: set $A := \emptyset$
2: **while** $A \neq V$ **do**
3:     Pick lexicographically smallest $u \in V \setminus A$ that does not have a valid certificate.
4:     Use Lemma 2 to try finding a certificate for u (where randomness might be used) using at most 2 uncharged consistent probes.
5:     **if** u turns out to be ideal **then**
6:         Use Lemma 1 with at most 1 uncharged consistent probe to check u.
7:         Add u to A (which might make some current certificates invalid)
8:     **end if**
9: **end while**

Lemma 3 (Probe Charging Analysis). *In Algorithm 1, suppose for each vertex u, there are X_u attempts to find a certificate for u. Then, the total number of probes made is at most $O(n + w + \sum_{u \in V} X_u)$.*

Proof. Observe that for each inconsistent probe, at most 1 consistent probe is charged to it, as in Lemma 2.

Hence, it suffices to bound the number of uncharged consistent probes. From the algorithm description, each vertex u incurs at most 1 uncharged consistent probes when it becomes ideal, or at most 2 uncharged consistent probes every time there is an attempt to find a certificate for it. Hence, the result follows.

Bounding the Number of Certificate Finding Attempts. According to Lemma 3, it suffices to bound the number of certificate finding attempts. One crucial observation made in [11] is that for each vertex, its pool of potentially valid certificate candidates changes in a deterministic fashion as a result of the following lemma.

Lemma 4 (Vertices Checked in a Deterministic Order). *Vertices are added to A in a deterministic order.*

Proof. This is because in Algorithm 1, we always pick the lexicographically smallest vertex u without a currently valid certificate. Moreover, according to Lemma 2, whether a certificate for u is found does not depend on the choices made in the process.

Certificate Candidates. For each vertex u without a currently valid certificate, the algorithm will first attempt to find a Type-1 certificate. Hence, initially all true p-neighbors S_u of u can potentially be a witness for a Type-1 certificate. However, as more vertices are added to A, a Type-1 certificate might become invalid.

Observe that the vertices in S_u are added to A in a deterministic order. At the moment when all vertices in S_u are added to A, there cannot be any valid Type-1 certificate for u. If u is still not in A, then the algorithm will start to find Type-2 certificate for u, where a witness is a distinct pair of vertices in S_u that are incomparable with respect to $<_A$. However, as more vertices are added to A, a Type-2 certificate may become invalid as the involved pair become comparable under $<_A$.

Uniform Certificate Sampling. In [11], the same technique is used to analyze the number of certificates found for a vertex u for each type. The idea is that there is initially a pool C_u of potential certificate candidates for u, and these candidates become invalid in some deterministic order. In the worst case scenario, we require that a valid certificate must be possessed by vertex u at all times until all potential candidates become invalid. When required, the intuition is to pick a certificate uniformly at random from the pool of remaining valid certificate candidates. To minimize the number of certificate finding attempts, it makes sense to pick a certificate that becomes invalid later. However, since the algorithm does not know the order in which the certificate candidates become invalid, a uniformly random one is picked to achieve a high probability bound. By considering all vertices together, we get a better measure concentration result than the previous analysis in [11].

Lemma 5 (New Measure Concentration Bound). *Suppose each vertex u has an initial pool C_u of certificate candidates that become invalid in a deterministic order. Suppose each vertex u independently samples a certificate uniformly at random from the remaining pool of valid certificates whenever needed. Then, except with probability at most $e^{-\Theta(n)}$, the total number of certificates sampled by all vertices is at most $O(\sum_{u \in V}(\log |C_u| + 1))$.*

Proof. In the original analysis [11], a high probability statement is proved to bound the number of certificates sampled for each vertex. Then, the union bound is used to achieve a high probability statement over all vertices. Our new insight is that a better measure concentration bound can be achieved if we directly consider the total number of sampled certificates for all vertices.

For each vertex u, it is equivalent to first sample a uniformly random priority on the collection C_u of candidates. Whenever a certificate is needed, a valid candidate with the highest priority is selected. Consider the candidate with the i-th highest priority; observe that this candidate is selected if among the i certificates with highest priority, it is the last one to become invalid. Let $X_i^u \in \{0, 1\}$ be the indicator random variable for this event.

A crucial observation is that $E[X_i^u] = \frac{1}{i}$ and the random variables X_i^u's are independent. This can be checked if we sample the random priority starting from the position with the least priority, i.e., $i = |C_u|$. In this case, $X_{|C_u|}^u = 1$ *iff* we pick the certificate that is the last to become invalid, which happens with probability $\frac{1}{|C_u|}$.

In general, when we pick a candidate for the i-th priority position, it does not matter what certificates have already been picked for the lower priority positions. This certificate will be possessed by u if among the i remaining candidates, the one that becomes invalid last is selected for the i-th position, which happens with probability $\frac{1}{i}$.

The above description assumes that certificates become invalid one by one, as in the case for Type-1 certificates. However, for Type-2 certificates, a block of certificates can become invalid at the same time. In this case, the above analysis is more pessimistic than reality, and so still serves as a correct upper bound analysis for the number of sampled certificates.

Therefore, the random variables $X_i^u \in \{0, 1\}$ are independent over different vertices u and i. Observe that the expectation satisfies:

$$n \leq E[\sum_{u \in V} \sum_{i=1}^{|C_u|} X_i^u] = \sum_{u \in V} \sum_{i=1}^{|C_u|} \frac{1}{i} \leq \sum_{u \in V}(\log |C_u| + 1).$$

For a sum Z of independent $\{0, 1\}$-random variables, the Chernoff Bound states for any $0 < \epsilon < 1$, $\Pr[Z \geq (1 + \epsilon)E[Z]] \leq \exp(-\frac{1}{3}\epsilon^2 E[Z])$. Using $\epsilon = \frac{1}{2}$ gives the required result.

Remark 2. Observe that the algorithm does not actually know the true p-neighbors. Hence, it can only do uniform sampling among plausible p-neighbors for a Type-1 certificate and among pairs of distinct p-neighbors for a Type-2 certificate. However, the proof in Lemma 2 shows that if any plausible p-neighbor turns out to be false, any consistent probes involved can be charged to inconsistent probes. Therefore, the procedure essentially samples uniformly over all remaining valid certificates.

Bounding the Pool of Certificate Candidates. In the original analysis [11], a trivial bound of n is used for the number of potential Type-1 certificate candidates and a bound of $\binom{n}{2}$ is used for that of Type-2. Our new insight is that these bounds can be improved to $O(w)$ and $O(w^2)$, respectively. From Lemmas 3 and 5, this implies that the total number of probes can be improved to $O(n \log w + w)$.

3   Path Decomposition

The main novel idea is that if there are only w number of mispredicted edges, then it is possible to partition the vertices into $O(w)$ number of disjoint *verified* directed paths using $O(n)$ number of probes, where a verified path means that edges in the path all belong to $\vec{E}$. Intuitively, for each node u, instead of considering each of its p-neighbors individually, we will consider a representative from each path, thereby reducing the number of consistent probes.

The first observation is that the predicted graph can be decomposed into disjoint paths or cycles in polynomial time.

Lemma 6 (Decomposition to Paths and Cycles in Prediction Graph).
There is a polynomial-time algorithm that partitions the vertices of the graph $(V, \vec{P})$ into at most $w+1$ number of paths and c number of cycles in $\vec{P}$, where w is the number of mispredicted edges in $\vec{P}$ and $c \leq w$. Observe that no edge probing is performed in this step.

Proof. Observe that in a directed graph where each vertex has in-degree and out-degree both at most 1, then each vertex lies on a directed path or cycle. Therefore, we achieve the desired decomposition by considering a matching problem in a bipartite graph constructed as follows.

The vertices in $V_B = L \cup R$ consists of $L := \{l_u : u \in V\}$ and $R = \{r_v : v \in V\}$. The bipartite graph $G_B = (V_B, E_B)$ is constructed such that an edge $(l_u, r_v) \in E_B$ iff $(u, v) \in \vec{P}$. Observe that a matching $M \subseteq E_B$ in the bipartite graph naturally induces a subset $E_M := \{(u, v) \in \vec{P} : (l_u, r_v) \in M\}$ of directed edges in $\vec{P}$ in which each vertex has in-degree and out-degree at most 1, i.e., E_M is a decomposition into disjoing paths or cycles.

In polynomial time, we can find a maximum matching M in (V_B, E_B). Let $\alpha = |E_M| = |M|$ be the size of the maximum matching. Observe that if $\vec{P}$ contains a Hamiltonian path, then this path induces a matching of size $n - 1$ in (V_B, E_B). However, since there are w mispredicted edges in $\vec{P}$, it follows that $\alpha \geq n - w - 1$.

We next want to give an upper bound on the number of weakly connected components in (V, E_M). Let c be the number of cycles in E_M. Since the cycles in E_M are disjoint and each cycle in $\vec{P}$ is due to a mispredicted edge, it follows that $c \leq w$.

Starting from the vertex set V, if we add edges in E_M one by one, every time we add an edge, the number of connected components drops by 1, unless the edge forms a cycle. Hence, the number of weakly connected components in E_M is $n - \alpha + c \leq w + 1 + c$, as required.

Lemma 7 (Verified Directed Paths). *Using at most n edge probes, the vertices in V can be partitioned into at most $2w + 1$ verified directed paths in $\vec{E} \cap \vec{P}$.*

Proof. According to Lemma 6, the vertices are partitioned to at most disjoint $w + 1$ number of directed paths and c number of cycles in $\vec{G}_P$, where $c \leq w$. We probe all the edges in these paths and cycles, and remove the mispedicted edges.

Observe that in each of the c cycles, there is at least one mispredicted edge whose removal does not increase the number of components. There can be at most $w - c$ remaining inconsistent probes, each of which may increase the number of components by 1.

Hence, after removing mispredicted edges, the number of components (directed paths) is at most $(w + 1 + c) + w - c = 2w + 1$.

Path Probing. We use Z to denote the collection of directed paths found in Lemma 7. Since all the edges in each such path have been probed, the vertices in each path are totally ordered by $<_{\vec{E}}$. Recall that in the framework described in Sect. 2, for each vertex u, all its p-neighbors in N_u are eventually revealed (to be either true or false). Observe that for each path $Q \in Z$, all vertices in $Q \cap N_u$ can be revealed using at most 1 consistent probe, because one can simply start from the largest plausible p-neighbor v in $Q \cap N_u$ and probe the edge (v, u) until the largest true p-neighbor (if any) in $Q \cap N_u$ is found. This observation directly recovers the result from [11] that $O(nw)$ probes are sufficient for a deterministic algorithm.

4   Modified Method for Finding Certificates

Under the path decomposition Z in Lemma 7, the framework in Algorithm 1 can still be applied, but now certificates are defined in terms of paths in Z.

Definition 5 (Modified Certificate). *The following is a valid certificate for u with respect to the current A.*

- *Type 1. A path $Q \in Z$ such that that contains a true p-neighbor of u that does not belong to A.*

- **Type 2.** *Suppose no Type 1 certificate exists. In this case, a Type 2 certificate is a pair of 2 distinct paths in Z that contain two true p-neighbors of u in A which are incomparable with respect to $<_A$. Observe those two incomparable true p-neighbors must lie on different paths in Z, because any two vertices in a path are comparable.*

Modified Global Probe Charging. We follow the same definition of consistent probe and inconsistent edge probes. There can be at most w inconsistent probes, which can be absorbed into the term $O(w)$. Hence, again we can focus on bounding the number of consistent probes. Observe that modified certificates are defined with respect to the paths in the decomposition Z.

At any time, for each node u, there can be at most $O(w)$ candidates for Type-1 certificates and at most $O(w^2)$ candidates for Type-2 certificates. We will prove in the following that at most 1 consistent probe is needed in each attempt to find a Type-1 certificate. For each attempt at finding Type-2 certificate, we will show after charging $O(1)$ consistent probe to each inconsistent probe, the number of uncharged probes would be at most 2. Therefore, if there are X valid certificates used during the execution of the algorithm, then $O(X)$ uncharged probes would be consumed, the total number of probes would be bounded by $O(X + w)$. However, in order to show that the X is small with high probability, we also need to prove that we can sample from the collection of valid potential certificates uniformly at random so that we can apply the measure concentration argument in Lemma 5.

Lemma 8 (Finding Type-1 Certificate). *For any vertex $u \notin A$ such that $\widetilde{T}_u \setminus A$ is non-empty, either a Type-1 certificate for u can be sampled uniformly at random over all valid ones or we can conclude that no Type-1 certificates can be found for u (in which case all p-neighbors in $\widetilde{T}_u \setminus A$ turns out to be false).*

If a Type-1 certificate is found, at most 1 consistent probe will be made; otherwise no consistent probes will be made.

Proof. We consider the paths $Z_u \subseteq Z$ that contain at least 1 plausible p-neighbor in A. However, at the beginning, the algorithm does not know whether these p-neighbors in A are true or false.

In order to sample a valid Type-1 certificate uniformly at random, the algorithm picks such a path Q in Z uniformly at random and verifies whether Q is a valid Type-1 certificate. Starting from the largest plausible p-neighbor v (which changes a more information is revealed) of u that does not belong to A, we probe (v, u). If v is a true p-neighbor, then Q is a valid Type-1 certificate; otherwise, we consider the next largest plausible p-neighbor in Q that is not contained in A. Observe that if it turns out that there is no more plausible p-neighbors from Q that is not in A, then Q cannot be a valid Type-1 certificate; in this case, no consistent probe is made, and we can sample uniformly at random from the remaining possible paths in $Z_u \setminus \{Q\}$.

Hence, the above process is actually doing uniform sampling among all valid Type-1 certificates (if any). Moreover, 1 consistent probe is incurred *iff* a valid Type-1 certificate is found.

Lemma 9 (Finding Type-2 Certificate). *For any vertex $u \notin A$ such that all plausible p-neighbors $\widetilde{T}_u \subseteq A$, either a Type-2 certificate for u can be sampled uniformly at random over all valid ones or we can conclude that u is ideal.*

If x inconsistent probes are made in the process, then at most $2x+2$ consistent probes would be made. Hence, $2x$ of these consistent probes can be charged to the x inconsistent probes, which means that there are at most 2 uncharged consistent probes.

Proof. Let $Z_u \subseteq Z$ be the set of paths where each path has at least one plausible p-neighbor of u. Let $\mathcal{L}$ be a random permutation of unordered pairs $\{Q_1, Q_2\}$ of paths that contain at least two incomparable p-neighbors of u with respect to A (but we do not know yet whether they are true or false). Observe that all valid Type-2 certificates are included in $\mathcal{L}$ and the relative order among them is uniformly at random.

Each pair $\{Q_1, Q_2\}$ in $\mathcal{L}$ is verified as follows. For $i \in \{1, 2\}$, starting from the largest plausible p-neighbor of u in Q_i, all p-neighbors in Q_i can be revealed as soon as the first true p-neighbor is found. After all p-neighbors of Q_1 and Q_2 are revealed, the pair forms a Type-2 certificate if they contain two incomparable p-neighbors with respect to A. (However, this pair may become invalid later as more vertices are included into A.) Since we give higher priority to Type-1 certificates, if no Type-2 certificate can be found for u, it means that node u is ideal.

The above procedure samples a valid Type-2 certificate uniformly at random if there is any. It remains to analyze probe charging. Observe that when each pair $\{Q_1, Q_2\}$ is verified, there will be at most 2 consistent probes, each coming from one of the paths. If the pair turns out to be a valid Type-2 certificate, we can keep these 2 consistent probes uncharged. If the pair turns out to be invalid, it means that at least 1 inconsistent probe is encountered, because at the beginning the two pairs contain two incomparable plausible p-neighbors (at least one of which is revealed to be false during verification). Hence, the 2 consistent probes can be charged to at least 1 inconsistent probe.

As before, the order in which vertices are included into A is deterministic.

Remark 3 (Becoming Ideal Independent of Choices). Observe that according to the process described in Lemma 8 and Lemma 9, the order in which we consider paths and pairs of paths for finding certificates cannot affect the outcome of whether u becomes ideal. Hence, the outcome of whether u becomes ideal does not depend on the choices made in the process.

Remark 4 (Predecessor in Verified Directed Paths). For a directed path $Q \in Z$, if a vertex $u \in Q$ belongs to A, then all vertices that are smaller than u in Q are all in A.

Modified Algorithm. The modified algorithm has the same framework as Algorithm 1, except that it uses Lemma 8 and Lemma 9 successively to try finding a certificate for the considered vertex u.

Lemma 10 (Identifying an Ideal Vertex). *An ideal vertex can always be identified in $V \setminus A$ if $A \neq V$.*

Proof. Let the Hamiltonian path in $\vec{G}$ be $(v_1, \cdots, v_n)$, and k the smallest index s.t. $v_k \notin A$, then $S_{v_k} \subseteq \{v_1, \cdots, v_{k-1}\} \subseteq A$. Hence the partial order $<_A$ restricted to S_{v_k} is a total order. By Lemma 8 and Lemma 9, if we follow the process of finding certificates on v_k, then v_k would be identified as an ideal vertex since wrong p-neighbors would all be removed from $\widetilde{T_u}$.

Lemma 11 (Correctness of the Algorithm). *The algorithm can always proceed and would terminate in finitely many steps.*

Proof. Applying the strategy described in Lemma 8 and Lemma 9 for a vertex that is not in A, we can either find a certificate for it or it becomes ideal. Moreover, according to Lemma 10, an ideal vertex can always be identified. Specifically, we can sequentially identify all ideal vertices in increasing order with respect to the true global ordering $<_V$. This implies that the algorithm will always terminate. Also, note that there are finitely many potential certificates for each vertex, and an invalid certificate would not be valid again. Hence there will only be a finite number of steps.

We recall Theorem 1:

Theorem 1 (Main Theorem). *There is a polynomial-time Monte Carlo algorithm which solves the generalized sorting with prediction problem using $O(n \log w + w)$ probes with probability at least $1 - e^{-\Theta(n)}$.*

Proof. Lemma 11 has shown the correctness of the modified algorithm. We now bound the number probes used.

First of all, we can use Lemma 7 to find in polynomial time $O(w)$ verified directed paths in $\vec{G}_P$ which can cover all vertices in V with at most n probes.

Lemma 4 still holds for the modified algorithm by the same proof since whether a certificate for the considered vertex is found does not depend on the choices made in the process in Lemma 8 and Lemma 9 neither.

The probe charging analysis for the modified algorithm is exactly the same as in Lemma 3, i.e., the total number of probes made is at most $O(n + w + \sum_{u \in V} X_u)$, where X_u is the number of attempts to find a certificate for u. Note that at most 2 uncharged consistent probes are incurred every time there is an attempt to find a certificate. Moreover, observe that the algorithm does not know the true p-neighbors. Hence it needs to do uniform sampling among all paths that contain plausible p-neighbors for a Type-1 certificate and among all pairs of distinct paths that contain at least one plausible p-neighbor for a Type-2 certificate. However, Lemma 8 and Lemma 9 show that if any plausible p-neighbor turns to be false, either no consistent probes would be consumed or any consistent probes involved can be charged to inconsistent probes. Lemma 8 and Lemma 9 are essentially to sample uniformly over all remaining valid certificates. Therefore, each vertex has an initial pool of $O(w)$ candidates for Type-1 certificate and $O(w^2)$ for Type-2 certificate.

Note that Lemma 5 will also hold for the modified algorithm. We now have that the initial pools C_u of Type-1 and Type-2 certificates are of size $O(w)$ and $O(w^2)$ separately. Therefore, the total number of certificates sampled by all vertices, i.e., $\sum_{u \in V} X_u$, is at most $O(\sum_{u \in V}(\log |C_u| + 1)) = O(n \log w)$ with probability at least $1 - e^{-\Theta(n)}$.

Combining all these together, we have that the modified algorithm uses at most $O(n \log w + w)$ probes with probability at least $1 - e^{-\Theta(n)}$ and it runs in polynomial running time.

5   Conclusion

In conclusion, this paper presented a novel approach to address the generalized sorting with predictions problem. The proposed algorithm partitions vertices into verified directed paths, which in turn helps to reduce the number of probes required for each vertex. This leads to a polynomial-time Monte Carlo algorithm that allows for $O(n \log w + w)$ probes with high probability, thus outperforming previous solutions when w is small. However, it remains an open problem whether the bound presented in our algorithm is tight, and if it is possible to remove the additive term w altogether. These are interesting directions for further research in this field.

References

1. Alon, N., Blum, M., Fiat, A., Kannan, S., Naor, M., Ostrovsky, R.: Matching nuts and bolts. In: SODA, pp. 690–696. ACM/SIAM (1994)
2. Banerjee, I., Richards, D.S.: Sorting under forbidden comparisons. In: SWAT. LIPIcs, vol. 53, pp. 22:1–22:13. Schloss Dagstuhl - Leibniz-Zentrum für Informatik (2016)
3. Biswas, A., Jayapaul, V., Raman, V.: Improved bounds for poset sorting in the forbidden-comparison regime. In: Gaur, D., Narayanaswamy, N.S. (eds.) CALDAM 2017. LNCS, vol. 10156, pp. 50–59. Springer, Cham (2017). https://doi.org/10.1007/978-3-319-53007-9_5
4. Blanc, G., Lange, J., Tan, L.: Query strategies for priced information, revisited. In: SODA, pp. 1638–1650. SIAM (2021)
5. Charikar, M., Fagin, R., Guruswami, V., Kleinberg, J.M., Raghavan, P., Sahai, A.: Query strategies for priced information. J. Comput. Syst. Sci. 64(4), 785–819 (2002)
6. Gupta, A., Kumar, A.: Sorting and selection with structured costs. In: FOCS, pp. 416–425. IEEE Computer Society (2001)
7. Huang, Z., Kannan, S., Khanna, S.: Algorithms for the generalized sorting problem. In: FOCS, pp. 738–747. IEEE Computer Society (2011)
8. Kannan, S., Khanna, S.: Selection with monotone comparison cost. In: SODA, pp. 10–17. ACM/SIAM (2003)
9. Komlós, J., Ma, Y., Szemerédi, E.: Matching nuts and bolts in o(n log n) time. SIAM J. Discret. Math. 11(3), 347–372 (1998)
10. Kuszmaul, W., Narayanan, S.: Stochastic and worst-case generalized sorting revisited. In: FOCS, pp. 1056–1067. IEEE (2021)
11. Lu, P., Ren, X., Sun, E., Zhang, Y.: Generalized sorting with predictions. In: SOSA, pp. 111–117. SIAM (2021)

Eliciting Truthful Reports with Partial Signals in Repeated Games

Yutong Wu[1(✉)], Ali Khodabakhsh[1], Bo Li[2], Evdokia Nikolova[1],
and Emmanouil Pountourakis[3]

[1] The University of Texas at Austin, Austin, TX 78712, USA
{yutong.wu,ali.kh}@utexas.edu, nikolova@austin.utexas.edu
[2] The Hong Kong Polytechnic University, Hung Hom, Hong Kong, China
comp-bo.li@polyu.edu.hk
[3] Drexel University, Philadelphia, PA 19104, USA
manolis@drexel.edu

Abstract. We consider a repeated game where a player self-reports her usage of a service and is charged a payment accordingly by a center. The center observes a partial signal, representing part of the player's true consumption, which is generated from a publicly known distribution. The player can report any value that does not contradict the signal and the center issues a payment based on the reported information. Such problems find application in net metering billing in the electricity market, where a customer's actual consumption of the electricity network is masked and complete verification is impractical. When the underlying true value is relatively constant, we propose a penalty mechanism that elicits truthful self-reports. Namely, besides charging the player the reported value, the mechanism charges a penalty proportional to her inconsistent reports. We show how fear of uncertainty in the future incentivizes the player to be truthful today. For Bernoulli distributions, we give the complete analysis and optimal strategies given any penalty. Since complete truthfulness is not possible for continuous distributions, we give approximate truthful results by a reduction from Bernoulli distributions. We also extend our mechanism to a multi-player cost-sharing setting and give equilibrium results.

Keywords: Energy·economics · Equilibrium analysis · Repeated games · Electricity market · Penalty mechanisms

1 Introduction

Consider the following repeated game where a center owns resources and one or more strategic players pay the center to consume the resources. In every round, a player self-reports their usage, which will then be used to determine their payment to the center. However, it is not always possible for the center to verify the submitted information from the players. Instead, only part of the actual consumption is revealed to the center based on some publicly known distribution.

© Springer Nature Switzerland AG 2023
M. Li et al. (Eds.): IJTCS-FAW 2023, LNCS 13933, pp. 42–57, 2023.
https://doi.org/10.1007/978-3-031-39344-0_4

A player can report any value that is at least the revealed amount. Without any external interference, a player will naturally report exactly the revealed amount (potentially lower than the true consumption) to minimize their payment. The center then needs to determine a payment mechanism such that each player is incentivized to report their true value.

The electricity market is facing precisely the described problem. As the number of electricity *prosumers* increases each year, new rate structures are designed to properly calculate the electricity bill for this special type of consumer while ensuring that every customer is still paying their fair share of the network costs. Prosumers are those who not only consume energy but also produce electricity via distributed energy resources such as rooftop solar panels. Among different rate structures, *net metering* is a popular billing mechanism that is currently adopted in more than 40 states in the US [22]. Net metering charges prosumers a payment proportional to their net consumption, i.e., gross consumption minus the production [26], demonstrated in Fig. 1. The payment includes the electricity usage as well as grid costs that are incurred by using the electricity network.

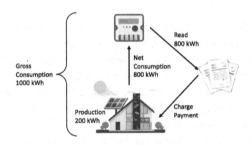

Fig. 1. Net metering for electricity prosumers.

The controversy in net metering lies in that prosumers fail to pay their share of the grid costs when they do not have local storage equipment [9]. In the United States, only 4% of the solar panel owners also own a battery to store the produced solar energy [18]. For those who do not own battery storage, the generated power has to be transmitted back to the grid. Accordingly, the daily consumption of power by these prosumers also needs to come from the grid instead of directly from the solar panels. In this way, most prosumers have under-paid their share of the network costs and become "free-riders" of the electricity grid. The grid is often subject to costly line upgrades and net metering unevenly shifts such costs to traditional consumers, who usually come from lower-income households [15]. Indeed, previous research works have suggested that prosumers should pay a part of the grid costs proportionally to their gross consumption, not net consumption [9,16]. However, the gross consumption is hidden from the utility companies since only net consumption can be observed from the meter. Meanwhile, there is no incentive for prosumers to voluntarily report their true consumption as it will only increase their electricity bills.

Fortunately, the production from solar panels usually follows some pattern while the gross consumption of electricity for a typical household stays relatively constant, which is especially true for industrial sites – the major consumers for utilities [21,27]. Thus, the observed consumption can be assumed to follow some natural distribution and the center is able to detect dishonesty when a player's report differs from their reporting history. With this idea, we propose a simple penalty mechanism, the *flux mechanism*, that elicits truthful reports from players in a repeated game setting when only partial verification is possible. Particularly, a player is charged their reported value as well as a penalty due to inconsistency in consecutive reports in each round. The main goal is to ensure that every player reports their true values and no penalty payment is collected. We show that the combining effect of (i) the penalty rate and (ii) the length of the game is sufficient for inducing truthful behavior from the player for the entire game. As the horizon of the game increases, the minimum penalty rate for truth-telling as an optimal strategy decreases. In other words, it is the fear of uncertainty in the future that incentivizes the player to be truthful today.

1.1 Our Contribution

We address the problem of eliciting truthful reports when the center is able to observe a part of the player's private value based on some publicly known distribution. The strategic player reports some value that is at least the publicly revealed value and is charged a payment accordingly. We propose a truth-eliciting mechanism, *flux mechanism*, that utilizes the player's fear of uncertainties to achieve truthfulness. In each round, the player is charged a "regular payment" proportional to the consumption they report. Starting from the second round, the player is charged an additional "penalty payment", which is r times the (absolute) difference between the reports in the current and the previous round, where the penalty rate r is set by the center before the game starts.

Intuitively, a player can save their regular payment by under-reporting their consumption, but they will then face the uncertainty of paying penalties in future rounds due to inconsistent reports. Under most settings, if r is set to be infinitely high, the players will be completely truthful to avoid any penalty payment. However, a severe punishment rule is undesirable and discourages players from participating. Therefore, we want to understand the following question.

What is the minimum penalty rate such that the player is willing to report their true value?

We observe that no finite penalty can achieve complete truthfulness for arbitrary distributions as a player's true consumption may never be revealed exactly. We can, however, obtain approximate truthfulness for a general distribution by analyzing complete truthfulness for a corresponding Bernoulli distribution. For $Ber(p)$, the partial signal equals the true consumption with probability p and 0 with probability $1-p$ for $p \in (0,1)$. We give results for Bernoulli distributions in Main Results 1 and 2. For arbitrary distributions, we redefine p as the probability of having a partial signal that is at least α times the true consumption, for $\alpha \in [0,1]$, to obtain α-truthfulness (Main Result 3).

Main Result 1 (Theorem 1). *For a T-round game with Bernoulli distribution Ber(p), the player is completely truthful if and only if the penalty rate is at least*

$$\frac{1-(1-p)^T}{p-p(1-p)^{T-1}}.$$

Main Result 1 gives the minimum penalty rate that guarantees complete truthfulness for $Ber(p)$ distributions. We also want to understand how players would behave if the penalty rate is not as high, which describes the situation when the center is willing to sacrifice some degree of truthfulness by lowering the penalty rate. Given any penalty rate, we show that a player's optimal strategies can be described as one or a combination of three basic strategies, *lying-till-end*, *lying-till-busted* and *honest-till-end*. Specifically, with a low penalty rate, the player is always untruthful to save regular payment, i.e., lying-till-end is optimal. As the penalty rate increases, the player's optimal strategy *gradually* moves to lying-till-busted, which is to be untruthful until the partial signal is revealed as the true consumption for the first time and then stays truthful for the rest of the game. When the penalty rate is sufficiently high, the player would avoid lying completely and reports the truth, i.e., she is honest-till-end.

Table 1. Optimal strategy given penalty rate r under $Ber(p)$ distributions

Bernoulli Prob.	Penalty Rate	Optimal Strategy
$p \geq 0.5$	$r \leq \frac{1}{2p}$	lying-till-end
	$\frac{1}{2p} < r \leq 1$	lying-till-busted + lying last round
	$1 < r < \frac{1-(1-p)^T}{p-p(1-p)^{T-1}}$	lying-till-busted
	$r \geq \frac{1-(1-p)^T}{p-p(1-p)^{T-1}}$	honest-till-end
$p < 0.5$	$r \leq 1$	lying-till-end
	$h(t-1) < r \leq h(t)$	lying-till-end first t rounds + lying-till-busted for rest
	$h(T-1) < r < \frac{1-(1-p)^T}{p-p(1-p)^{T-1}}$	lying-till-busted
	$r \geq \frac{1-(1-p)^T}{p-p(1-p)^{T-1}}$	honest-till-end

Main Result 2 (Theorems 1 and 2). *For a T-round game with Bernoulli distribution Ber(p), given any penalty rate r, the player's optimal strategy is summarized in Table 1, where*

$$h(t) = \frac{1-(1-p)^t}{2p-p(1-p)^{t-1}}, for 1 \leq t \leq T.$$

For arbitrary distributions, including uniform distributions, it is impossible to obtain complete truthfulness without setting the penalty to infinity. Main Result 3 gives a reduction from Bernoulli distributions to general distributions for approximate truthfulness.

Main Result 3 (Theorem 3). *Given $\alpha \in [0,1]$ and an arbitrary distribution with CDF F, if a penalty rate r achieves complete truthfulness for $Ber(p)$ where $p = 1 - F(\alpha D)$ and D is the player's true gross consumption, then the same r achieves α-approximate truthfulness for distribution F.*

Finally, we extend our results to multiple players. We note that if the players are charged independently, applying the flux mechanism to each individual elicits truthful reports. A more complicated and realistic setting is the cost-sharing problem where the players split an overhead cost based on their submitted reports. We propose the *multi-player flux mechanism* where the penalty payment is the same as before but the regular payment is now a share of some overhead cost. Again, if the penalty rate is sufficiently high, the players stay truthful, regardless of others' behavior, to avoid any penalty payment, i.e., the truthful report profile forms a *dominant strategy equilibrium*. As the penalty rate decreases, the truthfulness of a player may depend on other players' actions. That is, with a lower penalty rate, a truthful report profile forms a *Nash equilibrium*. For both equilibrium definitions, we are interested in the following question.

What is the minimum penalty rate for the truthful report profile to form a dominant strategy or Nash equilibrium?

We give exact penalty thresholds for both truthful equilibria under Bernoulli distributions and use a reduction to obtain approximate results under arbitrary distributions in Main Result 4.

Main Result 4 (Theorems 4, 5, 6 and 7). *For any T-round game with distribution $Ber(p)$, a truthful strategy profile is a dominant strategy equilibrium if and only if*

$$r \geq \frac{C}{nD} \frac{1 - (1-p)^{n-1}}{p} \frac{1 - (1-p)^T}{p - p(1-p)^{T-1}},$$

and a Nash equilibrium if and only if

$$r \geq \frac{C}{nD} \frac{1 - (1-p)^T}{p - p(1-p)^{T-1}}.$$

Given $\alpha \in [0,1]$ and any distribution with cumulative distribution function F, let $p = 1 - F(\alpha D)$, where D is the true gross consumption. Then α-approximate truthful profile is a Nash equilibrium if

$$r \geq \frac{1}{\alpha} \frac{C}{nD} \frac{1 - (1-p)^T}{p - p(1-p)^{T-1}},$$

and the α-approximate truthful profile is a dominant strategy equilibrium if

$$r \geq \frac{1}{\alpha} \frac{C}{nD} \frac{1 - (1-p)^n}{p} \frac{1 - (1-p)^T}{p - p(1-p)^{T-1}}.$$

1.2 Related Works

The economic effect of the net metering policy has been explored for different countries and regions [6,7,17,24,28]. It has been observed that net metering can cause inequality issues for traditional energy consumers [8,15,17,23]. Accordingly, alternative pricing mechanisms and tariff structures have been proposed to fairly compensate the energy production [4,9,10,16,25]. In particular, Gautier et al. [9] and Khodabakhsh et al. [16] proposed that individuals should be charged based on their true consumption, not net consumption. Our work is a continuation of [16], where a primitive version of the penalty mechanism is first proposed for promoting a fairer electricity rate structure. We formally define the mechanism and provide the corresponding theoretical analysis.

More broadly, fairness for the power grid has become an increasingly popular subject. First, Heylen et al. provided various indices to measure fairness and inequality in power system reliability [13]. Fairness is also explored for load shedding plans [14], electric vehicle charging schemes [1], demand response [2], etc. Moret and Pinson showed fairness can be improved with a "community-based electricity market", where prosumers are allowed to share their production on the community level [20]. Our model, on the other hand, addresses the fairness issues by modifying the current electricity structure, which is easier for utility companies to adopt.

Theoretically, our work is related to information elicitation with limited verification ability. Caragiannis et al. [5] and Ball et al. [3] worked on probabilistic verification where a lying player may be caught by a probability based on her type. Our work can be viewed as an extension of probabilistic verification under a repeated game setting where the verification is implicit and the main goal is to incentivize truthful reports. Another related problem is strategic classification, in which individuals can manipulate their input to obtain a better classification outcome [11,12,19,30]. Although we also consider the strategic behavior of players in a sequential game, our model is quite different. Strategic classification allows the center to learn the patterns of the players via a private classifier. In our setting, the scoring rule is transparent to the players and an additional measure (e.g., a penalty) has to be invoked to incentivize truthfulness.

2 Problem Statement

We formally define our problem under the single player setting and defer the extension to multiple players to Sect. 5. The player has a gross consumption $D \geq 0$, which is her private information. The game has T rounds where $T > 1$ as otherwise the flux mechanism becomes invalid. In each round t, the center observes a partial signal, $y_t \leq D$, which is randomly and independently drawn from a distribution F supported on $[0, D]$. We use $r \geq 0$ to denote the penalty rate. In a flux mechanism, a player cares more about the number of rounds left in the future rather than the number of rounds has passed. Thus we use $t = T, T - 1, \cdots, 1$ to denote the current round, where t means there are t rounds left, including the current round. For example, the first round is round

T, the last round is round 1, and the previous round of round t is round $t + 1$. For round $t \leq T$, the flux mechanism runs as follows.

- The center observes the player' net consumption $y_t \sim F$.
- The player submits their reported gross consumption which is at least the net consumption, $b_t \geq y_t$. The player may not be truthful, i.e., $b_t \leq D$.
- When $t < T$, the player's payment consists of regular payment b_t and penalty payment $r \cdot |b_{t+1} - b_t|$. When $t = T$, the player only pays the regular payment.

For $t < T$, we call b_{t+1} the *history* of round t. In each round t, the player wants to pay the lowest expected total payment by reporting b_t without knowing the partial signals for future rounds. We call a mechanism *truthful* if the player reports D for all rounds. When two reports bring the same expected payment, we break tie in favor of truthfulness. We adopt the assumption from Khodabakhsh *et al.* [16] that D does not vary with t. We explain in the full version of this paper an easy extension where D_t is drawn from a known range $[\underline{D}, \overline{D}]$.

3   Bernoulli Distributions

We start with the analysis of Bernoulli distribution as we show later a reduction from an arbitrary distribution to a Bernoulli distribution. We prove it is only optimal for a player to report zero or their true consumption in each round. The optimal strategies can then be characterized by three basic strategies (Definition 1). The penalty thresholds are computed by comparing the different combinations of the basic strategies. Due to space limit, we defer most proofs to the full version of this paper and focus on explaining the intuition in this section.

3.1   Basic Strategies

In a Bernoulli distribution setting, in each round t, the partial signal y_t is D with probability p and 0 with probability $1 - p$. When the partial signal equals to the private value, i.e., $y_t = D$, we say that the player is *"busted"* in round t. We first define three basic strategies.

Definition 1 (Basic Strategies). *For Bernoulli distributed net consumption $y_t \sim Ber(p)$, we define the following as the three basic strategies:*

- *lying-till-end: Report $b_t = 0$ when $y_t = 0$ and $b_t = D$ otherwise;*
- *lying-till-busted: Report $b_t = 0$ until $y_t = D$ for the first time, then report D for all future rounds;*
- *honest-till-end: Report $b_t = D$ for all rounds.*

We note that a player's optimal strategy for a given penalty rate r can be solved by backward induction. Let $OptCost(t, r, b_{t+1})$ denote the optimal expected cost for a player starting in round t with penalty rate r and report b_{t+1} for the previous round. Then

$$OptCost(t, r, b_{t+1}) = \min_{b_t} ExpCost(t, r, b_{t+1}, b_t),$$

where $ExpCost(t, r, b_{t+1}, b_t)$ is the expected cost for the player starting in round t and reporting b_t (if she is allowed to), with penalty rate r and history b_{t+1},

$$
\begin{aligned}
&ExpCost(t, r, b_{t+1}, b_t) \\
&= \mathbb{E}_{y_t}[\max\{y_t, b_t\} + r|\max\{y_t, b_t\} - b_{t+1}| + OptCost(t-1, r, \max\{y_t, b_t\})] \\
&= p(D + r(D - b_{t+1}) + OptCost(t-1, r, D)) \\
&\qquad + (1-p)(b_t + r|b_t - b_{t+1}| + OptCost(t-1, r, b_t)).
\end{aligned}
$$

The first term on the right side of the equation above refers to the cost when the partial signal is revealed as D and the player has to report D. The second term refers to the cost when the partial signal is 0 and the player chooses to report b_t. Let $OptCost(0, r, b_1) = 0$ for all b_1. When $t = T$, i.e., the first round, there is no history b_{T+1} and the player wants to minimize the following total cost,

$$
\begin{aligned}
OptCost(T, r) &= \min_{b_T} ExpCost(T, r, b_T) \\
&= p(D + OptCost(t-1, r, D)) + (1-p)(b_T + OptCost(t-1, r, b_T)).
\end{aligned}
$$

Solving the recursion will give the characterization of optimal strategies in Table 1, as we demonstrate in the full version of this paper. In what follows, we discuss a surprisingly simpler and more constructive proof by exploiting the properties of the flux mechanism, which may be of independent interest.

3.2 Main Theorems

We observe that there are two key elements that influence the decision making of the player.

(1) The player's history, b_{t+1} for $t < T$. The value of b_{t+1} directly affects the penalty payment in round t. Intuitively, a player is more reluctant to lie if b_{t+1} is high and better off lying if b_{t+1} is small.
(2) The number of rounds left to play, i.e., t, indirectly influences the probability and the number of times a player will be busted in the remaining rounds.

Via Lemmas 1–4, we show these are the *only two* elements that determine a rational player's action. The following lemma shows that it is not optimal for a player to report a value strictly between 0 and D. Moreover, if a player is untruthful in the previous round, it is better off to remain untruthful. With this lemma, we largely reduce the strategy space we need to consider.

Lemma 1. *For any round $t \leq T$, given $y_t = 0$, the optimal report in round t is $b_t \in \{0, D\}$. Moreover, if $t < T$ and $b_{t+1} = y_t = 0$, then the optimal report is $b_t = 0$.*

Next, we prove that in each round, the optimal strategy is determined by a penalty threshold such that a player will be truthful if and only if the penalty rate r is above the threshold. We call them *critical thresholds*.

Lemma 2 (Critical Thresholds). *For $t = T$, there is a threshold penalty rate $r_T^{(\emptyset)} \geq 0$ such that reporting D is optimal if and only if the penalty rate is at least $r_T^{(\emptyset)}$; For $t < T$, there is a threshold penalty rate $r_t^{(b_{t+1})} \geq 0$ such that reporting D is optimal for a player in round t with history b_{t+1} if and only if the penalty rate is at least $r_t^{(b_{t+1})}$.*

Lemmas 1 and 2 together imply that the optimal strategy can only be one or a combination of the basic strategies. In particular, by Lemma 1, $r_t^{(0)} = \infty$ for any t. Moreover, since b_{t+1} can only be 0 or D, by Lemma 2, we only need to determine the values of $r_T^{(\emptyset)}$ and $r_t^{(D)}$ for $t < T$ to complete the picture of optimal strategies. We now give some properties of these thresholds.

Lemma 3. $r_t^{(\emptyset)} \geq r_t^{(D)}$ *for $t \in \{1, \ldots, T\}$.*

Given the same t rounds left, Lemma 3 says a player is more inclined to lie without a history than with a truthful history. This is straightforward as lying with a truthful history results in an additional penalty payment.

Lemma 4. *Given $r_t^{(\emptyset)} \geq \frac{1}{p}$, $r_t^{(\emptyset)}$ decreases as t increases.*

Lemmas 3 and 4 together tell us the player is least incentivized to be truthful on the first round and $r_T^{(\emptyset)}$ is the penalty threshold that ensures truthfulness for the game. We give this important threshold in Theorem 1.

Theorem 1. *The minimum penalty for truthful reporting in a game of T rounds with $Ber(p)$ distribution is*

$$r_T^{(\emptyset)} = \frac{1 - (1 - p)^T}{p - p(1 - p)^{T-1}}. \tag{1}$$

We see $r_T^{(\emptyset)} \to 1/p$ as $T \to \infty$ and $r_T^{(\emptyset)}$ decreases as T increases. This implies the increasing length of the game incentivizes the player to speak the truth today, even when they do not have to. To understand Theorem 1, we observe that it is sufficient to compare lying-till-busted and honest-till-end since $r_T^{(\emptyset)}$ ensures the player to stay truthful after being busted. Before the player is busted for the first time, it is not optimal to oscillate between lying and truth-telling, as it is strictly dominated by lying completely. Therefore, the only viable strategies are lying-till-busted and honest-till-end, and the desired threshold sets the expected cost of these two strategies equal.

With a more involved argument, we get the exact values for the truthful threshold given a truthful history, i.e., the $r_t^{(D)}$'s. The values of $r_T^{(\emptyset)}$ and $r_t^{(D)}$ characterize the optimal strategies for a player and are an alternative representation of Table 1.

Theorem 2. *For $p \leq \frac{1}{2}$, $r_t^{(D)} = \frac{1-(1-p)^t}{2p-p(1-p)^{t-1}}$. For $p > \frac{1}{2}$, $r_t^{(D)} = 1$ for $t = 1$ and $r_t^{(D)} = \frac{1}{2p}$ for $t \geq 2$.*

The optimal strategy is visualized in Figs. 2a and 2b for $p = 0.3$ and $p = 0.7$, respectively. The x-axis is the number of rounds left (t), and the y-axis is the penalty thresholds for truthfulness. We give examples of penalties via the red dashed lines. For the first round, the player refers to the blue dot representing $r_T^{(\emptyset)}$ and is truthful if and only if the penalty is above the blue dot. Afterwards, given t rounds left and history D, the player looks at the green curve representing $r_t^{(D)}$ and is only truthful if the penalty is above the curve. If the history is 0, she remains untruthful and reports 0. Figures 2a and 2b visualize the optimal strategies given in Table 1. Both green curves are closely related to $\frac{1}{2p}$. An intuition is that in any round $t < T$, a player pays D if she is truthful and roughly $2prD$ if she lies, where the penalty payment rD comes from the previous and the next round, each with probability p. The penalty that sets these two costs equal is $\frac{1}{2p}$. The actual $r_t^{(D)}$ thresholds vary upon values of t and p.

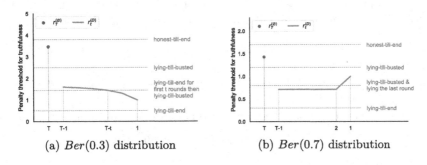

(a) $Ber(0.3)$ distribution   (b) $Ber(0.7)$ distribution

Fig. 2. Critical thresholds under two distributions with sample optimal strategies.

4   A Reduction for Arbitrary Distributions

As discussed in the introduction, only the infinite penalty rate will guarantee complete truthfulness under arbitrary distributions, yet there is still hope to obtain approximate results. The trick is to redefine being busted as having a partial signal that is less than α times the true consumption, for $\alpha \in [0,1]$. Then any arbitrary distribution is reduced to $Ber(p)$ where p is the probability that the partial signal is at least αD. For approximate truthfulness, we define being α-truthful as reporting at least αD. We reuse the arguments of comparing basic strategies from Sect. 3 to determine an upper bound for the penalty rate that guarantees α-truthfulness. We introduce the notion of approximate truthfulness in Definition 2 and give the reduction in Theorem 3. We demonstrate the reduction with uniform distributions in Example 1.

Definition 2 (α-truthfulness). *A reporting $\boldsymbol{b}$ is α-truthful when $b_t \geq \alpha D$ for all $t = 1, \ldots, T$.*

Theorem 3. *Given $\alpha \in [0,1]$ and an arbitrary distribution with CDF F, if a penalty rate r achieves complete truthfulness for $Ber(p)$ where $p = 1 - F(\alpha D)$, then the same r achieves α-approximate truthfulness for distribution F.*

Example 1. Assume partial signals follow a uniform distribution $U(0, D)$. Let r be the truthful threshold of $Ber(p)$ where $p = 1 - \alpha$, i.e. $r = \frac{1-\alpha^T}{(1-\alpha)(1-\alpha^{T-1})}$. Then using r ensures α-truthfulness for $U(0, D)$ by Theorem 3. For uniform distributions, it is impossible to obtain complete truthfulness unless $r = \infty$, which can be verified by setting $\alpha = 1$.

5   Extension: A Cost-Sharing Model

We extend the problem to the multi-player setting and focus on cost-sharing among homogeneous players. Let N be the set of players with $n = |N| \geq 1$. Each player $i \in N$ has a private value $x^i \geq 0$, and we assume all players are symmetric, i.e., $x^i = D$ for all $i \in N$ (see the full version of this paper for a relaxation). All players in N split an overhead cost C, which is at least the total gross consumption, i.e., $C \geq nD$. The game has T rounds in total. Given the penalty rate r, we analyze the following multi-player flux mechanism.

- The center observes a partial signal representing player i's net consumption $y_t^i \sim F$ for each player $i \in N$;
- Each player i submits their reported gross consumption that is at least their net consumption, $b_t^i \geq y_t^i$;
- If $t < T$, player i's pays regular payment $C \cdot \frac{b_t^i}{\sum_j b_t^j}$ and penalty payment $r \cdot |b_{t+1}^i - b_t^i|$. If $t = T$, the players only pay regular payments.

We call b_{t+1}^i the *history* for player i in round t and $\boldsymbol{b}_{t+1}$ the *group history*. If everyone lies in a round, the overhead cost is split evenly among all players. A mechanism is *truthful* if every player reports D for every round. We are interested in computing the minimum penalty rates such that truthful reports form a Nash equilibrium (NE) or a dominant strategy equilibrium (DSE). Informally, a strategy profile is a NE if no player wants to unilaterally deviate, and it is a DSE if no player wants to deviate no matter what the other players do. We show that approximate results for any arbitrary distribution can be deducted from an exact analysis for a Bernoulli distribution. Due to space limit, we defer all proofs to the full version of this paper.

Similar to the single-player setting, we avoid solving the recursion by exploiting the properties of the mechanism. Again, we start our analysis with F being a Bernoulli distribution and provide a reduction for approximate truthfulness when F is an arbitrary distribution. In the single-player model with Bernoulli-distributed F, we have shown that it is only optimal for a player to report 0 or her actual consumption D. We claim it is the same case for multiple players. Moreover, if a player lied yesterday and also has an observed consumption of 0 today, they will report 0 regardless of other players' actions.

Lemma 5. *For Bernoulli-distributed F, reporting anything strictly between 0 and D is sub-optimal in a multi-player flux mechanism. Moreover, if $b_{t+1}^i = y_t^i = 0$, it is optimal to report $b_t^i = 0$.*

Starting from this point, we assume that every player reports either 0 or D. When $n = 2$, we show that the multi-player model reduces to the single-player model with a multiplicative factor of $\frac{C}{2D}$. The reason for the reduction is that the savings of switching to lying from being truthful for a player are always $\frac{C}{2}$, regardless of what the other player does.

Lemma 6. *When $n = 2$, the multi-player model reduces to a single-player model. The truthful penalty threshold is $\frac{C}{2D}$ times (1).*

For general n, we show it is sufficient to analyze the maximum difference between lying and truth-telling for player i in round t given group history $\boldsymbol{b}_{t+1}$. In a DSE, a player achieves the biggest gain from lying if all players were lying in the previous round. We then use $\boldsymbol{b}_{t+1} = \boldsymbol{0}$ to compare lying and truth-telling for a player.

Theorem 4. *For the $Ber(p)$ distribution, a truthful strategy profile forms a dominant strategy equilibrium if and only if*

$$r \geq \frac{C}{nD} \frac{1 - (1-p)^{n-1}}{p} \frac{1 - (1-p)^T}{p - p(1-p)^{T-1}}. \tag{2}$$

If we slowly lower the penalty from (2), we will hit a threshold such that truth-telling is an NE. The difference between the truthful NE and the DSE is that now we can assume that every player $j \neq i$ is truthful in the first round and show that player i would not deviate unilaterally. However, we shall not assume that player $j \neq i$ remains truthful for the rest of the game. This is because if player i lies in the first round, player j can observe the report of i in the second round and deviate from truthful behavior. We first show that if $r \geq \frac{C}{nD}\frac{1}{p}$, players with truthful history stay truthful. Then we can safely assume player $j \neq i$ remains truthful throughout the game. In this way, truthful NE is reduced to the case where there is one strategic player and $n-1$ truthful players. It is not hard to see the threshold is precisely $\frac{C}{nD}\frac{1-(1-p)^T}{p-p(1-p)^{T-1}}$.

Theorem 5. *For the $Ber(p)$ distribution, a truthful strategy profile forms a Nash equilibrium if and only if*

$$r \geq \frac{C}{nD} \frac{1 - (1-p)^T}{p - p(1-p)^{T-1}}. \tag{3}$$

We visualize $Ber(p)$ penalty thresholds in Fig. 3 for different T's and p's. The x-axis is the total number of rounds for a game and the y-axis is the penalty rate that guarantees the specified equilibrium. The blue and orange lines are penalty thresholds for $p = \frac{1}{3}$ and $\frac{2}{3}$, respectively. The solid and dashed lines are thresholds for truthful DSE and NE, respectively. All four thresholds in Fig. 3 decrease as T increases, suggesting that the increasing length of the game promotes truthful equilibria. From expressions (2) and (3), we see that the DSE and NE thresholds tend to be the same as p approaches 1.

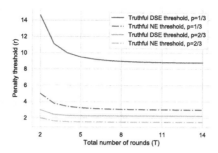

Fig. 3. Exact penalty thresholds for truthful DSE and NE, given the number of rounds T for $Ber(p)$ distributions. We assume $n = 20$, $D = 1$ and $C = n \cdot D = 20$.

Similar to the single-player model, we extend the results for Bernoulli distributions to approximate results for general distributions. Given $\alpha \in [0,1]$, we redefine being busted as having an observed consumption of at least αD. For the dominant strategy equilibrium, we find the threshold that being α-truthful is a dominant strategy. For Nash equilibrium, we first define the approximate truthful NE, a natural extension of the complete truthful NE.

Theorem 6. *Given $\alpha \in [0,1]$ and some general distribution F, let $p = 1 - F(\alpha D)$. The α-truthful strategy profile forms a dominant strategy equilibrium if*

$$r \geq \frac{1}{\alpha} \frac{C}{nD} \frac{1 - (1-p)^n}{p} \frac{1 - (1-p)^T}{p - p(1-p)^{T-1}}. \tag{4}$$

Definition 3 (α-truthful Nash equilibrium). *Given $\alpha \in [0,1]$, a reporting profile $b \in [0, D]^{n \times T}$ is an α-truthful Nash equilibrium if $b_t^i \geq \alpha D$ for all i, t and no player wants to deviate from being α-truthful in any round.*

Theorem 7. *Given $\alpha \in [0,1]$ and some general distribution F, let $p = 1 - F(\alpha D)$. The α-truthful strategy profile forms a Nash equilibrium if*

$$r \geq \frac{1}{\alpha} \frac{C}{nD} \frac{1 - (1-p)^T}{p - p(1-p)^{T-1}}. \tag{5}$$

We see that both the penalty thresholds, (4) and (5) are close to $\frac{1}{\alpha}$ times their Bernoulli thresholds, (2) and (3), for $p = 1 - F(\alpha D)$. Recall that in the single-player model, α-truthfulness can be obtained by directly using the Bernoulli threshold with $p = 1 - F(\alpha D)$. In the multi-player model, however, we have to multiply the Bernoulli threshold with a factor of $\frac{1}{\alpha}$, which suggests it is more difficult to get every player to speak the truth under the cost-sharing setting. We note that both penalty rates (4) and (5) are upper bounds for the actual thresholds. This is because we treat any report greater than αD as αD. We conjecture that the exact thresholds are not far from thresholds (4) and (5).

6 Conclusion and Open Problems

We propose a penalty mechanism for eliciting truthful self-reports when only partial signals are revealed in a repeated game. A player faces trade-off between under-reporting today and paying a penalty in the future due to the uncertainty of partial signals. We find that the length of the game naturally reduces the minimum penalty rate that incentivizes truth-telling. Given any penalty rate, we give a characterization of the optimal strategies under both single- and multiple-player settings for any distribution. We identify a penalty rate that achieves complete truthfulness for Bernoulli distributions, which can be used in a reduction to obtain approximate truthfulness for arbitrary distributions.

A possible future direction is to extend our results to asymmetric multi-player settings where players do not have the same gross consumption or the same distribution for partial signals. For heterogeneous players, we may then consider, in addition to truthfulness, the fairness of the mechanism. It would be interesting to develop a definition of fairness for the cost sharing model and compute the fairness ratios accordingly. It is also worthwhile to derive other truthful and fair mechanisms that do not involve penalty.

Acknowledgement. The full version of this paper can be found on arXiv [29]. The authors would like to thank the anonymous reviewers for their insightful and constructive comments. Part of this work was done when Bo Li was affiliated with the Department of Electrical and Computer Engineering at the University of Texas at Austin.

References

1. Aswantara, I.K.A., Ko, K.S., Sung, D.K.: A centralized EV charging scheme based on user satisfaction fairness and cost. In: 2013 IEEE Innovative Smart Grid Technologies-Asia (ISGT Asia), pp. 1–4. IEEE (2013)
2. Baharlouei, Z., Hashemi, M., Narimani, H., Mohsenian-Rad, H.: Achieving optimality and fairness in autonomous demand response: benchmarks and billing mechanisms. IEEE Trans. Smart Grid 4(2), 968–975 (2013)
3. Ball, I., Kattwinkel, D.: Probabilistic verification in mechanism design. In: Proceedings of the 2019 ACM Conference on Economics and Computation, pp. 389–390 (2019)
4. Burger, S., Schneider, I., Botterud, A., Pérez-Arriaga, I.: Fair, equitable, and efficient tariffs in the presence of distributed energy resources. Consumer, Prosumer, Prosumager: How Service Innovations will Disrupt the Utility Business Model, p. 155 (2019)
5. Caragiannis, I., Elkind, E., Szegedy, M., Yu, L.: Mechanism design: from partial to probabilistic verification. In: Proceedings of the 13th ACM Conference on Electronic Commerce, pp. 266–283 (2012)
6. Darghouth, N.R., Barbose, G., Wiser, R.: The impact of rate design and net metering on the bill savings from distributed PV for residential customers in California. Energy Policy 39(9), 5243–5253 (2011)
7. Dufo-López, R., Bernal-Agustín, J.L.: A comparative assessment of net metering and net billing policies. Study cases for Spain. Energy 84, 684–694 (2015)

8. Eid, C., Guillén, J.R., Marín, P.F., Hakvoort, R.: The economic effect of electricity net-metering with solar PV: consequences for network cost recovery, cross subsidies and policy objectives. Energy Policy **75**, 244–254 (2014)

9. Gautier, A., Jacqmin, J., Poudou, J.-C.: The prosumers and the grid. J. Regul. Econ. **53**(1), 100–126 (2018). https://doi.org/10.1007/s11149-018-9350-5

10. Glass, E., Glass, V.: Power to the prosumer: a transformative utility rate reform proposal that is fair and efficient. Electr. J. **34**(9), 107023 (2021)

11. Haghtalab, N., Immorlica, N., Lucier, B., Wang, J.Z.: Maximizing welfare with incentive-aware evaluation mechanisms. arXiv preprint arXiv:2011.01956 (2020)

12. Hardt, M., Megiddo, N., Papadimitriou, C., Wootters, M.: Strategic classification. In: Proceedings of the 2016 ACM Conference on Innovations in Theoretical Computer Science, pp. 111–122 (2016)

13. Heylen, E., Ovaere, M., Proost, S., Deconinck, G., Van Hertem, D.: Fairness and inequality in power system reliability: summarizing indices. Electr. Power Syst. Res. **168**, 313–323 (2019)

14. Heylen, E., Ovaere, M., Van Hertem, D., Deconinck, G.: Fairness of power system load-shedding plans. In: 2018 IEEE International Conference on Systems, Man, and Cybernetics (SMC), pp. 1404–1409. IEEE (2018)

15. Hoarau, Q., Perez, Y.: Network tariff design with prosumers and electromobility: who wins, who loses? Energy Econ. **83**, 26–39 (2019)

16. Khodabakhsh, A., Horn, J., Nikolova, E., Pountourakis, E.: Prosumer pricing, incentives and fairness. In: Proceedings of the Tenth ACM International Conference on Future Energy Systems, pp. 116–120 (2019)

17. Koumparou, I., Christoforidis, G.C., Efthymiou, V., Papagiannis, G.K., Georghiou, G.E.: Configuring residential PV net-metering policies–a focus on the Mediterranean region. Renew. Energy **113**, 795–812 (2017)

18. Leavitt, L.: Solar batteries: how renewable battery backups work (2021). https://www.cnet.com/home/energy-and-utilities/solar-batteries-how-renewable-battery-backups-work/. Accessed 13 Aug 2022

19. Liang, A., Madsen, E.: Data and incentives. In: Proceedings of the 21st ACM Conference on Economics and Computation, pp. 41–42 (2020)

20. Moret, F., Pinson, P.: Energy collectives: a community and fairness based approach to future electricity markets. IEEE Trans. Power Syst. **34**(5), 3994–4004 (2018)

21. Nadel, S., Young, R.: Why is electricity use no longer growing? In: American Council for an Energy-Efficient Economy Washington (2014)

22. National Conference of State Legislators: State net metering policies (2017). https://www.ncsl.org/research/energy/net-metering-policy-overview-and-state-legislative-updates.aspx. Accessed 13 Aug 2022

23. Negash, A.I., Kirschen, D.S.: Combined optimal retail rate restructuring and value of solar tariff. In: 2015 IEEE Power & Energy Society General Meeting, pp. 1–5. IEEE (2015)

24. Schelly, C., Louie, E.P., Pearce, J.M.: Examining interconnection and net metering policy for distributed generation in the United States. Renew. Energy Focus **22**, 10–19 (2017)

25. Singh, S.P., Scheller-Wolf, A.: That's not fair: tariff structures for electric utilities with rooftop solar. Manuf. Serv. Oper. Manag. **24**(1), 40–58 (2022)

26. Solar Energy Industry Associations: Net metering (2017). https://www.seia.org/initiatives/net-metering. Accessed 01 Sept 2021

27. United States Energy Information Administration: Hourly electricity consumption varies throughout the day and across seasons (2020). https://www.eia.gov/todayinenergy/detail.php?id=42915. Accessed 03 Nov 2021

28. Vieira, D., Shayani, R.A., De Oliveira, M.A.G.: Net metering in Brazil: regulation, opportunities and challenges. IEEE Lat. Am. Trans. **14**(8), 3687–3694 (2016)
29. Wu, Y., Khodabakhsh, A., Li, B., Nikolova, E., Pountourakis, E.: Eliciting information with partial signals in repeated games. CoRR abs/2109.04343 (2021)
30. Zhang, H., Conitzer, V.: Incentive-aware PAC learning. In: Proceedings of the AAAI Conference on Artificial Intelligence, vol. 35, pp. 5797–5804 (2021)

On the NP-Hardness of Two Scheduling Problems Under Linear Constraints

Kameng Nip$^{(\boxtimes)}$

School of Mathematical Sciences, Xiamen University, Xiamen, China
kmnip@xmu.edu.cn

Abstract. In this work, we investigate the computational complexity of two different scheduling problems under linear constraints, including single-machine scheduling problem with total completion time and no-wait two-machine flow shop scheduling problem. In these problems, a set of jobs must be scheduled one or more machines while the processing times of them are not fixed and known in advance, but are required to be determined by a system of given linear constraints. The objective is to determine the processing time of each job, and find the schedule that minimizes a specific criterion, e.g., makespan or total completion time among all the feasible choices. Although the original scheduling problems are polynomially solvable, we show that the problems under linear constraints become NP-hard. We also propose polynomial time exact or approximation algorithms for various special cases of them. Particularly, we show that when the total number of constraints is a fixed constant, both problems can be solved in polynomial time by utilizing the scheduling algorithms and the properties of linear programming.

Keywords: Scheduling · linear programming · computational complexity

1 Introduction

In the presented work, we study two different machine problems under linear constraints, including single-machine scheduling problem with minimizing total completion time $1||\sum_j C_j$ and no-wait two-machine flow shop scheduling problem $F2|no-wait|C_{\max}$. In scheduling problem under linear constraint problems, the processing times of jobs are not fixed and known beforehand, which distinguishes them from classic scheduling problems and adds an additional layer of flexibility to the decision-making process. Instead, the decision maker only knows the information that the processing times satisfy a system of given linear constraints. The goal is to determine the processing time of each job, and the

This research work is partially supported by the Natural Science Foundation of Fujian Province of China No. 2021J05011 and the Fundamental Research Funds for the Central Universities of Xiamen University No. 20720210033.

© Springer Nature Switzerland AG 2023
M. Li et al. (Eds.): IJTCS-FAW 2023, LNCS 13933, pp. 58–70, 2023.
https://doi.org/10.1007/978-3-031-39344-0_5

schedule to one or more machines such that certain objective, e.g., the makespan $C_{\max}$ or the total completion time $\sum_j C_j$ is minimized.

The scheduling problem under linear constraints (SLC in short) was first introduced in [18], in which the machine environment is identical parallel machine and the objective is to minimize the makespan, that is, $P||C_{\max}$ where $C_{\max}$ is the completion time of the last job. More specifically, there are k identical parallel machines and n jobs to be processed, with a matrix $A \in \mathbb{R}^{k \times n}$ and a vector $b \in \mathbb{R}^{k \times 1}$. The processing times $x = (x_1, ..., x_n) \in \mathbb{R}^n$ are also decision variables that have to satisfy $Ax \geq b$ and $x \geq 0$. The objective of the SLC problem is to find the values of processing times x as well as the schedule of jobs to the machines, which leads to the minimum makespan among all the feasible choices. Note that the SLC problem is a generalization of the original scheduling problem, since one can reduce the original one to the SLC problem by setting the linear constraints as $x_1 = p_1$, $x_2 = p_2$, ..., $x_n = p_n$, where p_i is the fixed processing times of the problem without linear constraints. Moreover, the authors showed that if the number of linear constraints k is a fixed constant, then the SLC problem can be solved in polynomial time through searching the basic feasible solutions of a series of linear programs. They also proposed several approximation algorithms for the general case where k is an input of problem instance. It is worth-noting that the original parallel machine scheduling problem $P||C_{\max}$ is a widely recognized NP-hard problem even for the case of two machines, and is strongly NP-hard in general [4]. Subsequently, several scheduling problems that involve processing times or other parameters (such as machine speeds) satisfying linear constraints have been investigated in this research direction. These problems includes uniformly parallel machines [27,28], the two-machine flow shop machine and other shop machine environments [16]. The above results illustrate a sharp difference in computational complexity between the original scheduling problems and those under linear constraints. For instance, the two-machine flow shop scheduling problem $F2||C_{\max}$ can be solved in polynomial time by Johnson's rule [10], while the two-machine flow shop scheduling problem under linear constraints (2-FLC in short) is NP-hard in the strong sense as shown in [15]. In comparison, the two-machine open shop scheduling problem under linear constraints (2-OLC in short) can be solved in polynomial time [16], which has the similar computational complexity as the original two-machine open shop scheduling problem $O2||C_{\max}$ [6]. Additionally, there have been recent research on different other combinatorial optimization problems with linear constraints, such as bin packing problem [24], knapsack problem [19], and various graph optimization problems [14,17]. The findings of these studies prompt further investigation into the computational complexity and algorithmic designs for problems under linear constraints. Such research could potentially offer theoretically benefits to the broader field combinatorial optimization as well.

As indicated in previous works, the scheduling problems under linear constraints could have potential practical applications beyond their theoretical interest. To motivate the two problems discussed in this work, we present some application scenarios as follows.

- *Consumer service.* The single machine scheduling problem, which minimizes total completion time, has a wide range of applications in service systems, such as hospitals, restaurants, and banks to enhance consumer satisfaction [21]. The processing time of each job can be viewed as a service time of each consumer, and the machine can be viewed as a service producer. The common objective to the service producer is to decide the serving schedule that can minimize the mean total waiting and service time of the consumers, which is closely related to the total completion time (also called mean flow time in the literature) of jobs in the machine. In the simplest and offline setting, the service producer possesses complete information regarding the service time required for each consumer, say, those make reservations for the upcoming business day. Therefore, it is optimal to schedule the service order according to the famous shortest processing time first rule [21], which is also known as Smith's rule [23], such that the mean total waiting time is minimized. However, in some situations, it is possible that the service time of each customer is not so accurately determined and could be flexible. For example, consider the scenario outlined in Table 1 for a concrete illustration. Assume

Table 1. Example for the application of consumer service.

Constraints	Consumer				
	1	2	$\cdots$	n	
revenue	$100x_1$	$200x_2$	$\cdots$	$120x_n$	≥ 10000
resource A	$3x_1$	$5x_2$	$\cdots$	$7x_n$	≤ 30
resource B	$5x_1$	$10x_2$	$\cdots$	$6x_n$	≤ 50
$\vdots$	$\vdots$	$\vdots$	$\vdots$	$\vdots$	$\vdots$
min. of 1	x_1	0	$\cdots$	0	≥ 10
max. of 1	x_1	0	$\cdots$	0	≤ 20
min. of 2	x_2	0	$\cdots$	0	≥ 15
max. of 2	x_2	0	$\cdots$	0	≤ 30
$\vdots$	$\vdots$	$\vdots$	$\vdots$	$\vdots$	$\vdots$
max. of n	0	0	$\cdots$	x_n	≤ 15

that we have n consumer to be served, and x_i is the service time of consumer i. Each consumer has an interval of her acceptable service time based on her own situation, e.g., $10 \leq x_1 \leq 20$ for consumer 1, which indicates that the appropriate time to serve this consumer would be between 10 and 20 unit times. Moreover, each unit of service time will generate a specific amount of profit or utility to the service producer. For example, if the service time of consumer 1, 2, ..., n are x_1, ..., n, respectively, then the service producer would receive revenues of $100x_1$, $200x_2$, ..., $120x_n$, respectively. The service producer aims to gain a total revenue of at least 10000, which naturally leads to $100x_1 + 200x_2 + \cdots 120x_n \geq 10000$. Moreover, the service typically requires the consumption of various resources, including human labor, electricity, computational resources, and more. Particularly in Table 1, each unit service time of consumer 1 requires 3 unit amounts of resource A, 5 unit amounts for consumer 2, and so forth. In other words, the processing time of jobs should also satisfy the linear constraint $3x_1 + 5x_2 + \cdots 7x_n \leq 30$ for the limit of resource A, and so for the other resources. The service provider needs to assign service times to consumers based on linear constraints and schedule their service to minimize the total completion time.

- *Industrial Production.* Scheduling problems have a wide range of applications in industrial production. In previous works [15,16,18], the authors presented various practical scenarios regarding the scheduling problems under linear constraints. For instance, the 2-FLC problem is motivated by steel manufacturing [15,16]. The decision maker wants to obtain specific quantities of several raw metals by extracting them from multiple types of steel. Each type of steel (iron, copper, aluminum, and etc.) corresponds to a job. The restrictions such as requirements of the raw metals corresponds to the linear constraints of the processing times. There are different essential steps in the flow-shop production process, such as wire-drawing and annealing, and must be completed in that order for each job. The decision maker aims to determine the processing time of each job while consider linear constraints, and to schedule the jobs on the flow-shop machines to minimize the makespan.
In certain scenarios, such as chemical processing, food processing, automobile assembly and working planning [3,8], it is necessary for a job to start processing on the second machine immediately after completing its operations on the first machine. For instance, if the materials are not cooled down quickly, it can cause undesirable issue such as transforming into some other substances. Therefore, it is practical to extend the above 2-FLC problem to the machine environment that the jobs satisfy no-wait restrictions.

In this work, we focus on investigating the computational complexity of the two scheduling problems under linear constraints mentioned above, namely, single-machine scheduling problem under linear constraints (with minimizing total completion time) and no-wait two-machine flow shop scheduling problem under linear constraints. We show that both two problems are NP-hard. It is worth noting that the original problems $1 || \sum_j C_j$ and $F2 | no - wait | C_{\max}$ can be solved in polynomial time, as we will review in later sections. Our findings indicate that solving the problems under linear constraints is more computationally difficult than solving their original counterparts. Moreover, we propose approximation algorithms for the general cases, which are obtained by solving some specific linear programs. Then we consider several nontrivial special cases of them. In particular, we show that if the total number of fixed constraints is a constant, then both problems can be solved in polynomial time. The algorithms are based on the properties of basic feasible solutions for linear programming and their scheduling algorithms.

The remainder of this work is organized as follows. In Sect. 2, we study the single machine scheduling problem under linear constraints, while in Sect. 3, we study the no-wait two-machine flow shop scheduling problem under linear constraints. For each section, we first formally define the corresponding problem and briefly review the related literature. Then, we analyze the computational complexity, and present polynomial-time optimal or approximation algorithms for various cases. Finally, in Sect. 4, we provide some concluding remarks.

2   Single Machine Scheduling Under Linear Constraints

2.1   Problem Definition and Literature Review

The single machine scheduling problem under linear constraints (SSLC problem in short) is formally defined as follows.

Definition 1. *Given n jobs and a single machine. Each job i has a processing time x_i, which are determined by k linear constraints $Ax \geq b$ with $A \in \mathbb{R}^{k \times n}$ and $b \in \mathbb{R}^{k \times 1}$. The goal of the SSLC problem is to determine the processing times of the jobs such that they satisfy the linear constraints and to schedule the jobs to the machines to minimize the total completion time $\sum_j C_j$.*

The problem $1||\sum_j C_j$ is perhaps the simplest scheduling model studied in the field of scheduling. In [23], it is shown that the optimal schedule is to assign the jobs in a non-decreasing order of their processing jobs. Such rule is referred to Smith's rule or the shortest processing time first rule [21], and can be implemented in $O(n \log n)$ time. In other words, let $p_1 \leq p_2 \leq \dots \leq p_n$ the (fixed) processing times of the n jobs and OPT be the optimal value, then the optimal total completion time for $1||\sum_j C_j$ is given by

$$OPT = np_1 + (n-1)p_2 + \cdots + 2p_{n-1} + p_n. \tag{1}$$

Smith [23] also extended the idea of Simth's rule to solve a more general model $1||\sum_j w_j C_j$, in which each job has a nonnegative weight w_j and the objective is to minimize the total weighted completion time. Later, [1] showed that a more complicated extension to the unrelated parallel machine environments $R||\sum_j C_j$ can be solved in polynomial time, by reducing it to the transportation problem. However, many slight extensions of single machine scheduling problem turn out to be NP-hard. For instance, [1] showed that $P2||\sum_j w_j C_j$ is NP-hard, and [12] proved that $1|r_j|\sum_j C_j$ is strongly NP-hard in which r_j is the arrival time of job j. For more details on the machine scheduling problem with minimizing total completion time, we refer to the textbook [21] and some recent works [9,11]. For the scheduling problem under the linear constraints, [18] showed that the problem $1||C_{\max}$ with linear constraints is also polynomially solvable, by solving a linear program that minimizes the total processing time $\sum_{j=1}^{n} p_j$. Conversely, the SSLC problem turns out to be more difficult than the original scheduling problem $1||\sum_j C_j$, as we will demonstrate that it is indeed NP-hard.

2.2   Computational Complexity

By (1), the optimal schedule of $1||\sum_j C_j$ is to assign the jobs in a non-decreasing order of their processing times. Therefore, we can reformulate the SSLC problem as the following mathematical optimization problem (2):

$$\min \quad n\Theta_1(\boldsymbol{x}) + (n-1)\Theta_2(\boldsymbol{x}) + \cdots + 2\Theta_{n-1}(\boldsymbol{x}) + \Theta_n(\boldsymbol{x}) \qquad (2a)$$
$$\text{s.t.} \quad A\boldsymbol{x} \geq \boldsymbol{b} \qquad (2b)$$
$$\boldsymbol{x} \geq 0, \qquad (2c)$$

where $\Theta_k : \mathbb{R}^n \to \mathbb{R}$ maps $\boldsymbol{x}$ to its k-largest element. In other words, we have $\Theta_1(\boldsymbol{x}) = \min_{i=1,\dots,n}\{x_i\}$, $\Theta_n(\boldsymbol{x}) = \max_{i=1,\dots,n}\{x_i\}$ and $\Theta_1(\boldsymbol{x}) \leq \Theta_2(\boldsymbol{x}) \leq \cdots \leq \Theta_n(\boldsymbol{x})$ where $\Theta_k(\boldsymbol{x})$ is the k-largest element of $\boldsymbol{x}$. We remark that the objective function is considered as a weighted average function in the literature [20, 25, 26]. Particularly, let $f(\boldsymbol{x}) = w_1\Theta_1(\boldsymbol{x}) + w_2\Theta_2(\boldsymbol{x}) \cdots + w_{n-1}\Theta_{n-1}(\boldsymbol{x}) + w_n\Theta_n(\boldsymbol{x})$. If the weights satisfy $w_1 \leq w_2 \leq \cdots \leq w_n$, then it has been shown that the nonlinear optimization problem $\min f(\boldsymbol{x})$ subject to (2b) and (2b) can be equivalently reformulated as a linear programming problem [20]. In other words, the corresponding problem can be solvable in polynomial time, e.g., by some ellipsoid method or interior point method [13]. However, the idea of transforming (2) into an equivalent linear programming formulation may not be a viable option when the weights follow the condition $w_1 \geq w_2 \geq \cdots \geq w_n$. To the best of our knowledge, the computational complexity of such linear programming problem with a non-increasing weighted average objective function is unclear. In the following, we show that the SSLC problem (2) is NP-hard, which significantly differs from the original scheduling problem $1||\sum_j C_j$, as well as other optimization problems involving a weighted average function.

Theorem 1. *The SSLC problem (2) is NP-hard.*

Proof. We reduce the independent set problem to the decision problem of the SSLC problem (2). The decision problem of independent set problem is given a graph $G = (V, E)$ and an integer K, to decide whether there is a vertex set $V' \in V$ with size at least K in which no two vertices are adjacent. Let $n = |V|$ and $m = |E|$, we construct an instance of SSLC with $2n$ variables that has processing times denoted by $x_1, \dots, x_n$ and $y_1, \dots, y_n$, where x_j and y_j correspond to vertex v_j in V, and $m + n + 1$ linear constraints:

$$\sum_{i=1}^{n} x_i \geq K \qquad (3a)$$
$$x_u + x_v \leq 1, \quad \forall (u, v) \in E \qquad (3b)$$
$$x_j + y_j = 1, \quad \forall j = 1, \dots, n \qquad (3c)$$
$$\boldsymbol{x}, \boldsymbol{y} \geq 0. \qquad (3d)$$

It suffices to show that there is an independent set with size at least K if and only if there is a setting of processing times for the jobs that are feasible to (3) of the SSLC problem, and a corresponding schedule with a total completion time of no more than $\frac{n(n+1)}{2}$.

On one hand, if there is an independent set V' with size at least K, then for each vertex $v_j \in V'$, we set $x_j = 1$ and $y_j = 0$; for $v_j \notin V'$, we set $x_j = 0$ and $y_j = 1$. The processing times of jobs are feasible to (3), since the jobs correspond to V' form an independent set. Moreover, there are exactly n jobs that have processing times 1, and the other n jobs have processing times 0. The total completion time of this schedule of this instance is exactly $n + (n-1) + \cdots + 1 = \frac{n(n+1)}{2}$.

On the other hand, assume that there is an instance of the SSLC problem, in which the processing times x and y are feasible to (3), and a corresponding schedule S with total completion time no more than $\frac{n(n+1)}{2}$. Let $p_1 \leq p_2 \leq \cdots \leq p_{n-1} \leq p_n$ be the n largest values among the $2n$ variables x and y in this solution, where p_j could be certain x_i or y_i for some i. Accordingly, for each $j = 1, ..., n$, we denote $\tilde{p}_j = y_i$ if $p_j = x_i$, and $\tilde{p}_j = x_i$ if $p_j = y_i$. Note that we have $x_j + y_j = 1$ for all j by constraint (3c), and $1 - p_n \geq \cdots \geq 1 - p_1$ by the definition of p_j. In other words, we relabel the variables $x_1, ..., x_n$ and $y_1, ..., y_n$ by $p_1, ..., p_n$ and $\tilde{p}_1, ..., \tilde{p}_n$, which satisfies

$$\tilde{p}_n \leq \cdots \leq \tilde{p}_1 \leq p_1 \leq \cdots \leq p_n. \tag{4}$$

Next we claim that any feasible schedule of this instance must have total completion time at least $\frac{n(n+1)}{2}$, and the equality holds only if $p_1 = \cdots = p_n = 1$ and $\tilde{p}_1 = \cdots = \tilde{p}_n = 0$. To see this, we consider the best possible schedule of this instance, which is scheduled by the shortest processing time first/Smith's rule. Then from (1) and (4), its total completion time is given by

$$2n\tilde{p}_n + (2n-1)\tilde{p}_{n-1} + \cdots + (n+1)\tilde{p}_1 + np_1 + \cdots + 2p_{n-1} + p_n$$
$$= 2n(1-p_n) + p_n + (2n-1)(1-p_{n-1}) + 2p_{n-1} + \cdots + (n+1)(1-p_1) + np_1$$
$$= 2n + (2n-1) + \cdots + (n+1) + (1-2n)p_n + (3-2n)p_{n-1} + \cdots + (-1)p_1.$$

Consider the following linear programming problem:

$$\begin{aligned} \min \ &(1-2n)p_n + (3-2n)p_{n-1} + \cdots + (-1)p_1 \\ \text{s.t.} \quad &0 \leq p_j \leq 1, \qquad \forall j = 1, ..., n \end{aligned} \tag{5}$$

Since all coefficients of its objective function are negative, we can verify that $p_1 = \cdots = p_n = 1$ is the unique optimal solution to (5). By definition, it follows that $\tilde{p}_1 = \cdots = \tilde{p}_n = 0$ and the claim is proved. Therefore, the total completion time of any feasible schedule to this instance is no less than the optiaml value to (5), namely, $n + (n-1) + \cdots + 1 = \frac{n(n+1)}{2}$. By assumption, the total completion time of the schedule in this SSLC instance is exactly $\frac{n(n+1)}{2}$, in which exactly n variables among x and y are 1, and the other n variables are 0. We can just select the vertex v_j with $x_j = 1$ into V', which constitutes an independent set with size at least K by (3a) and (3b). It finishes the proof of the theorem. □

2.3   Algorithms

The hardness result in Theorem 1 indicates that it is impossible to find an optimal solution to (2) in polynomial time unless $P = NP$. In this subsection, we develop algorithms for solving the SSLC problem. For the general case, we can obtain an n-approximation algorithm as described in Theorem 2. Due to the limitation of space, we omit the details of the subsequent lemmas and theorems. The details will be provided in the full version.

Theorem 2. *The SSLC problem has an n-approximation algorithm.*

It should be noted that the number of constraints k in the instance of the hardness reduction in Theorem 1 is not a fixed constant. In the following, we consider a special case when k is not fixed. We show that this case can be solved in polynomial time, which depends on the following property.

Lemma 1. *The SSLC problem has an optimal solution in which at most k jobs have nonzero processing time.*

By Lemma 1, there exists an optimal solution that contains a constant number k of nonzero processing time. Therefore, we can find the optimal solution by first enumerating all the nonzero processing time jobs and the optimal schedule, then solve a specific linear program to obtain the best processing times. The detail is summarized in Algorithm 1 and Theorem 3.

Algorithm 1. Enumeration algorithm for the SSLC problem with fixed k

---

1: **for** each subset J' of J with k jobs **do**
2:    **for** each possible permutation of J' **do**
3:       Let $(\sigma(1), ..., \sigma(k))$ be the permutation, solve the following linear program while setting $x_i = 0$ for $i \notin J'$:

$$\min k x_{\sigma(1)} + (k-1)x_{\sigma(2)} + \cdots + x_{\sigma(k)}$$
$$\text{s.t.} \qquad Ax \geq b \qquad\qquad (6)$$
$$x \geq 0.$$

4:    **if** (6) is feasible **then**
5:       Let the processing times of jobs be the optimal solution to (6), and record the schedule and the makespan.
6: **return** the schedule with the smallest total completion time among all these iterations and its corresponding processing times.

Theorem 3. *Algorithm 1 returns an optimal solution to the SLC problem and has time complexity $O(n^k L)$, where the parameter L is the input length of (6).*

3 No-Wait Two-Machine Flow Shop Scheduling Problem Under Linear Constraints

3.1 Problem Definition and Literature Review

The no-wait two-machine flow shop scheduling problem under the linear constraints (no-wait 2FLC problem in short) is formally defined as follows.

Definition 2. *Given n jobs and two flow-shop machines. Each job has to be processed on the first machine and then on the second machine, and the process in the second machine must be start immediately after its finish in the first machine (no-wait restriction). The processing times of job i on the first machine and the second machine are x_i and y_i respectively, which are determined by k linear constraints, $A\boldsymbol{x} + C\boldsymbol{y} \geq \boldsymbol{b}$ with $A, C \in \mathbb{R}^{k \times n}$ and $\boldsymbol{b} \in \mathbb{R}^{k \times 1}$. The goal of the no-wait 2FLC problem is to determine the processing times of the jobs such that they satisfy the linear constraints and to schedule the jobs to the two flow-shop machines to minimize the makespan.*

Flow shop scheduling is one of the three basic models (open shop, flow shop, job shop) of multi-stage scheduling problems. Flow shop scheduling with minimizing the makespan is usually denoted by $Fm||C_{\max}$, where m is the number of machines. Garey et al. [5] proved that $Fm||C_{\max}$ is strongly NP-hard for $m \geq 3$, and Hall [7] proposed a PTAS algorithm for $Fm||_{\max}$. Particularly, the two-machine flow shop scheduling problem $F2||C_{\max}$ can be solved by Johnson's algorithm in $O(n \log n)$ time [10]. If all the jobs are processed in the same order, then we call this schedule a permutation schedule. It is known that $F2||C_{max}$ or $F3||C_{max}$ has an optimal permutation schedule [2]. The no-wait flow shop scheduling, which is denoted by $Fm|no - wait|C_{\max}$. In $Fm|no - wait|C_{\max}$, once a job has been processed, each stage must be started after its completion of the previous stage without any delay. In the traditional setting, each job has exactly two distinct operations and must be scheduled fulfilling the no-wait restrictions. In other words, each of the two processes of jobs must be scheduled even if it has zero processing time (see, e.g., [3, Section 6.3]). We note that the no-wait 2-FLC problem studied in this paper correspond to this traditional setting of no-wait restriction. Gilmore and Gomory [6] proposed an $O(n \log n)$ algorithm to solve this problem in polynomial time, which is by relating it to a polynomially solvable case of the traveling salesman problem. Researchers also concerned on the problem in which some job may have only one stage, namely, with missing operations. Surprisingly, [22] showed that the problem with missing operations is NP-hard in the strong sense and hence is much harder than that without missing operations. Furthermore, it is observed that any feasible schedule of $F2|no - wait|C_{\max}$ must be a permutation schedule. We refer to the literature [3,8] for more discussion of no-wait scheduling and its applications.

For 2-FLC problem, [15,16] showed that the problem under linear constraints is NP-hard in the strong sense, which sharply differs from the computational complexity of the original $F2||C_{\max}$. In the following, we will show that the no-wait 2-FLC problem is also NP-hard, which is more difficult than its original version. We remark that the hardness result of 2-FLC problem cannot be directly applied to no-wait 2-FLC problem, since the jobs used in the reduction [15,16] do not satisfy no-wait restriction.

To close this section, we state a classic result of $F2|no - wait|C_{\max}$ (see, e.g., [3, Section 6.3]), which will be frequently used in the subsequent analysis. Let $(\sigma(1), \sigma(2), ..., \sigma(n))$ be an arbitrary feasible schedule of $F2|no - wait|C_{\max}$. Then the makespan of this schedule is given by

$$C_{\max} = x_{\sigma(1)} + \sum_{i=2}^{n} \max(y_{\sigma(i-1)}, x_{\sigma(i)}) + y_{\sigma(n)}. \tag{7}$$

3.2   Computational Complexity

In this section, we show that the no-wait 2-FLC problem is NP-hard. As mentioned above, we consider the flow-shop scheduling problem without any missing operations. For the original problem without linear constraints, on can use Gilmore and Gomory's algorithm [6] to find an optimal schedule in polynomial time. Our hardness result indicates that the problem under linear constraints is much more difficult. We summarize the result in Theorem 4.

Theorem 4. *The no-wait 2-FLC problem is NP-hard, even if each job has strictly positive processing times on both stages.*

3.3   Algorithms

In the general case, it can be observed that a straightforward 2-approximation algorithm exists for the no-wait 2-FLC problem, which is based on solving a linear program that minimizes the total processing time of the jobs.

Theorem 5. *The no-wait 2-FLC problem has a 2-approximation algorithm.*

Next we study a special case where the no-wait 2-FLC problem can be solved in polynomial time, in which the number of constraints k is fixed. The key is to prove a similar result as the SSLC problem, which shows that the no-wait 2-FLC problem has an optimal solution with a fixed number of nonzero processing time jobs if k is fixed. We remark that the case of no-wait 2-FLC problem is more complicated than Lemma 2, since the objective function (7) is not linear in general even when the optimal schedule σ is known.

Lemma 2. *The no-wait 2-FLC problem has an optimal solution, in which each machine has at most k jobs with nonzero processing time.*

Based on Lemma 2, we propose an enumeration algorithm for 2-FLC problem with a fixed number of constraints k. We summarize it in Algorithm 2 and Theorem 6.

Theorem 6. *Algorithm 2 returns an optimal solution to the no-wait 2FLC problem and has time complexity $O(n^{2k}L)$, where the parameter L is the input length of (8).*

Algorithm 2. Enumeration algorithm for 2-FLC problem with fixed k

---
1: **for** each subset J' of J with k jobs **do**
2:    **for** each possible permutation of the jobs in J' **do**
3:        Let $(\sigma(1), ..., \sigma(k))$ be the permutation, i.e., the schedule of these k jobs. Solve the following LP while setting $y_i = 0$ for $i \notin J'$.

$$\min \quad x_{\sigma(1)} + \frac{1}{2}\sum_{i=2}^{n}(y_{\sigma(i-1)} + x_{\sigma(i)} + z_i^+ + z_i^-) + y_{\sigma(n)} \tag{8a}$$

$$\text{s.t.} \quad \sum_{i=1}^{n} a_{ji}x_i + c_{ji}y_i \geq b_j \qquad \forall\, j = 1,\ldots,k \tag{8b}$$

$$x_{\sigma(i)} + z_i^+ = y_{\sigma(i-1)} + z_i^- \qquad \forall\, i = 2,\ldots,n \tag{8c}$$

$$z_i^+, z_i^-, x_i, y_i \geq 0 \qquad \forall\, i = 1,\ldots,n,$$

4:    **if** (8) is feasible **then**
5:        Let the processing times of jobs be the optimal solution to (8), and record the schedule and the makespan.
6: **return** the schedule with the smallest makespan among all these iterations and its corresponding processing times.

---

4 Conclusions

In this work, we investigate the computational complexity of two different scheduling problems under linear constraints. We show that the problems with linear constraints are NP-hard, while the original versions of both problems are polynomially solvable. Additionally, we propose polynomial time exact or approximation algorithms for various cases of the problems. One potential research direction is to develop improved approximation algorithms for the general cases of these problems, or to show the possibility of inapproximability. Moreover, it is interesting to explore other types of objectives (e.g., lateness or tardiness) or machine restrictions (e.g., release date, due dates or job unavailability) for the machine scheduling problems under linear constraints.

References

1. Bruno, J.L., Coffman, E.G., Jr., Sethi, R.: Scheduling independent tasks to reduce mean finishing time. Commun. ACM **17**(7), 382–387 (1974)
2. Conway, R.W., Maxwell, W.L., Miller, L.W.: Theory of Scheduling. Reading (1967)
3. Emmons, H., Vairaktarakis, G.: Flow Shop Scheduling: Theoretical Results, Algorithms, and Applications. Springer, New York (2013). https://doi.org/10.1007/978-1-4614-5152-5
4. Garey, M.R., Johnson, D.S.: Computers and Intractability: A Guide to the Theory of NP-Completeness. Freeman, New York (1979)

5. Garey, M.R., Johnson, D.S., Sethi, R.: The complexity of flowshop and jobshop scheduling. Math. Oper. Res. **1**, 117–129 (1976)
6. Gilmore, C., Gomory, R.E.: Sequencing a one state-variable machine: a solvable case of the travelling salesman problem. Oper. Res. **12**, 655–679 (1964)
7. Hall, L.A.: Approximability of flow shop scheduling. Math. Program. **82**, 175–190 (1998)
8. Hall, N.G., Sriskandarajah, C.: A survey of machine scheduling problems with blocking and no-wait in process. Oper. Res. **44**(3), 510–525 (1996)
9. Jansen, K., Lassota, A., Maack, M., Pikies, T.: Total completion time minimization for scheduling with incompatibility cliques. In: Proceedings of the International Conference on Automated Planning and Scheduling, vol. 31, no. 1, pp. 192–200 (2021)
10. Johnson, S.M.: Optimal two- and three-stage production schedules with setup times included. Naval Res. Logist. Q. **1**, 61–68 (1954)
11. Knop, D., Koutecký, M.: Scheduling meets n-fold integer programming. J. Sched. **21**(5), 493–503 (2018)
12. Lawler, J.L., Johnson, E.L., Lenstra, J.K., Rinnooy Kan, A.H.G., Shmoys, D.B.: The complexity of machine scheduling problems. Ann. Discrete Math. **1**, 343–362 (1977)
13. Luenberger, D.G., Ye, Y.: Linear and Nonlinear Programming, 4th edn. Springer, Cham (2016)
14. Nip, K., Shi, T., Wang, Z.: Some graph optimization problems with weights satisfying linear constraints. J. Comb. Optim. **43**, 200–225 (2022)
15. Nip, K., Wang, Z.: Two-machine flow shop scheduling problem under linear constraints. In: Li, Y., Cardei, M., Huang, Y. (eds.) COCOA 2019. LNCS, vol. 11949, pp. 400–411. Springer, Cham (2019). https://doi.org/10.1007/978-3-030-36412-0_32
16. Nip, K., Wang, Z.: A complexity analysis and algorithms for two-machine shop scheduling problems under linear constraints. J. Sched. (2021)
17. Nip, K., Wang, Z., Shi, T.: Some graph optimization problems with weights satisfying linear constraints. In: Li, Y., Cardei, M., Huang, Y. (eds.) COCOA 2019. LNCS, vol. 11949, pp. 412–424. Springer, Cham (2019). https://doi.org/10.1007/978-3-030-36412-0_33
18. Nip, K., Wang, Z., Wang, Z.: Scheduling under linear constraints. Eur. J. Oper. Res. **253**(2), 290–297 (2016)
19. Nip, K., Wang, Z., Wang, Z.: Knapsack with variable weights satisfying linear constraints. J. Global Optim. **69**(3), 713–725 (2017). https://doi.org/10.1007/s10898-017-0540-y
20. Ogryczak, W., Śliwiński, T.: On solving linear programs with the ordered weighted averaging objective. Eur. J. Oper. Res. **148**(1), 80–91 (2003)
21. Pinedo, M.: Scheduling: Theory, Algorithms, and Systems. Springer, New York (2016)
22. Sahni, S., Cho, Y.: Complexity of scheduling shops with no wait in process. Math. Oper. Res. **4**(4), 448–457 (1979)
23. Smith, W.E.: Various optimizers for single-stage production. Naval Res. Logist. Q. **3**, 59–66 (1956)
24. Wang, Z., Nip, K.: Bin packing under linear constraints. J. Comb. Optim. **34**(4), 1198–1209 (2017). https://doi.org/10.1007/s10878-017-0140-2
25. Yager, R.R.: On ordered weighted averaging aggregation operators in multicriteria decisionmaking. IEEE Trans. Syst. Man Cybern. **18**(1), 183–190 (1988)

26. Yager, R.R.: Constrained OWA aggregation. Fuzzy Sets Syst. **81**(1), 89–101 (1996)
27. Zhang, S., Nip, K., Wang, Z.: Related machine scheduling with machine speeds satisfying linear constraints. In: Kim, D., Uma, R.N., Zelikovsky, A. (eds.) COCOA 2018. LNCS, vol. 11346, pp. 314–328. Springer, Cham (2018). https://doi.org/10.1007/978-3-030-04651-4_21
28. Zhang, S., Nip, K., Wang, Z.: Related machine scheduling with machine speeds satisfying linear constraints. J. Comb. Optim. **44**(3), 1724–1740 (2022)

On the Matching Number of k-Uniform Connected Hypergraphs with Maximum Degree

Zhongzheng Tang[1], Haoyang Zou[2], and Zhuo Diao[2(✉)]

[1] School of Science, Beijing University of Posts and Telecommunications,
Beijing 100876, China
tangzhongzheng@amss.ac.cn
[2] School of Statistics and Mathematics, Central University of Finance
and Economics, Beijing 100081, China
diaozhuo@amss.ac.cn

Abstract. For $k \geq 2$, let $H(V, E)$ be a k-uniform connected hypergraph with maximum degree Δ on n vertices and m edges. A set of edges $A \subseteq E$ is a matching if every two distinct edges in A have no common vertices. The matching number is the maximum cardinality of a matching, denoted by $\nu(H)$. In this paper, we prove the following inequality: $\nu(H) \geq \frac{n-(k-2)m-1}{\Delta}$ and characterize the extremal hypergraphs with equality holds. A class of hypergraphs called Δ-star hypertrees are introduced, which are exactly the extremal hypergraphs. These results is a generalization of the theorems by Tang and Diao in IJTCS-FAW 2022.

Keywords: k-uniform hypergraphs · maximum degree · matching number · extremal hypergraphs

1 Introduction

A hypergraph is a generalization of a graph in which an edge can join any number of vertices. A simple hypergraph is a hypergraph without multiple edges. Let $H = (V, E)$ be a simple hypergraph with vertex set V and edge set E. As for a graph, the order of H, denoted by n, is the number of vertices. The number of edges is denoted by m.

For each vertex $v \in V$, the degree $d(v)$ is the number of edges in E that contains v. We say v is an isolated vertex of H if $d(v) = 0$. Hypergraph H is k-regular if each vertex's degree is k ($d(v) = k, \forall v \in V$). The maximum degree of H is $\Delta(H) = \max_{v \in V} d(v)$. Hypergraph H is k-uniform if each edge contains exactly k vertices ($|e| = k, \forall e \in E$). Hypergraph H is called linear if any two distinct edges have at most one common vertex.

Supported by National Natural Science Foundation of China under Grant No. 11901605, No. 12101069, the disciplinary funding of Central University of Finance and Economics, the Emerging Interdisciplinary Project of CUFE.

M. Li et al. (Eds.): IJTCS-FAW 2023, LNCS 13933, pp. 71–84, 2023.
https://doi.org/10.1007/978-3-031-39344-0_6

Let $k \geq 2$ be an integer. A cycle of length k, denoted as k-cycle, is a vertex-edge sequence $C = v_1 e_1 v_2 e_2 \cdots v_k e_k v_1$ with: (1)$\{e_1, e_2, \ldots, e_k\}$ are distinct edges of H. (2)$\{v_1, v_2, \ldots, v_k\}$ are distinct vertices of H. (3)$\{v_i, v_{i+1}\} \subseteq e_i$ for each $i \in [k]$, here $v_{k+1} = v_1$. We consider the cycle C as a sub-hypergraph of H with vertex set $\{v_i, i \in [k]\}$ and edge set $\{e_j, j \in [k]\}$. For any vertex set $S \subseteq V$, we write $H \backslash S$ for the sub-hypergraph of H obtained from H by deleting all vertices in S and all edges incident with some vertices in S. For any edge set $A \subseteq E$, we write $H \backslash A$ for the sub-hypergraph of H obtained from H by deleting all edges in A and keeping vertices. If S is a singleton set $\{s\}$, we write $H \backslash s$ instead of $H \backslash \{s\}$.

A hypertree $T(V, E)$ is a connected hypergraph with no cycle. In a hypertree $T(V, E)$, for any two distinct vertices $\{u, v\} \subseteq V$, there is a unique $u - v$ path P in T. Given a hypergraph $H(V, E)$, a set of edges $A \subseteq E$ is a feedback edge set (FES) if $H \backslash A$ is acyclic. The feedback edge set number is the minimum cardinality of a feedback edge set, denoted by $\tau'_c(H)$. Given a hypergraph $H(V, E)$, a set of edges $A \subseteq E$ is a matching if every two distinct edges have no common vertex. The matching number is the maximum cardinality of a matching, denoted by $\nu(H)$. In this paper, we consider the matching number in k-uniform connected hypergraphs.

1.1 Related Works

The problem of finding a maximum matching is fundamental in both practical and theoretical computer science, and has numerous applications. Maximum matching in bipartite graphs is significantly simpler than in general graphs, as computations of augmenting paths do not encounter odd cycles. The seminal algorithm of Hopcroft and Karp [14] solves maximum bipartite matching in $O(\sqrt{n}m)$ time, for graphs with n vertices and m edges. The first polynomial time algorithm for finding a maximum matching in a general graph was obtained by Edmonds [5]. The currently fastest deterministic algorithm for this problem, obtained by Micali and Vazirani, runs in $O(\sqrt{n}m)$ time (see [8,18,23]). Faster algorithms are known for several important special classes of graphs.

Various lower bounds on the matching number for regular graphs have appeared in the literature. For example, Biedl et al. [2] proved that if G is a cubic graph, then $\nu(G) \geq \frac{4n-1}{9}$. This result was generalized to regular graphs of higher degree by Henning and Yeo [11], see also O and West [19]. O and West [21] established lower bounds on the matching number with given edge-connectivity in regular graphs. Cioaba et al. [3,4,9] studied matchings in regular graphs from eigenvalues. Suil [20] studied spectral radius and matchings in graphs. Lower bounds on the matching number in subcubic graphs (graphs with maximum degree at most 3) were studied by, among others, Henning et al. [13] and Haxell et al. [10]. Lower bounds on the matching number for general graphs and bipartite graphs were obtained by Jahanbekam et al. [17] and Henning et al. [12].

Yet although the graph matching problem is fairly well understood, and solvable in polynomial time, most of the problems related to hypergraph matching tend to be very difficult and remain unsolved. Indeed, the hypergraph matching problem is known to be NP-hard even for 3-uniform hypergraphs, without any good approximation algorithm.

One of the most basic open questions in this area was raised in 1965 by Erdös [6], who asked for the determination of the maximum possible number of edges that can appear in any k-uniform hypergraph with matching number $\nu(H) < t \leq \frac{n}{k}$. See [1,7,15,16] for latest developments on the problem.

In IJTCS-FAW 2022, for simple graphs, Tang and Diao [22] have established a sharp lower bound on the matching number with given maximum degree, as stated in Theorem 1. In this paper, we generalize the results to k-uniform hypergraphs.

Theorem 1. *For every connected graph $G(V, E)$ with maximum degree Δ on n vertices and m edges, $\nu(G) \geq \frac{n-1}{\Delta}$ holds. Furthermore, the equality $\nu(G) = \frac{n-1}{\Delta}$ holds if and only if $G(V, E)$ is a Δ-star tree when $\Delta \geq 3$.*

1.2   Our Results

In this paper, for every k-uniform connected hypergraph, we give a sharp lower bound for the matching number and characterize the extremal hypergraphs with lower bound attained. The result is stated in Theorem 2.

Theorem 2. *For every k-uniform connected hypergraph $H(V, E)$ with maximum degree Δ on n vertices and m edges, $\nu(H) \geq \frac{n-(k-2)m-1}{\Delta}$ holds. Furthermore, the equality $\nu(H) = \frac{n-(k-2)m-1}{\Delta}$ holds if and only if $H(V, E)$ is a Δ-star hypertree when $\Delta \geq 3$.*

The main content of the article is organized as follows:

- In Sect. 2, we prove the lower bound for the matching number.
- In Sect. 3, we characterize the extremal hypergraphs with lower bound attained.

2   The Lower Bound of Matching Number with Maximum Degree

In this section, for every k-uniform connected hypergraph, a lower bound of the matching number with maximum degree is proved, as stated in Theorem 3. The content is organized as follows:

- In Lemma 1, for every k-uniform hypertree, a lower bound of the matching number with maximum degree is proved.
- In Lemma 2, for every k-uniform connected hypergraph, an upper bound of the feedback edge number is proved.
- In Theorem 3, for every k-uniform connected hypergraph, a lower bound of the matching number with maximum degree is proved.

Theorem 3. *For every k-uniform connected hypergraph $H(V, E)$ with maximum degree Δ on n vertices and m edges, $\nu(H) \geq \frac{n-(k-2)m-1}{\Delta}$.*

Lemma 1. *For every k-uniform hypertree $T(V, E)$ with maximum degree Δ on m edges, $\nu(T) \geq \frac{m}{\Delta}$.*

Proof. It is equivalent to prove the next proposition:

Proposition. *For every k-uniform hypertree $T(V, E)$ with $d(v) \leq p, \forall v \in V$, $\nu(T) \geq \frac{m}{p}$.*

We prove this proposition by contradiction. Let us take out the counterexample $T(V, E)$ with minimum edges. Thus $d(v) \leq p, \forall v \in V$, $\nu(T) < \frac{m}{p}$. Obviously $T(V, E)$ has at least three edges. The longest path in T is $P = v_1 e_1 v_2 \cdots v_t e_t v_{t+1}$, which connects one leaf v_1 to another leaf v_{t+1} and v_2 is the only one vertex in e_1 with degree more than one, as shown in Fig. 1. The degree of v_2 is $d(v_2)$ and $T \setminus v_2$ has $(k-1)d(v_2)$ components, denoted as $\{T_i, 1 \leq i \leq (k-1)d(v_2)\}$. Assume that T_i contains m_i edges.

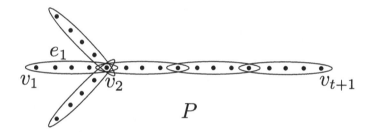

Fig. 1. The longest path P between leaves v_1 and v_{t+1}

Claim 1. $\nu(T \setminus v_2) \geq \frac{m-p}{p}$.

T is the counterexample with minimum edges, thus $\nu(T_i) \geq \frac{m_i}{p}$. Combined with $d(v_2) \leq p$, we have the following inequalities:

$$\nu(T \setminus v_2) = \sum_{1 \leq i \leq (k-1)d(v_2)} \nu(T_i) \geq \sum_{1 \leq i \leq (k-1)d(v_2)} \frac{m_i}{p} = \frac{m - d(v_2)}{p} \geq \frac{m-p}{p}. \quad \blacksquare$$

Claim 2. $\nu(T) \geq \nu(T \setminus v_2) + 1$.

The longest path in T is $P = v_1 e_1 v_2 \cdots v_t e_t v_{t+1}$ thus v_2 is the only one vertex in e_1 with degree more than one. For every matching M in $T \setminus v_2$, $M \cup \{e_1\}$ is a matching in T. $\blacksquare$

According to these claims, we have the following inequalities:

$$\nu(T) \geq \nu(T \setminus v_2) + 1 \geq \frac{m-p}{p} + 1 = \frac{m}{p},$$

which is a contradiction with $\nu(T) < \frac{m}{p}$. $\square$

Next for every k-uniform connected hypergraph, we give an upper bound of the feedback edge number.

Lemma 2. *Let $H(V, E)$ be a k-uniform connected hypergraph, $\tau_c'(H) \leq (k-1)m - n + 1$.*

Before proving the lemma above, we will prove a series of lemmas which are very useful.

Lemma 3. *For every k-uniform connected hypergraph $H(V, E)$, $n \leq (k-1)m+1$ holds on.*

Proof. We prove this lemma by induction on m. When $m = 0$, $H(V, E)$ is an isolate vertex, $n \leq (k-1)m+1$ holds on. Assume this lemma holds on for $m \leq t$. When $m = t + 1$, take arbitrarily one edge e and consider the subgraph $H \setminus e$. obviously, $H \setminus e$ has at most k components. Assume $H \setminus e$ has p components $H_i(V_i, E_i)$, n_i is the vertex number and m_i is the edge number for each $i \in \{1, \ldots, p\}$. Then by our induction, $n_i \leq (k-1)m_i + 1$ holds on. So we have

$$n = n_1 + \cdots + n_p \leq (k-1)m_1 + \cdots + (k-1)m_p + p$$

$$= (k-1)(m-1) + p = (k-1)m + p - k + 1 \leq (k-1)m + 1.$$

By induction, we finish our proof. □

Lemma 4. *For every k-uniform connected hypergraph $H(V, E)$, $n = (k-1)m+1$ if and only if H is a hypertree.*

Proof. Sufficiency: if H is a hypertree, we prove $n = (k-1)m+1$ by induction on m. When $m = 0$, $H(V, E)$ is an isolate vertex, $n = (k-1)m+1$ holds on. Assume this lemma holds on for $m \leq t$. When $m = t + 1$, take arbitrarily one edge e and consider the subhypergraph $H \setminus e$. Because H is a k-uniform hypertree, $H \setminus e$ has exactly k components, denoted by $H_i(V_i, E_i)$, n_i is the vertex number and m_i is the edge number for each $i \in \{1, \ldots, p\}$. Because every component is a hypertree, thus by our induction, $n_i = (k-1)m_i + 1$ holds on. So we have

$$n = n_1 + \cdots + n_k = (k-1)m_1 + \cdots + (k-1)m_k + k = (k-1)(m-1) + k = (k-1)m+1.$$

By induction, we finish the sufficiency proof.

Necessity: We prove by contradiction. If H is not a hypertree, H contain a cycle C. Take arbitrarily one edge e in C and consider the subgraph $H \setminus e$. obviously, $H \setminus e$ has at most $k - 1$ components. Assume $H \setminus e$ has p components $H_i(V_i, E_i)$, n_i is the vertex number and m_i is the edge number for each $i \in \{1, \ldots, p\}$. Then by Lemma 3, $n_i \leq (k-1)m_i + 1$ holds on. So we have

$$n = n_1 + \cdots + n_p \leq (k-1)m_1 + \cdots + (k-1)m_p + p$$

$$= (k-1)(m-1) + p = (k-1)m + p - k + 1 \leq (k-1)m < (k-1)m + 1,$$

which is a contradiction with $n = (k-1)m + 1$. Thus H is a hypertree and we finish our necessity proof. □

Next we will prove a generalization of Lemma 2:

Lemma 5. *Let $H = (V, E)$ be a k-uniform hypergraph with p components, then $\tau_c'(H) \leq (k-1)m - n + p$.*

Proof. Pick arbitrarily a minimum FES $A \subseteq E$. Suppose that $H \setminus A$ contains exactly t components $H_i = (V_i, E_i)$ with n_i vertices and m_i edges, $i = 1, \ldots, t$. It follows from Lemma 4 that $n_i = (k-1)m_i + 1$ for each $i \in [t]$. Thus $n = \sum_{i \in [t]} n_i = (k-1)\sum_{i \in [k]} m_i + t = (k-1)(m - \tau_c'(H)) + t$, which means $(k-1)\tau_c'(H) = (k-1)m - n + t$. To establish the lemma, it suffices to prove $t \leq (k-2)\tau_c'(H) + p$.

In case of $\tau_c'(H) = 0$, we have $A = \varnothing$ and $t = p = (k-2)\tau_c'(H) + p$. In case of $\tau_c'(H) \geq 1$, suppose that $A = \{e_1, \ldots, e_q\}$. Because A is a minimum FES of H, for each $i \in [q]$, there is a cycle C_i in $H \setminus (A \setminus \{e_i\})$ such that $e_i \in C_i$. Considering $H \setminus A$ being obtained from H by removing $e_1, e_2, \ldots, e_q$ sequentially, for $i = 1, \ldots, q$, since e_i have k vertices, the presence of C_i implies that the removal of e_i can create at most $k - 2$ more components. Therefore we have $t \leq (k-2)\tau_c'(H) + p$ as desired. □

Corollary 1. *Let $H = (V, E)$ be a k-uniform hypergraph with p components and $\tau_c'(H) = (k-1)m - n + p$. $A \subseteq E$ is a minimum feedback edge set. For any $e \in A$, $C(H \setminus A) = C(H \setminus \{A \setminus e\}) + k - 2$ holds, here $C(H \setminus A)$ is the number of components in $H \setminus A$ and $C(H \setminus \{A \setminus e\})$ is the number of components in $H \setminus \{A \setminus e\}$.*

Now we will prove the main theorem in this section, for every k-uniform connected hypergraph, a lower bound of the matching number with maximum degree is proved.

Theorem 3. *For every k-uniform connected hypergraph $H(V, E)$ with maximum degree Δ on n vertices and m edges, $\nu(H) \geq \frac{n-(k-2)m-1}{\Delta}$.*

Proof. It is equivalent to prove the next proposition:

Proposition. *For every k-uniform connected hypergraph $H(V, E)$ on n vertices and m edges, $d(v) \leq p, \forall v \in V$, $\nu(H) \geq \frac{n-(k-2)m-1}{p}$.*

Pick arbitrarily a minimum FES $A \subseteq E$. Consider the hypergraph $H' = H \setminus A$ and m' is the edge number. According to Lemma 5,

$$m' = m - \tau_c'(H) \geq m - [(k-1)m - n + 1] = n - (k-2)m - 1.$$

H' is acyclic and by Lemma 1, we have $\nu(H') \geq \frac{m'}{p}$. Combining the above inequalities, there is

$$\nu(H) \geq \nu(H') \geq \frac{m'}{p} \geq \frac{n - (k-2)m - 1}{p}. \qquad \square$$

Remark 1. For every k-uniform connected hypergraph $H(V, E)$ with maximum degree Δ on n vertices and m edges, Theorem 3 implies a polynomial-time algorithm for computing a matching with cardinality at least $\frac{n-(k-2)m-1}{\Delta}$.

3   The Extremal Hypergraphs on Matching Number with Maximimum Degree

In this section, we characterize the extremal hypergraphs H with $\Delta \geq 3$ and $\nu(H) = \frac{n-(k-2)m-1}{\Delta}$. A class of hypergraphs called Δ-star hypertrees are introduced and we prove Δ-star hypertrees are exactly the extremal hypergraphs, as stated in Theorem 4.

Theorem 4. *For every k-uniform connected hypergraph $H(V, E)$ with maximum degree $\Delta \geq 3$ on n vertices and m edges, $\nu(H) = \frac{n-(k-2)m-1}{\Delta}$ if and only if $H(V, E)$ is a Δ-star hypertree.*

The content is organized as follows:

- In Definitions 1, 2 and 3, a class of hypergraphs called p-star hypertrees are introduced.
- In Lemmas 6 and 7, some useful results about p-star hypertrees are proved.
- In Lemma 8, we prove Δ-star hypertrees are exactly the extremal hypertrees with $\nu(T) = \frac{m}{\Delta}$.
- In Theorem 4, we prove Δ-star hypertrees are exactly the extremal hypergraphs with $\nu(H) = \frac{n-(k-2)m-1}{\Delta}$.

Define that k-uniform p-star is a $(kp-p+1)$-vertex tree with $(kp-p)$ vertices with degree 1 and the central vertex of a p-star is a p-degree vertex, an example is shown as Fig. 2. Then we introduce the definition of p-star hypertree.

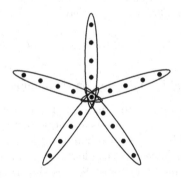

Fig. 2. 5-star

Definition 1. *A k-uniform hypertree $T(V, E)$ is called a p-star tree if it satisfies:*

- *Each vertex's degree is no more than p.*
- *The edges of T can be decomposed into several p-stars.*

Definition 2. *For a p-star hypertree $T(V, E)$, the central vertices of p-stars are called* central vertices *and other vertices are called* noncentral vertices. *The vertices connecting different p-stars are called* adjacent vertices. *Noncentral vertices are formed by adjacent vertices and leaves.*

Definition 3. *For a p-star hypertree $T(V, E)$, the* structure tree *describes the structure of T as formed by its p-stars. Let A denote the set of adjacent vertices of T, and B the set of its p-stars. Then we have a natural tree $T'(A \cup B, E')$ on $A \cup B$ formed by the edges $e'(a, b) \in E'$ with $a \in A, b \in B, a \in b$, which means that the adjacent vertex a belongs to the p-star b in T. An example is shown in Fig. 3.*

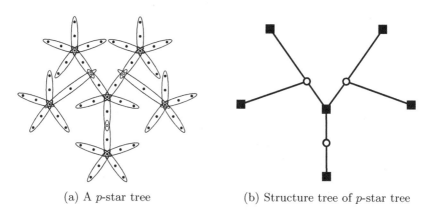

(a) A p-star tree   (b) Structure tree of p-star tree

Fig. 3. A p-star tree, where the squares, the hollow dots and the solid dots are central vertices, adjacent vertices and leaves, respectively.

Lemma 6. *For a p-star tree $T(V, E)$, there is a unique p-star decomposition of T.*

Proof. This can be finished by induction on the number t of p-stars in T.

– When $t = 0$, T is an isolated vertex and there is a unique p-star decomposition.
– When $t = 1$, T is a p-star and there is a unique p-star decomposition.
– Assume the lemma holds for $t \leq q$. When $t = q + 1$, let us consider the structure tree T' of the p-star tree T. v' is a leaf in T' and $S_{v'}$ is the corresponding p-star in T. v is the central vertex of $S_{v'}$, as shown in Fig. 4. $T \setminus v$ is also a p-star tree and the number of p-stars in $T \setminus v$ is exactly q. By our induction, there is a unique p-star decomposition D in $T \setminus v$. Thus $D \cup S_{v'}$ is the unique q-star decomposition in T. The lemma holds for $t = q + 1$. By induction, the lemma holds for every p-star tree.   □

Lemma 7. *$T(V, E)$ is a p-star hypertree.*

1. *$u \in V, e \in E, u \in e$, u is a noncentral vertex, $C_u(T \setminus e)$ is the number of edges in the component of $T \setminus e$ containing u. Then $C_u(T \setminus e)$ is congruent to 0 modulo p, that is $C_u(T \setminus e) \equiv 0 \pmod{p}$.*

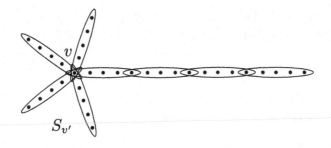

Fig. 4. The central vertex v of $S_{v'}$

2. $u \in V, e \in E, u \in e$, u is a central vertex, $C_u(T \setminus e)$ is the number of edges in the component of $T \setminus e$ containing u. Then $C_u(T \setminus e)$ is congruent to $p-1$ modulo p, that is $C_u(T \setminus e) \equiv (p-1) \pmod{p}$.

3. $u \in V, \{e_1, e_2\} \subseteq E, u \in e_1, u \in e_2$, u is a central vertex, $C_u(T \setminus \{e_1, e_2\})$ is the number of edges in the component of $T \setminus \{e_1, e_2\}$ containing u. Then $C_u(T \setminus \{e_1, e_2\})$ is congruent to $p-2$ modulo p, that is $C_u(T \setminus \{e_1, e_2\}) \equiv (p-2) \pmod{p}$.

Proof. This can be finished by induction on the number t of p-stars in T.

- When $t = 0$, T is an isolated vertex and the three propositions hold.
- When $t = 1$, T is a p-star and the three propositions hold.
- Assume the three propositions hold for $t \leq q$. When $t = q+1$, let us consider the structure tree T' of the p-star tree T. v' is a leaf in T' and $S_{v'}$ is the corresponding p-star in T. v is the central vertex of $S_{v'}$, as shown in Fig. 4. $T \setminus v$ is also a p-star hypertree and the number of p-stars in $T \setminus v$ is exactly q. By our induction, the three propositions hold in $T \setminus v$. T is formed by adding a p-star to $T \setminus v$ and the operation keeps the same remainder modulo p. Thus the three propositions also hold in T. The lemma holds for $t = q + 1$. By induction, the lemma holds for every p-star hypertree. $\qquad\square$

Next we will prove Δ-star hypertrees are exactly the extremal hypertrees with $\nu(T) = \frac{m}{\Delta}$, as stated in Lemma 8.

Lemma 8. *For every k-uniform hypertree $T(V, E)$ with maximum degree Δ on n vertices and m edges, $\nu(T) = \frac{m}{\Delta}$ if and only if $T(V, E)$ is a Δ-star hypertree.*

Proof. It is equivalent to prove the next proposition:

Proposition. *For every k-uniform connected hypertree $T(V, E)$ on n vertices and m edges, $d(v) \leq p, \forall v \in V$, $\nu(T) = \frac{m}{p}$ if and only if T is a p-star hypertree.*

Necessity: $T(V, E)$ is a hypertree with $d(v) \leq p, \forall v \in V$ and $\nu(T) = \frac{m}{p}$ holds. It needs to show T is a p-star hypertree. Obviously $T(V, E)$ has at least two edges. The longest path in T is $P = v_1 e_1 v_2 \cdots v_t e_t v_{t+1}$, which connects one leaf v_1 to another leaf v_{t+1} and v_2 is the only one vertex in e_1 with degree more than one, as shown in Fig. 1. The degree of v_2 is $d(v_2)$ and $T \setminus v_2$ has $(k-1)d(v_2)$ components, denoted as $\{T_i, 1 \le i \le (k-1)d(v_2)\}$. Assume that T_i contains m_i edges.

Claim 3. $\nu(T \setminus v_2) \ge \frac{m-p}{p}$.

According to Lemma 1, $\nu(T_i) \ge \frac{m_i}{p}$. Combined with $d(v_2) \le p$, we have the following inequalities:

$$\nu(T \setminus v_2) = \sum_{1 \le i \le (k-1)d(v_2)} \nu(T_i) \ge \sum_{1 \le i \le (k-1)d(v_2)} \frac{m_i}{p} = \frac{m-d(v_2)}{p} \ge \frac{m-p}{p}. \quad \blacksquare$$

Claim 4. $\nu(T) \ge \nu(T \setminus v_2) + 1$.

The longest path in T is $P = v_1 e_1 v_2 \cdots v_t e_t v_{t+1}$ thus v_2 is the only one vertex in e_1 with degree more than one. For every matching M in $T \setminus v_2$, $M \cup \{e_1\}$ is a matching in T. $\quad \blacksquare$

According to these claims, we have the following inequalities:

$$\nu(T) \ge \nu(T \setminus v_2) + 1 \ge \frac{m-p}{p} + 1 = \frac{m}{p}$$

Combined with $\nu(T) = \frac{m}{p}$, we have $\nu(T \setminus v_2) = \frac{m-p}{p}$. This means the degree of v_2 is exactly p and in $T \setminus v_2$, each component $\{T_i, 1 \le i \le (k-1)d(v_2)\}$ satisfies $\nu(T_i) = \frac{m_i}{p}$ holds. Take out $\{T_i, 1 \le i \le (k-1)d(v_2)\}$ as T and repeat the above analysis process. Finally, there are some isolated vertices. Denote the deleted vertices as $\{v_j, 1 \le j \le q\}$. A p-star S_j is deleted when v_j is deleted. Thus the edges of T can be decomposed into several p-stars. According to Definition 1, T is a p-star tree.

Sufficiency: T is a p-star tree. It needs to show $\nu(T) = \frac{m}{p}$. This can be finished by induction on the number t of p-stars in T.

- When $t = 0$, T is an isolated vertex and $\nu(T) = \frac{m}{p}$ holds.
- When $t = 1$, T is a p-star and $\nu(T) = \frac{m}{p}$ holds.
- Assume the sufficiency holds for $t \le q$. When $t = q + 1$, let us consider the structure tree T' of the p-star tree T. v' is a leaf in T' and $S_{v'}$ is the corresponding p-star in T. v is the central vertex of $S_{v'}$, as shown in Fig. 4. $T \setminus v$ is also a p-star hypertree and the number of p-stars in $T \setminus v$ is exactly q. By our induction, $\nu(T \setminus v) = \frac{m-p}{p}$ holds. Thus we have

$$\nu(T) = \nu(T \setminus v) + 1 = \frac{m-p}{p} + 1 = \frac{m}{p}$$

The sufficiency holds for $t = q + 1$. By induction, the sufficiency holds for every p-star tree. $\quad \square$

Remark 2. The above proof also demonstrates a polynomial-time algorithm to decide whether a tree T is a p-star hypertree. In addition, If T is a p-star hypertree, the algorithm gives the unique p-star decomposition.

Next we will prove our main theorem: Δ-star hypertrees are exactly the extremal hypergraphs with $\nu(H) = \frac{n-(k-2)m-1}{\Delta}$ when $\Delta \geq 3$.

Theorem 4. *For every k-uniform connected hypergraph $H(V, E)$ with maximum degree $\Delta \geq 3$ on n vertices and m edges, $\nu(H) = \frac{n-(k-2)m-1}{\Delta}$ if and only if $H(V, E)$ is a Δ-star hypertree.*

Proof. Sufficiency: $H(V, E)$ is a Δ-star hypertree. According to Lemmas 4 and 8, there is

$$n = (k-1)m + 1, \ \nu(H) = \frac{m}{\Delta} \Rightarrow \nu(H) = \frac{n-(k-2)m-1}{\Delta}.$$

Necessity: For every k-uniform connected hypergraph $H(V, E)$ with maximum degree Δ on n vertices and m edges, $\nu(H) = \frac{n-(k-2)m-1}{\Delta}$. It needs to prove $H(V, E)$ is a Δ-star hypertree. We prove by contradiction. Suppose $H(V, E)$ is not a hypertree. Take out arbitrarily a minimum feedback edge set $A \subseteq E$, then $T = H \setminus A$ is acyclic.

Claim 5. *Each component of T is a Δ-star hypertree.*

Denote $m(T)$ as the edge number of T and $\Delta(T)$ as the maximum degree of vertices in T. The components of T are T_i with m_i edges for $1 \leq i \leq p$. According to Lemmas 1 and 2, there is

$$m(T) = m - \tau'_c(H), \ \tau'_c(H) \leq (k-1)m - n + 1, \ \nu(T) \geq \frac{m(T)}{\Delta(T)}.$$

$$\Rightarrow \nu(H) \geq \nu(T) = \sum_{i=1}^{p} \nu(T_i) \geq \sum_{i=1}^{p} \frac{m_i}{\Delta} = \frac{m(T)}{\Delta} \geq \frac{n-(k-2)m-1}{\Delta}.$$

Combining with $\nu(H) = \frac{n-(k-2)m-1}{\Delta}$, there is $\nu(T_i) = \frac{m_i}{\Delta}$. According to Lemma 8, each component of T is a Δ-star hypertree. ∎

Now take out arbitrarily an edge $e \in A$. e is an edge in the minimum feedback edge set, thus there is a cycle C containing e in $H \setminus \{A \setminus e\}$. Assume the cycle $C = v_1 e_1 v_2 \cdots v_t e_t v_{t+1} e v_1$ and $P = v_1 e_1 v_2 \cdots v_t e_t v_{t+1}$ is a path from v_1 to v_{t+1} in $T = H \setminus A$, as shown in Fig. 5(a)(b).

Claim 6. *For each $e_i \in P, 1 \leq i \leq t$, $A \cup \{e_i\} \setminus e$ is a minimum feedback edge set of H.*

Because A and $A \cup \{e_i\} \setminus e$ have the same cardinality and A is a minimum feedback edge set of H, it is suffice to prove $A \cup \{e_i\} \setminus e$ is also a feedback edge set of H. Suppose $A \cup \{e_i\} \setminus e$ is not a feedback edge set, $H \setminus \{A \cup \{e_i\} \setminus e\}$ has a cycle C'. Because $H \setminus A$ is acyclic, C' must contain e. According to Corollary 1, the number of components in $H \setminus A$ is $k - 2$ more than the number of components in $H \setminus \{A \setminus e\}$. This means v_1 and v_{t+1} are two adjacent vertices of e in the cycle C'. But this is impossible because the unique $v_1 - v_{t+1}$ path $P = v_1 e_1 v_2 \cdots v_t e_t v_{t+1}$ in $H \setminus A$ is cut off in $H \setminus \{A \cup \{e_i\} \setminus e\}$. ∎

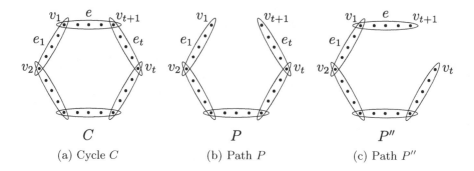

Fig. 5. Schematic diagrams in the proof of Theorem 4

According to Claim 5, each component of $T = H \setminus A$ is a Δ-star hypertree. $P = v_1 e_1 v_2 \cdots v_t e_t v_{t+1}$ is a path from v_1 to v_{t+1} in $T = H \setminus A$. Next let us consider the roles of v_1 and v_{t+1} in the Δ-star hypertree. v_1 and v_{t+1} are noncentral vertices of the Δ-star hypertree. $T = H \setminus A$, $\{v_1, v_{t+1}\} \subseteq e$, if one of v_1 and v_{t+1} is a central vertex of the Δ-star hypertree, the degree of v_1 and v_{t+1} is $\Delta + 1$ in H, this is impossible.

Claim 7. *In the path $P = v_1 e_1 v_2 \cdots v_t e_t v_{t+1}$ of $T = H \setminus A$, v_1 and v_{t+1} are noncentral vertices. Furthermore, the central vertices and noncentral vertices are alternate in the path P.*

The component of $T = H \setminus A$ is a Δ-star hypertree. In the Δ-star hypertree, two central vertices are not adjacent. Suppose in the path $P = v_1 e_1 v_2 \cdots v_t e_t v_{t+1}$, there are two noncentral vertices v_i and v_{i+1}. We consider the edge set $A \cup \{e_i\} \setminus e$. According to Claim 6, $A \cup \{e_i\} \setminus e$ is also a minimum feedback edge set of H.

On one hand, according to Claim 5, each component of $T' = H \setminus \{A \cup \{e_i\} \setminus e\}$ is also a Δ-star hypertree, thus the edge number of each component is congruent to 0 modulo Δ.

On the other hand, each component of $T = H \setminus A$ is a Δ-star hypertree, v_i and v_{i+1} are noncentral vertices and $\{v_i, v_{i+1}\} \subseteq e_i$. According to Lemma 7, in $T \setminus e_i$, the edge number of component containing v_i is congruent to 0 modulo Δ, the edge number of component containing v_{i+1} is congruent to 0 modulo Δ. This means the edge number of component containing e in $T' = H \setminus \{A \cup \{e_i\} \setminus e\}$ is congruent to 1 modulo Δ, which contradicts with T' is a Δ-star hypertree. ∎

Claim 8. *$A \cup \{e_t\} \setminus e$ is a minimum feedback edge set of H. Each component of $T'' = H \setminus \{A \cup \{e_t\} \setminus e\}$ is a Δ-star hypertree. In the path $P'' = v_{t+1} e v_1 e_1 v_2 \cdots v_{t-1} e_{t-1} v_t$ of T'' as shown in Fig. 5(c), v_t and v_{t+1} are noncentral vertices. Furthermore, the central vertices and noncentral vertices are alternate in the path P''.*

According to Claims 5 and 6, $A \cup \{e_t\} \setminus e$ is a minimum feedback edge set of H. Each component of $T'' = H \setminus \{A \cup \{e_t\} \setminus e\}$ is a Δ-star hypertree. A similar analysis as Claim 7 proves in the path $P'' = v_{t+1}ev_1e_1v_2\cdots v_{t-1}e_{t-1}v_t$ of T'', v_t and v_{t+1} are noncentral vertices. The central vertices and noncentral vertices are alternate in the path P''. ∎

On one hand, each component of $T = H \setminus A$ is a Δ-star hypertree. According to Claim 7, in the path $P = v_1e_1v_2\cdots v_te_tv_{t+1}$ of T, v_1 and v_{t+1} are noncentral vertices. The central vertices and noncentral vertices are alternate in the path P. This means v_t is a central vertex in the Δ-star hypertree of T.

On the other hand, each component of $T'' = H \setminus \{A \cup \{e_t\} \setminus e\}$ is a Δ-star hypertree. According to Claim 8, in the path $P'' = v_{t+1}ev_1e_1v_2\cdots v_{t-1}e_{t-1}v_t$ of T'', v_t and v_{t+1} are noncentral vertices. The central vertices and noncentral vertices are alternate in the path P''. This means v_t is a noncentral vertex in the Δ-star hypertree of T''.

According to Claims 7 and 8, v_1 is a noncentral vertex in P and v_{t+1} is a noncentral vertex in P'', so v_1 is a central vertex in P''. Thus, the roles of v_1 are different in P and P''.

On one hand, in the Δ-star hypertree of $T = H \setminus A$, v_1 is a noncentral vertex in P and $v_1 \in e_1$. According to Lemma 7, in $T \setminus e_1$, the edge number of component containing v_1 is congruent to 0 modulo Δ.

On the other hand, in the Δ-star hypertree of $T'' = H \setminus \{A \cup \{e_t\} \setminus e\}$, v_1 is a central vertex in P'' and $v_1 \in e_1, v_1 \in e$. According to Lemma 7, in $T'' \setminus \{e_1, e\}$, the edge number of component containing v_1 is congruent to $\Delta - 2$ modulo Δ.

The components containing v_1 in $T \setminus e_1$ and $T'' \setminus \{e_1, e\}$ are same, which is a contradiction due to $\Delta \geq 3$. Thus our initial assumption doesn't hold. $H(V, E)$ is a hypertree. According to Lemma 8, $H(V, E)$ is a Δ-star hypertree. The necessity proof is finished. □

Remark 3. The condition of Theorem 4 is that $\Delta \geq 3$. For $\Delta = 2$, the result does not hold since an odd cycle is a simple example. It is interesting to characterize the structure of H with $\Delta(H) = 2$ and $\nu(H) = \frac{n-(k-2)m-1}{2}$.

References

1. Alon, N., Frankl, P., Huang, H., Rödl, V., Ruciński, A., Sudakov, B.: Large matchings in uniform hypergraphs and the conjectures of Erdős and Samuels. J. Comb. Theory Ser. A **119**(6), 1200–1215 (2012)
2. Biedl, T., Demaine, E.D., Duncan, C.A., Fleischer, R., Kobourov, S.G.: Tight bounds on maximal and maximum matchings. Discret. Math. **285**, 7–15 (2004)
3. Cioabă, S.M., Gu, X.: Connectivity, toughness, spanning trees of bounded degree, and the spectrum of regular graphs. Czechoslov. Math. J. **66**(3), 913–924 (2016). https://doi.org/10.1007/s10587-016-0300-z
4. Cioaba, S.M., Gregory, D.A., Haemers, W.H.: Matchings in regular graphs from eigenvalues. J. Comb. Theory Ser. B **99**(2), 287–297 (2009)
5. Edmonds, J.: Path, trees, and flowers. Can. J. Math. **17**(3), 449–467 (1965)
6. Erdos, P.: A problem on independent r-tuples. Ann. Univ. Sci. Budapest. Eötvös Sect. Math. **8**, 93–95 (1965)
7. Frankl, P.: On the maximum number of edges in a hypergraph with given matching number. Discret. Appl. Math. **216**, 562–581 (2017)

8. Gabow, H.N., Tarjan, R.E.: Faster scaling algorithms for general graph matching problems. J. ACM **38**(4), 815–853 (1991)
9. Gu, X.: Regular factors and eigenvalues of regular graphs. Eur. J. Comb. **42**, 15–25 (2014)
10. Haxell, P., Scott, A.: On lower bounds for the matching number of subcubic graphs. J. Graph Theory **85**, 336–348 (2017)
11. Henning, M.A., Yeo, A.: Tight lower bounds on the size of a maximum matching in a regular graph. Graphs Comb. **23**(6), 647–657 (2007)
12. Henning, M., Yeo, A.: Tight lower bounds on the matching number in a graph with given maximum degree. J. Graph Theory **89**, 115–149 (2018)
13. Henning, M.A., Lwenstein, C., Rautenbach, D.: Independent sets and matchings in subcubic graphs. Discret. Math. **312**(11), 1900–1910 (2012)
14. Hopcroft, J.E., Karp, R.M.: A $n^{5/2}$ algorithm for maximum matchings in bipartite graphs. SIAM J. Comput. **2**, 225 (1973)
15. Huang, H., Loh, P.S., Sudakov, B.: The size of a hypergraph and its matching number. Comb. Probab. Comput. **21**(3), 442–450 (2012)
16. Huang, H., Zhao, Y.: Degree versions of the Erdős-Ko-Rado theorem and Erdős hypergraph matching conjecture. J. Comb. Theory Ser. A **150**, 233–247 (2017)
17. Jahanbekam, S., West, D.B.: New lower bounds on matching numbers of general and bipartite graphs. Congr. Numer. **218**, 57–59 (2013)
18. Micali, S., Vazirani, V.: An $O(\sqrt{n}m)$ algorithm for finding maximum matchings in general graphs. In: Proceedings of the 21st Annual Symposium on Foundations of Computer Science (FOCS), pp. 17–27 (1980)
19. Suil, O., Douglas, B.: West: balloons, cut-edges, matchings, and total domination in regular graphs of odd degree. J. Graph Theory **64**, 116–131 (2010)
20. Suil, O.: Spectral radius and matchings in graphs. Linear Algebra Appl. **614**(2), 316–324 (2021)
21. Suil, O., West, D.B.: Matching and edge-connectivity in regular graphs. Eur. J. Comb. **32**(2), 324–329 (2011)
22. Tang, Z., Diao, Z.: On the transversal number of rank k hypergraphs. In: Li, M., Sun, X. (eds.) IJTCS-FAW 2022. LNCS, vol. 13461, pp. 162–175. Springer, Cham (2022). https://doi.org/10.1007/978-3-031-20796-9_12
23. Vazirani, V.V.: A theory of alternating paths and blossoms for proving correctness of the general graph matching algorithm. Combinatorica **14**(1), 71–109 (1994)

Max-Min Greedy Matching Problem: Hardness for the Adversary and Fractional Variant

T.-H. Hubert Chan[1], Zhihao Gavin Tang[2], and Quan Xue[1(✉)]

[1] The University of Hong Kong, Hong Kong, China
`csxuequan@connect.hku.hk`
[2] ITCS, Shanghai University of Finance and Economics, Shanghai, China

Abstract. Eden, Feige, and Feldman considered the max-min greedy matching problem can be viewed as a game between an *algorithm* and an *adversary*. A bipartite graph between items and players is given to both parties upfront. The algorithm first chooses a priority order on the items, and then depending on the algorithm's choice, the adversary chooses a priority order on the players. Then, the two priority orders are used in a greedy process to produce a matching between the items and the players; specifically, when it is a player's turn, the highest priority item among its still available neighbors will be matched. The goal of the algorithm is to maximize the size of the resulting matching, while the goal of the adversary is to minimize its size. The previous work shows that the algorithm has a polynomial-time strategy to ensure a competitive ratio of strictly greater than $\frac{1}{2}$.

In this work, we show that from the adversary's perspective, the adversarial order minimum matching problem is NP-hard to approximate with a ratio better than $\frac{6}{5}$, assuming the small set expansion (SSE) hypothesis. On the other hand, we propose a fractional variant of the problem and examine the interplay between the algorithm and the adversary when one or both parties may use fractional permutations. An interesting result is that if the algorithm uses only integral item permutations, then an optimal response for the adversary can also be an integral player permutation. Moreover, we also show that in a fractional variant, the algorithm can use a round-robin strategy to achieve a competitive ratio of at least $1 - \frac{1}{e}$ for input graphs with large enough granularity parameter m. Furthermore, we show that the analysis for the round-robin strategy is tight even for regular graphs.

Keywords: max-min greedy matching · adversarial hardness · fractional matching

T.-H. H. Chan—This research was partially supported by the Hong Kong RGC grants 17201220, 17202121 and 17203122.

M. Li et al. (Eds.): IJTCS-FAW 2023, LNCS 13933, pp. 85–104, 2023.
https://doi.org/10.1007/978-3-031-39344-0_7

1 Introduction

Karp, Vazirani, and Vazirani [10] introduced the *ranking* algorithm for the online bipartite matching problem, which can be described as a greedy matching process as follows. The input of the problem is a bipartite graph $G = (U, V; E)$ between players U and items V. Based on G, the adversary chooses an arrival order of the players and the algorithm picks a uniformly random priority order on the items. When a player arrives, it is matched to the highest priority item among its still unmatched (if any) neighbors. Observe that even though the algorithm's strategy is independent of G, the resulting matching has an expected size that achieves a competitive ratio of at least $1 - \frac{1}{e}$.

Eden, Feige, and Feldman [5] studied a related variant of the bipartite matching problem in which the rules for how each party makes its decision are modified. In the *max-min greedy matching* problem, the algorithm may refer to the input graph and first picks a priority order on the items, after which the adversary can adaptively pick the arrival order of the players depending on the input graph and also the algorithm's choice. For clarity, we describe the matching process in more details as follows.

Bipartite Matching with Priorities. Suppose $G(U, V; E)$ is a bipartite graph with n vertices on each side. As above, we will think of U as players and V as items. For every vertex $w \in U \cup V$, let $N(w)$ denote the set of neighbors of w in G. We use $\mathcal{S}(A)$ to denote the collection of permutations over a set A, which represents a priority order on A. Given a player permutation $\sigma \in \mathcal{S}(U)$ and an item permutation $\pi \in \mathcal{S}(V)$, we use $M_G[\sigma, \pi]$ to denote the matching achieved by the following greedy process. Players in U arrive in the order given by σ, and an arriving vertex $u \in U$ is matched to the unmatched vertex (if any) in $N(u)$ with the highest priority under π. If all vertices in $N(u)$ are matched, then u is left unmatched. The competitive ratio is the size of $M_G[\sigma, \pi]$ divided by that of a maximum matching in G.

Max-Min Greedy Matching Problem. This is a game between an *algorithm* and an *adversary*. Given an input graph $G(U, V; E)$, the algorithm first picks $\pi \in \mathcal{S}(V)$ and then the adversary picks $\sigma \in \mathcal{S}(U)$, which may depend on the algorithm's choice. The algorithm's objective is to maximize the size of the matching $M_G[\sigma, \pi]$, while the adversary's objective is to minimize its size. Hence, formally, the game can be described by the following max-min problem:

$$\max_{\pi \in \mathcal{S}(V)} \min_{\sigma \in \mathcal{S}(U)} |M_G[\sigma, \pi]|.$$

The aforementioned previous work [5] showed that the algorithm has a polynomial-time strategy that achieves a competitive ratio of larger than 0.51. Moreover, they showed that for certain classes of regular graphs with arbitrarily large degrees, no strategy by the algorithm can achieve a competitive ratio of larger than $\frac{8}{9}$. While they conjectured that finding the optimal algorithm strategy is computationally hard, they showed that there is a polynomial-time procedure that decides whether the optimal algorithm strategy has competitive ratio of exactly 1, and if "yes", can also return a corresponding optimal item permutation.

1.1 Our Results

We summarize the contribution of this work and the structure of the paper as follows.

- **Section 3: Hardness of the adversary side.** Given a bipartite graph $G = (U, V; E)$ and an item permutation π over V, we refer to the strategy made by the adversary as the *adversarial order minimum matching* problem, which we show (in Theorem 1) is NP-hard to approximate with ratio smaller than $\frac{6}{5}$, assuming the small set expansion hypothesis.
- **Section 4: Fractional Variant.** We utilize the concept of fractional permutations, which can be interpreted as breaking a vertex in smaller pieces that we refer to as *atoms*. An interesting question is for each party, whether utilizing a fractional permutation would offer any advantage as opposed to utilizing only an integral one. We examine the interplay between the algorithm and the adversary when one or both parties implement fractional permutations. Among several cases, we think that the most interesting result is that if the algorithm uses only integral item permutations, then the adversary may respond optimally using only an integral player permutation (Theorem 2).
- **Section 5: Tight Analysis of the Round-Robin Algorithm.** Inspired by the ranking algorithm, the algorithm uses an oblivious strategy for the fractional variant, in which atoms for distinct items are chosen in a round-robin fashion to form the priority order; the formal procedure is described in Sect. 4.

  We show that when the granularity of the atoms is fine enough, the round-robin fractional algorithm can achieve a competitive ratio approaching $1 - \frac{1}{e}$ in Theorem 3.

  On the other hand, we also show that for certain classes of regular graphs, the round-robin algorithm cannot achieve a competitive ratio of strictly larger than $1 - \frac{1}{e}$ in Theorem 4.

1.2 Related Work

Since we study the same problem as in the most relevant aforementioned work [5], we will summarize some of the related works and contrast them with our results.

Online Bipartite Matching. As mentioned above, the *ranking* algorithm proposed by Karp et al. [10] and its subsequent analysis [8] have given us the inspiration to analyze the round-robin algorithm for the fractional variant to achieve the same competitive ratio of $1 - \frac{1}{e}$.

For regular graphs with large degrees d, there are randomized algorithms that can achieve ratios close to 1. For instance, Cohen and Wajc [2] presented a randomized algorithm that can achieve an expected value of $1 - O(\frac{\sqrt{\log d}}{\sqrt{d}})$ for d-regular graphs and has a lower bound of $1 - O(\frac{1}{\sqrt{d}})$. In contrast, our construction in Theorem 4 shows that the analysis of the round-robin algorithm is tight for regular graphs.

The case when the adversary chooses a random player arrival order has also been analyzed. The deterministic greedy algorithm attains $1-\frac{1}{e}$ and no deterministic algorithm can attain more than $\frac{3}{4}$ [2]. For random player arrival order, the ranking algorithm obtains a competitive ratio of between 0.696 and 0.727 [9,12], while no randomized algorithm can achieve more than 0.823 [13].

More details on online bipartite matching are given in the survey [15] and experimental results are given in the recent paper [1].

Minimum Maximal Matching. Since the greedy matching process always produces a maximal matching, the hardness for the adversary side is closely related to the *minimum maximal matching problem* studied by Dudycz et al. [3], where for a given graph, the goal is to return a maximal matching with minimum size. Assuming the *Unique Games Conjecture* (UGC), the hardness of approximation is 2 for general graphs. For bipartite graphs, the hardness result is $\frac{4}{3}$ also under UGC; however, under the stronger small set expansion (SSE) hypothesis the hardness result for bipartite graphs can be improved to $\frac{3}{2}$. The techniques used in achieving our hardness results are also inspired by [3].

Pricing Mechanisms. As mentioned in [5], the max-min greedy matching problem is also related to pricing mechanisms [7], where setting prices of items corresponds to an item permutation. Furthermore, various forms of posted price mechanisms have been proposed for different combinatorial settings [4,6,7,11]. Hence, it is not surprising that our analysis of the round-robin algorithm in Theorem 3 is also based on a pricing strategy.

2   Preliminaries

Recall that max-min greedy matching is a game between an *algorithm* and an *adversary*. The aim of the algorithm is to maximize the matching, while the aim of the adversary is to minimize the matching. The bipartite graph $G(U, V; E)$ is given upfront. Upon seeing G the algorithm chooses a linear order π over V. Upon seeing G and π, the adversary chooses a linear order σ over U. The combination of G, π and σ defines a unique matching $M_G[\sigma; \pi]$ that we refer to as the greedy matching. It is the matching produced by the greedy matching algorithm in which vertices of U arrive in order σ and each vertex $u \in U$ is matched to the highest (under π) yet unmatched $v \in N(u)$ (or left unmatched, if all of $N(u)$ has already been matched).

The following observation states that any upper bound on the competitive ratio can be attained by instances admitting perfect matchings.

Observation 1 (Instances with Perfect Matching are Hardest for Algorithm). *If there is an algorithm that achieves competitive ratio at least r for bipartite graphs with perfect matching, then the algorithm can be modified to achieve ratio at least r for general bipartite graphs.*

Proof. Suppose algorithm *ALG* achieves a competitive ratio of r for graphs with perfect matching. Consider a general graph $G = (U, V; E)$ where the maximum matching is M. We provide *ALG* with the induced subgraph $G[M]$ on the matched vertices in M as input, which returns a permutation π_M of matched items.

Then, the modified algorithm returns π_M followed by items unmatched in M in any arbitrary order. Because ALG achieves competitive ratio r, any permutation σ_M on the matched players will lead to at least $r \cdot |M|$ matched items. Finally, observe that interleaving unmatched players among σ_M cannot decrease the size of the matching. Hence, the result follows. □

(Discrete) Fractional Variant. Given a bipartite graph instance G and a granularity parameter $m \in \mathbb{Z}$, we define a fractional version of G, denoted as G_m. To construct G_m, each vertex w in G is subdivided into m *atoms* (each having a weight of $\frac{1}{m}$). Edges in G_m are induced by edges in G in the natural way: every edge $(u, v) \in G$ induces a complete bipartite graph between atoms derived from u with atoms derived from v. The game between the *algorithm* and the *adversary* on the new graph G_m is the same as before, except that each party chooses a permutation of the atoms on its side. The weight of the corresponding (fractional) matching $M_{G_m}[\sigma, \pi]$ refers to the total weights of matched atoms on each side.

Fractional Permutations. A fractional permutation refers to a permutation of atoms (derived with respect to some granularity parameter m).

Integral Permutations. An integral permutation refers to a permutation of atoms in which atoms derived from the same original vertex are located consecutively. This naturally corresponds to a permutation of the original vertices.

Continuous Variant. It is possible to define a continuous variant directly (with $m = \infty$), but complicated mathematical notions in measure theory are needed to formally define a continuous permutation and describe the matching procedure given two continuous permutations on both sides of the bipartite graphs. However, since we only consider a relatively simple algorithm (known as *Round Robin* in Sect. 5) for the continuous variant, it is simpler to treat the continuous variant as the limiting behavior as m tends to infinity.

3 Hardness for the Adversary Side

In previous work [5], it is shown that there is a polynomial-time algorithm that achieves a competitive ratio of strictly larger than $\frac{1}{2}$, even against an adversary with unlimited computational power. To the best of our knowledge, the complexity of the max-min greedy matching problem on the adversary side has not been investigated. We show that the problem faced by the adversary side is actually hard to approximate given some standard hardness assumption. Our proofs borrow techniques that are used to achieve hardness results for the similar *minimum maximal matching problem* [3]. We first define formally the problem for the adversary side.

Adversarial Order (AO) Minimum Matching Problem. Given a bipartite graph $G = (U, V; E)$ and an item permutation $\pi \in \mathcal{S}(V)$, the adversary returns a player permutation $\sigma \in \mathcal{S}(U)$ that minimizes the size of the matching $M_G[\sigma; \pi]$. The approximation ratio (which is at least 1) for the adversary is compared with respect to the optimal $\min_{\sigma \in \mathcal{S}(U)} |M_G[\sigma, \pi]|$.

As in previous works [3,14], our inapproximability result is based on the following hardness assumption.

Small Set Expansion (SSE) Hypothesis [16, Conjecture 1.3]. The edge expansion of a set X of vertices in a graph G is defined as: $\frac{|\partial X|}{|X|}$, where ∂X denotes the subset of edges that have exactly one endpoint in X. The small set expansion of a graph with n vertices is defined to be the minimum edge expansion among its subsets of at most $\frac{n}{\log_2 n}$ vertices. The small set expansion hypothesis asserts that, for every $\epsilon > 0$, it is NP-hard to distinguish for d-regular graphs between the two cases of whether the small set expansion is at least $(1 - \epsilon)d$, or at most ϵd.

Similar to the minimum maximal matching problem [3], our reduction is also from the decision version of the *maximum balanced biclique* (MBB) problem.

Fact 1 (Decisional MBB is SSE-Hard [3]). *Assuming the SSE hypothesis, for every $\epsilon > 0$, it is NP-hard to distinguish, given a bipartite graph $G = (A, B; E)$, with $|A| = |B|$, between the following two cases:*

- Yes *case:* $\exists K_A \subset A, K_B \subset B, |K_A| = |K_B| = (\frac{1}{2} - \epsilon)n$ *such that* $K_A \times K_B \subseteq G$. *Namely, there is a balanced biclique in G with almost half of the vertices.*
- No *case:* $\forall K_A \subset A, K_B \subset B, K_A = K_B > \epsilon|A| \implies \exists a \in K_A, b \in K_B, (a, b) \notin E(G)$. *Namely, there is no balanced biclique with more than ϵ-fraction of vertices.*

The following inapproximability result is the main result of this section.

Theorem 1 (Hardness of AO Minimum Matching). *The AO minimum matching problem is NP-hard to approximate with ratio strictly smaller than $\frac{6}{5}$ under the SSE hypothesis.*

Following the proof strategy in [3], we show that if there exists a polynomial-time algorithm for AO minimum matching problem with an approximation ratio smaller than $\frac{6}{5}$, then we can use this algorithm to construct a randomized algorithm that solves the MBB problem with a success probability greater than some positive constant. This would imply that the MBB problem is in RP, which contradicts Fact 1 under the SSE hypothesis.

3.1   Proof of Theorem 1: Reduction from the MBB Problem

Assuming that we have an adversary Adv that can approximate the AO minimum matching problem with some small ratio ρ, we construct a randomized algorithm ALG for the decisional MBB problem.

Given an instance G of the MBB problem, we first describe how to transform it into an instance of the AO minimum matching problem.

Transformation from MBB to AO Minimum Matching. Let $G = (A, B; E)$ (together with the distinguishing parameter ϵ) be an instance of the MBB problem. Our transformation will return a bipartite graph G' together with a collection S of item permutations.

Augmented Bipartite Graph. Suppose $n = |A| = |B|$. We use the same construction of G' as in [3, Section 7], where $G' := (A \cup A', B \cup B'; \overline{E} \cup E')$. The construction takes the following steps.

1. Start from the complement bipartite graph of G, i.e., add the edge $(a, b) \in A \times B$ iff $(a, b) \notin E$.
2. Add $n(\frac{1}{2} + \epsilon)$ new vertices to each side, i.e., $|A'| = |B'| = n(\frac{1}{2} + \epsilon)$.
3. Each newly added vertex is connected to all vertices (both old and new) on the other side, i.e., $E' = (A' \times B) \cup (A \times B') \cup (A' \times B')$.

Recall that in the AO minimum matching problem, $A \cup A'$ are the players and $B \cup B'$ are the items.

Collection S of Item Permutations. Let $\pi_{B'}$ be an arbitrary permutation of B'. Let $S = \{\pi_{B'} \oplus \pi_B : \pi_B \in \mathcal{S}(B)\}$, i.e., the items in B' have some highest fixed priorities, followed by items in B arranged uniformly at random.

Randomized Algorithm ALG for MBB. We generate $N = n^2$ independent instances of AO minimum matching. Each instance uses the same augmented G' and samples an item permutation uniformly at random from S. Instance $i \in [N]$ is passed to the adversary Adv that returns some player permutation, and let $\mathcal{M}_i$ be the size of the corresponding matching. Define $\overline{\mathcal{M}} := \frac{1}{N} \sum_{i \in [N]} \mathcal{M}_i$. If $\overline{\mathcal{M}} < (\frac{3}{2} - \epsilon)n$, ALG returns "yes"; otherwise, ALG returns "no".

To complete the inapproximability proof, it suffices to consider the behavior of ALG for "yes" and "no" instances G of MBB. For "no" instances, we can readily use the result in [3, Lemma 25].

Fact 2 (No Instances of MBB). *If G has no biclique $K_{\epsilon n, \epsilon n}$, every maximal matching in G' contains at least $(\frac{3}{2} - \epsilon)n$ edges.*

Lemma 1. *If G has no biclique $K_{\epsilon n, \epsilon n}$, then ALG outputs "no" with probability 1.*

Proof. Observe that every player permutation returned by Adv will lead to a maximal matching in G'. Hence, by Fact 2, for every instance $i \in [N]$, $\mathcal{M}_i \geq (\frac{3}{2} - \epsilon)n$. This implies that $\overline{\mathcal{M}} \geq (\frac{3}{2} - \epsilon)n$ with probability 1. $\square$

For "yes" instances G of MBB, we show that ALG returns "yes" with some constant probability. The idea is that if the adversary ideally knew a $K_{n(\frac{1}{2} - \epsilon), n(\frac{1}{2} - \epsilon)}$ biclique $K_A \times K_B$ in G, then it can return a player permutation such that the expected size of the resulting matching is small, where the randomness is over the random item permutation sampled from S. However, our polynomial-time adversary can only achieve some approximation ratio ρ, which inflates the expectation of the ideal case by a factor of at most ρ. Finally, the repetition with N independent instances allows us to use standard measure concentration techniques to achieve a constant probability statement.

Lemma 2. *If G has a $K_{n(\frac{1}{2}-\epsilon),n(\frac{1}{2}-\epsilon)}$ biclique $K_A \times K_B$, then in the transformed instance G' for AO minimum matching, if an item permutation is sampled uniformly from S, an (unbounded) adversary has a strategy such that the size $\mathcal{M}$ of the resulting matching satisfies:*
$$E[\mathcal{M}] \leq n(\tfrac{5}{4} + 3\epsilon - \epsilon^2).$$

Proof. Since the adversary is unbounded, we may assume that it knows the players K_A in the biclique. Then, the adversary picks a permutation of the form: $\sigma = (A - K_A, A', K_A)$, which consists of three blocks. The first block includes players in $A - K_A$, the second block includes players in A', and the third block includes players in K_A. The order of players within each block can be arbitrary.

Consider a permutation π sampled uniformly at random from S, i.e., the $(\frac{1}{2}+\epsilon)n$ items in B' have the highest priorities, followed by a random permutation of the n items in B.

Observe that for the player permutation σ, the $(\frac{1}{2} + \epsilon)n$ players in $A - K_A$ will be matched to the items in B'. Then, the top $(\frac{1}{2} - \epsilon)n$ items in B according to π will be matched to the players in A'. Since K_A and K_B have no edges in G', any item in K_B falling into the last $(\frac{1}{2} - \epsilon)n$ positions will remain unmatched. Observe that an item in K_B falls into the last $(\frac{1}{2} - \epsilon)n$ positions in π with probability $\frac{1}{2} - \epsilon$.

Hence, the expected number of such items is $|K_B| \cdot (\frac{1}{2} - \epsilon)$. This implies that the size $\mathcal{M}$ of the resulting matching has expectation $E[\mathcal{M}] \leq n(\frac{3}{2} + \epsilon) - |K_B| \cdot (\frac{1}{2} - \epsilon) = n(\frac{5}{4} + 3\epsilon - \epsilon^2)$. □

Lemma 3. *Suppose an adversary Adv for AO minimum matching can achieve approximation ratio $\rho < \frac{6}{5}$. For small enough $\epsilon > 0$ and large enough n, if G has a $K_{n(\frac{1}{2}-\epsilon),n(\frac{1}{2}-\epsilon)}$ biclique, then ALG outputs "yes" with probability at least $1 - e^{-2}$.*

Proof. Lemma 2 and the ρ-approximation of the adversary Adv implies that for each instance $i \in [N]$, the size $\mathcal{M}_i$ of the resulting matching has expectation $E[\mathcal{M}_i] \leq \rho \cdot n(\frac{5}{4} + 3\epsilon - \epsilon^2)$.

Observe that it is guaranteed that all items in B' will be matched. Hence, it follows that with probability 1, $\mathcal{M}_i \in [(\frac{1}{2} + \epsilon)n, (\frac{3}{2} + \epsilon)n]$, which is an interval of width n. By Hoeffding's inequality, for any $t > 0$, we have

$$P(N \cdot \overline{\mathcal{M}} - E[N \cdot \overline{\mathcal{M}}] \geq t) \leq e^{-\frac{2t^2}{Nn^2}}.$$

We pick $t = n^2$ and $N = n^2$. It suffices to check that if $\rho < \frac{6}{5}$, then for small enough $\epsilon > 0$ and large enough n, we have:
$\frac{t}{N} + E[\overline{\mathcal{M}}] = 1 + \rho \cdot n(\frac{5}{4} + 3\epsilon - \epsilon^2) < (\frac{3}{2} - \epsilon)n$.
This implies that $\Pr[\overline{\mathcal{M}} < (\frac{3}{2} - \epsilon)n] \geq 1 - e^{-2}$, as required. □

Lemmas 1 and 3 together complete the proof of Theorem 1.

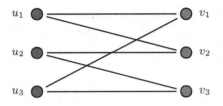

Fig. 1. The C_6 Example

4   The Power of Fractional Permutations

In Sect. 2, we introduce the notion of (discrete) fractional permutations. A natural question is whether a party can gain some advantage if it uses a fractional permutation instead of only an integral one. In this section, we investigate the interplay between the algorithm and the adversary when one or both parties use fractional permutations. Here is a summary of scenarios that we will consider.

- We use the simple C_6 example as in [5] and consider the competitive ratio when both parties use only integral permutations as the baseline.
- In the same example, if the algorithm is allowed to use a fractional permutation, while the adversary must still use an integral permutation, we show that the competitive ratio can strictly increase.
- However, we show that if the adversary is also allowed to use a fractional permutation in this example, then it can cause a slightly lower competitive ratio, but it is still larger than when both parties use integral permutations.
- Therefore, a natural question is whether a better competitive ratio can be achieved when both sides use fractional permutations, as opposed to both sides using integral permutations. Obviously, for some examples such as a single edge, there is no difference.

  The non-trivial result is that in general, allowing both parties to use fractional permutations will not decrease the competitive ratio of the max-min greedy matching problem, comparing to the case that both parties use integral permutations.

  Another way to interpret this result is that if the algorithm only uses integral permutations, then the adversary does not have an extra advantage to use fractional permutations.

The C_6 Example [5] **in Fig.** 1. The underlying bipartite graph is a cycle with 6 vertices. The set of items is $V = \{v_1, v_2, v_3\}$ and the set of players is $U = \{u_1, u_2, u_3\}$. It is simple to analyze this graph because all items and all users interact in a symmetric way. As noted in [5], all item permutations picked by the algorithm are equivalent, and the adversary can always cause one item being unmatched. For instance, if the algorithm picks $\pi = (v_1, v_2, v_3)$, the adversary can pick $\sigma = (u_1, u_3, u_2)$ such that both u_2 and v_3 are left unmatched.

Fact 3 (Both Parties Use Integral Permutations). *For C_6, when both parties act optimally using only integral permutations, the competitive ratio for the max-min greedy problem is $\frac{2}{3}$.*

Fractional Item Permutations. In this paper, we consider a special type of fractional item permutations that is oblivious to the input graph structure.

Round-Robin Permutation. Recall that each vertex is divided into m atoms in the fractional variant. A round-robin permutation consists of m blocks, where each block is a permutation of n atoms derived from distinct vertices in the original graph. Note that the permutation of the n atoms within each block can be arbitrary. Note that when m is large, the choice of the permutation within each block becomes less relevant.

Item Fractional-Player Integral. If we allow item permutation to be fractional, but restrict player permutation to be integral, by Lemma 4, it's possible that the best item permutation can only be fractional.

Lemma 4 (Fractional Item Permutations vs Integral Player Permutations). *For C_6 with granularity parameter m which is a multiple of 4, if the algorithm chooses a round-robin (fractional) item permutation, then any integral player permutation chosen by the adversary will lead to a competitive ratio $\frac{11}{12}$.*

Proof. Let $m = 4k$. Because of symmetry in C_6, we can assume without loss of generality that the adversary picks $\pi = (u_1, u_2, u_3)$. After all $4k$ atoms from u_1 arrived, they are matched to $2k$ atoms from each of v_1 and v_2. When atoms from u_2 arrive, $3k$ of them are matched to atoms from v_3, while k of them are matched to atoms from v_2. After all atoms from u_3 arrived, $2k$ of them are matched to atoms from v_1, while k of them are matched to v_3. Since at this point, all atoms of v_1 and v_3 are fully matched. Therefore, k atoms from u_3 left unmatched. Thus, the resulting ratio is $\frac{11}{12}$. $\square$

Remark 1. For large enough m, even if it is not a multiple of 4, similar analysis shows that the resulting ratio approaches $\frac{11}{12}$ as m tends to infinity.

Both Parties Use Fractional Permutations. In the context of regular graphs, the (discrete) fractional variant of C_6 has been analyzed. We paraphrase their result [5, Theorem 2] as follows.

Fact 4 (Adversary's Strategy for C_6 [5]). *For C_6, even if the algorithm is allowed to use fractional item permutations, the adversary has a strategy to also use fractional player permutations such that competitive ratio is at most $\frac{8}{9}$.*

Lemma 5 demonstrates that the algorithm can use a round-robin item permutation to achieve the competitive ratio $\frac{8}{9}$. Hence, for the fractional variant on C_6, we see that when both parties act optimally, the achieved competitive ratio is $\frac{8}{9}$, which is strictly greater than $\frac{2}{3}$ for the case when both parties may use only integral permutations.

Lemma 5 (Round-Robin Item Permutation). *For the graph C_6 and some large enough granularity parameter m, the algorithm can use a round-robin item permutation to achieve competitive ratio $\frac{8}{9}$ (no matter what fractional player permutation the adversary chooses).*

Proof. Suppose in the fractional variant, when the algorithm picks a round-robin item permutation, there is some fractional player permutation such that the competitive ratio is strictly less than 1.

Without loss of generality, suppose there is at least one player u_1-atom that is unmatched. Because the matching is maximal, this implies that all item v_1- and v_2-atoms are matched, which means that only v_3-atoms may be unmatched. Furthermore, this implies that all player u_2- and u_3-atoms are matched. We will restrict our attention to (fractional) player permutations that satisfy these conditions.

For some player permutation, consider the moment just after the last player u_2- or u_3-atom arrives. Observe that if β is the fraction of remaining unmatched v_3-atoms at this moment, they will always remain unmatched. Hence, the competitive ratio is $1 - \frac{\beta}{3}$.

Hence, the idea is to make β as large as possible, i.e., when the last u_2- or u_3-atom arrives, the fraction of matched v_3-atoms should be as small as possible. Since the algorithm uses a round-robin permutation, the adversary should keep all player u_1-atoms arriving at the end.

Moreover, since v_3 is connected to both u_2 and u_3, a round-robin item permutation implies that among the three items, the fraction of matched v_3-atoms is close to the maximum fraction for the other two types of item atoms. This implies that to maximize the fraction β of unmatched v_3-atoms, all three types of item atoms should be matched for roughly the same fraction (to player u_2- and u_3-atoms). Hence, it follows that $\beta \leq \frac{1}{3} + \frac{1}{m}$. When m is a multiple of 3, $\beta = \frac{1}{3}$.

Hence, when m is large enough, the competitive ratio is at least $1 - \frac{\beta}{3} \geq \frac{8}{9}$, as required. □

4.1 Integral Item Permutation vs Fractional Player Permutation

Fact 3 and Lemma 4 together show that if the adversary only uses integral player permutation, then it is possible for the algorithm to have an advantage by using fractional item permutations. In this section, we show an interesting result that the reverse is not true. In general, if the algorithm only uses integral item permutations, then the adversary does not lose anything by also using only integral player permutations. The formal description for this result is provided in Theorem 2.

Theorem 2. *In the fractional AO minimum matching problem, if the algorithm picks an integral item permutation, then the adversary can act optimally using also an integral player permutation.*

Proof Overview of Theorem 2. We first construct a program P which is a relaxation for the AO minimum matching problem (Lemma 6). Then, we show in Lemma 7 that any integral solution of P corresponds to an integral player permutation with exactly the same objective value.

Then, it suffices to show that P does not have an integrality gap. The main idea of this step is similar to the proof of no integrality gap of LP relaxation P_0 of maximum bipartite matching problem: all extreme points are integral. However, unlike P_0, P is not a linear programming problem. Therefore, we need extra steps to show that the feasible region R is closed and bounded in Lemma 8 and P has an optimal solution which is also an extreme point of R in Lemma 9. Then, the proof is completed by showing that any fractional solution is not an extreme point in Lemma 10, because this implies that there is an integral optimal solution to the program P. The main idea of this step is also similar to the proof in P_0: any non-integral solution can be written as a convex combination of two other points in the feasible region. But the process to find these two points is more sophisticated.

Program Relaxation P. We prove Theorem 2 with the help of a minimization program P. Given any integral item permutation, we rename the items as $V = [n]$, where $i < j$ means that item i has a higher priority than j to be matched. For player u and item v, the variable $x_{uv} \in [0, 1]$ represents the fraction of u-atoms that are matched to v-atoms.

$$min \sum_{(u,v)\in E} x_{uv}$$

$$s.t \quad \forall (u,v) \in E, \quad \sum_l x_{lv} < 1 \implies \sum_{s \leq v} x_{us} = 1 \tag{1}$$

$$\forall v \in V, \sum_l x_{lv} \leq 1$$

$$\forall u \in U, \sum_k x_{uk} \leq 1$$

$$\forall (u,v) \in E, x_{uv} \geq 0$$

The constraints for the program P consist of the usual matching constraints for a bipartite graph, but with an extra condition (1) that essentially says that if not all item v-atoms are matched, but there is a player u such that $(u, v) \in E$, then it must be the case that all player u-atoms are matched to item atoms with priorities higher or equal to v. This immediately implies that P is a relaxation of the AO minimum matching problem. However, note that because of the extra constraint, P is not exactly an LP.

Lemma 6 (Program Relaxation). *Any fractional player permutation corresponds to a feasible fractional solution to the program P, whose objective equals the resulting weight of the matching.*

Proof. For any fractional player permutation, consider the greedy matching procedure in which for any $(u, v) \in E$ in the original graph, x_{uv} is the fraction of player u-atoms matched to item v-atoms. The objective is exactly the weight of the (fractional) matching.

Then, it is clear that the variables x satisfy the usual matching constraints. Moreover, the above discussion states that x must also satisfy constraint (1). This is because if $(u, v) \in E$ and $\sum_l x_{lv} < 1$, then some item v-atom is not matched. By the max-min greedy matching property, all player u-atoms must be fully matched. Additionally, since there are some available item v-atoms left unmatched, no player u-atom can match to any item atom derived from an item with lower priority than v.   □

In general, it is not clear if every feasible solution of the program P would correspond to a (fractional) player permutation. However, we can show that every integral feasible solution corresponds to an integral player permutation.

Lemma 7 (Integral Feasible Solution). *For any integral solution x of P, there is some integral player permutation that produces a matching that agrees with x.*

Proof. We construct a player permutation σ based the feasible integral solution x. Since x is an integral solution that is feasible, x corresponds to an integral matching M between items and players. Constraint (1) implies that the matching M is maximal

In σ, we let matched players have the highest priorities. The priorities among matched players are resolved by the priorities of the corresponding matched items, i.e., a player matched to an item with higher priority will have a higher priority in σ. Unmatched players have the lowest priorities and can be arranged arbitrarily in σ.

Then, it follows that if players arrive according to σ, then the matching M will be produced.   □

Proof of Theorem 2. In view of Lemmas 6 and 7, it suffices to show that there exists an integral optimal solution to program P. This is achieved in the following two steps.

1. An optimal solution can be attained by an extreme point of the collection R of feasible solutions. Recall that an extreme point of R is one that cannot be expressed as a convex combination of two other different points in R.
2. Only integral solutions can be extreme points of R.

Lemma 8 (Feasible Region is Closed and Bounded). *The feasible region R of program P is closed and bounded.*

Proof. The matching constraints show that any feasible x must satisfy $\|x\|_\infty \leq 1$. Hence, the feasible region R is bounded.

Next, it suffices to check that R is closed. Consider an arbitrary sequence $\{x^{(n)}\}$ in R that converges to y, i.e., for all $(u,v) \in E$, $\lim_{n\to\infty} x_{uv}^{(n)} = y_{uv}$. We check that y is a feasible solution to P.

By the property of limits, it is immediate that y satisfies the matching constraints. We next check that y satisfies the constraint (1). Consider any $(u,v) \in E$. If $\sum_l y_{lv} = 1$, then there is nothing more to check; hence, we consider the case that $\sum_l y_{lv} < 1$.

By the property of limits, there exists some N such that for all $n \geq N$, $\sum_l x_{lv}^{(n)} < 1$, which implies that $\sum_{s\leq v} x_{us}^{(n)} = 1$, since $x^{(n)}$ is feasible. Hence, as n tends to infinity, we have $\sum_{s\leq v} y_{us} = 1$. Therefore, y is also a feasible solution.

Hence, the feasible region $\bar{R}$ is closed.   □

Lemma 9 (Existence of Optimal Solution that is an Extreme Point).
The program P has an optimal solution that is an extreme point.

Proof. Consider the hyperplane H such that the objective value of P equals the optimal value. By Lemma 8, the feasible region R is closed and bounded. Hence, the collection $R \cap H$ of optimal solutions is also closed and bounded. Since the ℓ_2-norm is a continuous function, let $c \in R \cap H$ be an optimal solution with the maximum ℓ_2 norm.

We show that c is an extreme point of R. Otherwise, there exists distinct $a, b \in R$ such that all three points $\{a, b, c\}$ are distinct and $c = \lambda \cdot a + (1 - \lambda) \cdot b$ for some $0 < \lambda < 1$.

Because the objective function of P is linear, the optimality of c and the feasibility of both a and b imply that both a and b are in $R \cap H$.

However, since the ℓ_2-norm is a strongly convex function, it follows that either a or b has a strictly larger ℓ_2-norm than c. In other words, when there are three distinct points on a straight line, only one of the two extreme points can have maximum ℓ_2-norm among the three points. Hence, we have reached a contradiction.   □

Lemma 10 (Only Integral Solutions Can be Extreme Points). *Any (strictly) fractional solution x of P is not an extreme point of the feasible region R.*

Proof. A strictly fractional solution x is one such that there is some $(u,v) \in E$ such that $0 < x_{uv} < 1$. In this case, we show that it is possible to create two distinct different feasible points, denoted as x^+ and x^-, such that $x = \frac{1}{2}(x^+ + x^-)$. The process for constructing these points is outlined as follows.

We say that an edge (u,v) is fractional if $0 < x_{uv} < 1$. Define G_x to be the collection of these fractional edges. We say a vertex w is saturated if it is fully matched in x, i.e., its weighted degree with respect to x is 1.

There are two possible cases for G_x. Either there exists a cycle $C \subseteq G_x$ (case 1), or there's no cycle (case 2).

1. **Case 1: there exists a cycle $C \subseteq G_x$.** Let a and b be the minimum and maximum values of x_e for $e \in C$, respectively. Define $\epsilon \leq \frac{1}{2} \cdot min(1 -$

b, a). Obtain a new (fractional) matching x^+ from x by alternatively adding and subtracting ϵ from weights on edges along C. By performing the same operations on x but reverse additions and subtractions, we obtain x^-.

Now we argue that x^+ is still feasible. A similar argument can be used to demonstrate the feasibility of x^-. Observe that the following features do not change as the solution changes from x to x^+.

- The weighted degree of every vertex does not change.
- The fractional set G_x of edges does not change. Integral edges in x also have the same value for x^+.

In particular, this means that if the weighted degree for a vertex is due to non-zero weights in x for a subset F of edges, then its weighted degree in x^+ is also derived from exactly F.

Hence, all constraints in P are still satisfied.

2. **Case 2: there's no cycle in G_x.** Since G_x contains at least one edge, let $Q \in G_x$ be a maximal path in G_x. Let s_1, s_2 be the two end points of Q. Note that s_1 and s_2 must be non-saturated as otherwise, Q is not maximal. Moreover, because of constraint (1), both s_1 and s_2 must be item vertices. Define a, b in the same way as in case 1. Let $c = \sum_l x_{s_1 l}$, $d = \sum_l x_{s_2 l}$. Define $\epsilon = \frac{1}{2} \cdot \min(1 - b, a, c, d, 1 - c, 1 - d)$. Obtain x^+ and x^- similarly as case 1. Note that apart from the weighted degree of s_1 or s_2, all other quantities remain unchanged as in case 1. However, observe that even if the weighted degree of these two vertices may change, they will remain strictly fractional. Hence, no constraint in P can be violated.

Therefore, we can conclude that both x^+ and x^- are feasible.

For both cases, since we can construct two feasible points x^+ and x^- such that $x = \frac{1}{2}(x^+ + x^-)$. Therefore, x is not an extreme point. □

5 Tight Analysis of the Fractional Round-Robin Algorithm

In Sect. 4, we showed in Lemma 4 that for some graphs, the algorithm can achieve a strictly larger competitive ratio by using fractional item permutations. Specifically, the round-robin algorithm uses a fractional item permutation that is oblivious to the input graph. In this section, we will give a tight analysis for the round-robin algorithm.

Theorem 3. *For any input bipartite graph G, when the granularity parameter m is large enough, the round-robin algorithm achieves competitive ratio approaching $1 - \frac{1}{e}$.*

Proof. To prove the lemma, we assign a price to each atom of the items to facilitate our analysis. Note that this assignment of prices can result in a (possibly fractional) permutation of the items where a smaller price corresponds to a higher priority.

100 T.-H. H. Chan et al.

Regarding the matching between an item atom a and a player atom b, we define the contribution of a to be its price divided by m, which is equal to $\frac{p_a}{m}$. Similarly, we define the contribution of b to be $\frac{1-p_a}{m}$. Then, the size of the resulting matching is the sum of the contributions of all atoms.

Let M be a maximum matching of G. To further investigate the matching, we analyze the total contribution R of all atoms of an item v and its partner in M.

Assume that for an item $v \in V$, all of its m atoms are priced from the smallest to the largest as $p_1, p_2, \ldots, p_m$. Suppose among these m atoms, the first i of them are matched, corresponding respectively to the prices $p_1, p_2, \ldots, p_i$. It holds that $R \geq 1 - p_{i+1} + \sum_{s=1}^{i} p_s \cdot \frac{1}{m}$, where we let $p_{m+1} = 1$.

Let $f(x)$ be a function such that $f(\frac{s}{m}) = p_s$. Then, let $t = \frac{i+1}{m}$. For a large number m, the expression $1 - p_{i+1} + \sum_{s=1}^{i} p_s \cdot \frac{1}{m}$ is approximately equal to $1 - f(t) + \int_0^t f(x)dx$.

When we choose the function $f(x) = e^{x-1}$, it follows that $1 - f(t) + \int_0^t f(x)dx = 1 - \frac{1}{e}$. Thus, we have shown that $R \geq 1 - \frac{1}{e}$. $\square$

In the proof of Theorem 4, we give a construction of regular graphs for which the performance of the round-robin algorithm is tight.

Theorem 4. *There exists a class of regular bipartite graphs such that when the granularity parameter m is large enough, the adversary has a strategy against the round-robin algorithm to achieve a competitive ratio approaching $1 - \frac{1}{e}$.*

Proof. To prove the theorem, we will construct a regular graph G with an integral player permutation π such that if the item permutation is round-robin, the player permutation yields a ratio of approximately $1 - \frac{1}{e}$. G is a d-regular graph with n items that are divisible by d, with $k = \frac{n}{d}$, and where d is divisible by an integer t with $z = \frac{d}{t}$. Since π is integral, we will indicate the set of items with which each player connects in player's arrival order in π to describe G and π. Lastly, we will show that when z and k are large, the competitive ratio is approximately $1 - \frac{1}{e}$.

The set of items is divided into k groups with a size of d items per group, and each group of items is further partitioned into z subgroups with a size of t items per subgroup. We will index each subgroup arbitrarily as $\alpha_1, \alpha_2, \ldots, \alpha_{kz}$. Then the ith group of items is $\bigcup_{j=1}^{z} \alpha_{(i-1)z+j}$.

The player permutation π consists of k consecutive blocks of players, and we will specify each block in descending priority order, starting with the highest. We will use μ_ℓ and h to describe the players, defining them as follows:

$$\mu_\ell = \begin{cases} \frac{z}{k(z-1)+1}, & \ell = 1 \\ \frac{z}{(k-(\ell-1))(z-1)+1} \cdot \prod_{i=0}^{\ell-2} \frac{(k-i)(z-1)}{(k-i)(z-1)+1}, & \ell \geq 2 \end{cases}$$

h is the largest integer such that $\sum_{i=1}^{h} \mu_i \leq 1$. Additionally, t should be such that $t \cdot \mu_\ell$ is an integer for all $1 \leq \ell \leq h$. Given that μ_ℓ is a rational number, such a t exists.

The ℓth Block

- $1 \leq \ell \leq h$. The block starts with $t \cdot \mu_\ell \cdot z$ identical players, each of whom is incident to all items in the ℓth group of items.
  The set of all subsets of the ℓth group that contain $z - 1$ subgroups (of items) can be arbitrarily indexed as $\{s_1, s_2, \ldots, s_z\}$. There are then $(k - \ell) \cdot z$ sub-blocks of players that will arrive, each consisting of $t \cdot \mu_\ell$ players. Each player in the ith sub-block likes all items in $s_{i \bmod z} \cup \alpha_i$.
  After all players in the ℓth block have arrived, note that the degree of all items in the ℓth group is d. Additionally, for groups (of items) with an index $\geq \ell$, the proportion of items that have been matched, denoted as x_v, is equal to $\sum_{i=1}^\ell \mu_i$.
- $h < \ell \leq k$. After the arrival of the hth block of players, every item in a group (of items) with an index of $\leq h$ has a degree of d.
  For the ℓth block of players, $t \cdot (z - \sum_{i=1}^h \mu_i)$ players arrive, each liking all the items in the ℓth group. Once all players in the ℓth block have arrived, all the items in the ℓth group will have a degree of d. As a result, the resulting graph will be d-regular.

Now that we have finished describing the graph G and player permutation π, we will show that if the item permutation is round-robin, then π yields an approximate ratio of $1 - \frac{1}{e}$ for graph G.

Competitive Ratio Analysis

By the definition of h, $\mu_{h+1} + \sum_{i=1}^h \mu_i > 1$, while $\sum_{i=1}^h \mu_i \leq 1$.

$$\mu_{h+1} = \frac{z}{(k-h)(z-1)+1} \cdot \prod_{i=0}^{h-1} \frac{(k-i)(z-1)}{(k-i)(z-1)+1}$$

$$= \frac{z}{(k-h)(z-1)+1} \cdot \prod_{j=k-h+1}^{k} \frac{j(z-1)}{j(z-1)+1}$$

$$= \frac{z}{(k-h)(z-1)+1} \cdot \prod_{j=k-h+1}^{k} 1 - \frac{1}{j(z-1)+1}$$

Take ln on both sides:

$$\ln(\mu_{h+1}) = \ln(\frac{z}{(k-h)(z-1)+1}) + \sum_{j=k-h+1}^{k} \ln(1 - \frac{1}{j(z-1)+1})$$

When $k \to \infty$, $z \to \infty$:

$$\lim_{z,k\to\infty} \ln(\mu_{h+1}) = \lim_{z\to\infty} \ln(\frac{1+\frac{1}{(z-1)}}{(k-h)+\frac{1}{(z-1)}}) + \sum_{j=k-h+1}^{k} \lim_{z\to\infty} \ln(1 - \frac{1}{j(z-1)+1})$$

$$= \ln(\frac{1}{k-h}) - \sum_{j=k-h+1}^{k} \frac{1}{j(z-1)+1}$$

$$= \ln(\frac{1}{k-h}) - \frac{1}{(z-1)} \cdot \sum_{j=k-h+1}^{k} \lim_{z\to\infty} \frac{1}{j+\frac{1}{(z-1)}}$$

$$= \ln(\frac{1}{k-h}) - \frac{1}{(z-1)} \cdot \sum_{j=k-h+1}^{k} \frac{1}{j}$$

$$= \ln(\frac{1}{k-h}) - \frac{1}{(z-1)} \cdot \lim_{k\to\infty} \sum_{j=k-h+1}^{k} \frac{1}{\frac{j}{k}} \cdot \frac{1}{k}$$

$$= \ln(\frac{1}{k-h}) - \frac{1}{(z-1)} \cdot \int_{1-\frac{h}{k}}^{1} \frac{1}{x} \, dx$$

$$= \ln(\frac{1}{k-h}) - \frac{1}{(z-1)} \cdot \ln(\frac{k}{k-h})$$

$$= \ln(\frac{1}{k-h} \cdot (\frac{k-h}{k})^{\frac{1}{(z-1)}}) \implies$$

$$\lim_{z,k\to\infty} \mu_{h+1} = \frac{1}{k-h} \cdot (\frac{k-h}{k})^{\frac{1}{(z-1)}}$$

Let's now analyze the relationship between h and k. Define $\rho_\ell = z - \sum_{i=1}^{\ell-1} \mu_\ell$. It can be proved by induction that:

$$\rho_\ell = \begin{cases} z, & \ell = 1 \\ z \cdot \prod_{i=0}^{\ell-2} \frac{(k-i)(z-1)}{(k-i)(z-1)+1}, & \ell \geq 2 \end{cases}$$

Using similar argument as above we can show that $\lim_{z,k\to\infty} \rho_{h+1} = z \cdot e^{-\frac{1}{z-1}\cdot\ln(\frac{k}{k-h})}$.

If we assume that $\frac{h}{k} = \epsilon$ where $0 < \epsilon < 1$. Then $\lim_{z,k\to\infty} \mu_{h+1} = \lim_{z,k\to\infty} \frac{1}{(1-\epsilon)k} \cdot 1 = 0$, which implies that $\lim_{z,k\to\infty} \sum_{i=1}^{h} \mu_i = 1$. Since $\rho_{h+1} = z - \sum_{i=1}^{h} \mu_i$, therefore, it holds that:

$$\lim_{z,k\to\infty} \frac{\rho_{h+1}}{z} = 1 - \frac{1}{z} = e^{-\frac{1}{z-1}\cdot\log(\frac{k}{k-h})}$$

$$\implies \lim_{z,k\to\infty} \frac{1}{z} = \frac{1}{z-1} \cdot \log(\frac{k}{k-h}) \implies \log(\frac{k}{k-h}) = \lim_{z\to\infty} \frac{z-1}{z} = 1$$

$$\implies \frac{k}{k-h} = e$$

This implies that $\epsilon = 1 - \frac{1}{e}$, which does not contradict our assumption. Since after the hth iteration, all groups of items with index $> h$ will not be matched. Therefore, the competitive ratio is:

$$\lim_{z,k \to \infty} 1 - \frac{(z-1)t(k-h-1)}{zkt} = 1 - \frac{k-h}{k} = 1 - \frac{1}{e}$$

$\square$

References

1. Borodin, A., Karavasilis, C., Pankratov, D.: An experimental study of algorithms for online bipartite matching. J. Exp. Algorithmics (JEA) **25**, 1–37 (2020)
2. Cohen, I.R., Wajc, D.: Randomized online matching in regular graphs. In: Proceedings of the Twenty-Ninth Annual ACM-SIAM Symposium on Discrete Algorithms, pp. 960–979. SIAM (2018)
3. Dudycz, S., Lewandowski, M., Marcinkowski, J.: Tight approximation ratio for minimum maximal matching. In: Lodi, A., Nagarajan, V. (eds.) IPCO 2019. LNCS, vol. 11480, pp. 181–193. Springer, Cham (2019). https://doi.org/10.1007/978-3-030-17953-3_14
4. Dutting, P., Feldman, M., Kesselheim, T., Lucier, B.: Prophet inequalities made easy: stochastic optimization by pricing nonstochastic inputs. SIAM J. Comput. **49**(3), 540–582 (2020)
5. Eden, A., Feige, U., Feldman, M.: Max-min greedy matching. Theory Comput. **18**(6), 1–33 (2022). https://doi.org/10.4086/toc.2022.v018a006, https://theoryofcomputing.org/articles/v018a006
6. Ezra, T., Feldman, M., Roughgarden, T., Suksompong, W.: Pricing multi-unit markets. ACM Trans. Econ. Comput. (TEAC) **7**(4), 1–29 (2020)
7. Feldman, M., Gravin, N., Lucier, B.: Combinatorial auctions via posted prices. In: Proceedings of the Twenty-Sixth Annual ACM-SIAM Symposium on Discrete Algorithms, pp. 123–135. SIAM (2014)
8. Goel, G., Mehta, A.: Online budgeted matching in random input models with applications to adwords. In: SODA, vol. 8, pp. 982–991 (2008)
9. Karande, C., Mehta, A., Tripathi, P.: Online bipartite matching with unknown distributions. In: Proceedings of the Forty-Third Annual ACM Symposium on Theory of Computing, pp. 587–596 (2011)
10. Karp, R.M., Vazirani, U.V., Vazirani, V.V.: An optimal algorithm for on-line bipartite matching. In: Proceedings of the Twenty-Second Annual ACM Symposium on Theory of Computing, pp. 352–358 (1990)
11. Lucier, B.: An economic view of prophet inequalities. ACM SIGecom Exchanges **16**(1), 24–47 (2017)
12. Mahdian, M., Yan, Q.: Online bipartite matching with random arrivals: an approach based on strongly factor-revealing LPS. In: Proceedings of the Forty-Third Annual ACM Symposium on Theory of Computing, pp. 597–606 (2011)
13. Manshadi, V.H., Gharan, S.O., Saberi, A.: Online stochastic matching: online actions based on offline statistics. Math. Oper. Res. **37**(4), 559–573 (2012)
14. Manurangsi, P.: Inapproximability of maximum biclique problems, minimum k-cut and densest at-least-k-subgraph from the small set expansion hypothesis. Algorithms **11**(1), 10 (2018). https://doi.org/10.3390/a11010010

15. Mehta, A., et al.: Online matching and ad allocation. Found. Trends® Theor. Comput. Sci. **8**(4), 265–368 (2013)
16. Raghavendra, P., Steurer, D.: Graph expansion and the unique games conjecture. In: Proceedings of the Forty-Second ACM Symposium on Theory of Computing, pp. 755–764 (2010)

Approximate Core Allocations for Edge Cover Games

Tianhang Lu[ID], Han Xiao[⊠][ID], and Qizhi Fang

School of Mathematical Sciences, Ocean University of China, Qingdao, China
tlu@stu.ouc.edu.cn, {hxiao,qfang}@ouc.edu.cn

Abstract. Edge cover games are cooperative cost games arising from edge cover problems, where each player controls a vertex and the cost of a coalition is the minimum weight of edge covers in the subgraph induced by the coalition. In this paper, we study the approximate core for edge cover games. We show that the ratio of approximate core depends on the shortest odd cycle of underlying graphs and the $\frac{3}{4}$-core is always non-empty. We also show that the approximate ratio $\frac{3}{4}$ is tight, since it coincides with the integrality gap of the natural LP for edge cover problems.

Keywords: Edge cover game · approximate core · linear program duality · integrality gap

1 Introduction

Game theory studies the decision-making of rational, self-interested agents in strategic environments [22]. Cooperative game theory is the branch of game theory which studies situations where players are able to making binding agreements about the distribution of payoffs outside the rules of the game [24]. One central problem in cooperative game theory is to distribute the total cost of cooperation to its participants. There are many criteria for evaluating allocations [3,24], such as stability, fairness, and satisfaction. Emphases on different criteria lead to different allocations, e.g., the core, the stable set, the Shapley value, the nucleon and the nucleolus.

The core [12], which addresses the issue of stability, is one of the most attractive solution concepts in cooperative game theory. The allocations in core are stable in the sense that no subset of players has an incentive to deviate from the grand coalition. The approximate core, which is introduced by Faigle and Kern [9], provides an alternative solution for stability. Unlike the core which can be empty, it offers an approximation to the core and is always existent, as long as the approximation ratio is bad enough. Moreover, the approximate

This work is supported in part by the National Natural Science Foundation of China (Nos. 12001507, 11871442, 11971447 and 12171444) and Natural Science Foundation of Shandong (No. ZR2020QA024).

M. Li et al. (Eds.): IJTCS-FAW 2023, LNCS 13933, pp. 105–115, 2023.
https://doi.org/10.1007/978-3-031-39344-0_8

core captures a wider range of solution concepts compared to the core, and it eventually reduces to the core when the approximation ratio equals one. In the study of approximate core, a central problem is to determine the best ratio so that the approximation to the core is as close as possible. This problem has been widely discussed for a number of cooperative games, such as matching games [9,28,30], TSP games [8,25,27], bin packing games [10,17,26] and facility location games [13,19].

Edge cover games was first studied by Deng et al. [6] to model the cost allocation problem arising from edge cover problems. From Gallai's Theorem, the core of edge cover game can be represented by the core of matching games. In this sense, the core of the edge cover game may be empty by using of the fact that the core of the matching games may be empty. On the other hand, the convex of the core can not be represented by the set of maximum independent sets on the underlying graph.

In another work, Liu and Fang [20] studied a variant of edge cover games and provided a complete characterization for the core and a sufficient condition for verifying the non-emptiness of the core. In a follow-up work, Kim [18] studied rigid fractional edge cover games and its relaxed games. They showed that a characterization of the cores of both games and found relationships between them. Park et al. [23] also studied different variants of edge cover games, including rigid k-edge cover games and its relaxed games. They gave a characterization of the cores of both games, found relationships between them, and gave necessary and sufficient conditions for the balancedness of both of them.

In this work, we study edge cover games and present a characterization for the approximate core by employing the integrality gap of the underlying problem. Our analysis demonstrates that the best ratio of the approximate core is upper bounded by the reciprocal of integrality gap. Consequently, the most promising ratio for guaranteeing the non-emptiness of approximate core in edge cover games is $\frac{3}{4}$. Additionally, we illustrate that it is always feasible to construct an allocation in the $\frac{3}{4}$-core of edge cover games efficiently.

The rest of this work is organized as follows. In Sect. 2, some notions and notations used in this paper are introduced. Section 3 is devoted to a characterization for the approximate core of edge cover games with the integrality gap. Section 4 gives the concluding remark and some possible future research for the edge cover games.

2   Preliminaries

A cooperative cost game $\Gamma(N, c)$ consists of a *player set* $N = \{1, 2, \ldots, n\}$ and a *characteristic function* $c : 2^N \to \mathbb{R}$, where for each *coalition* $S \subseteq N$, $c(S)$ represents the cost incurred by the players in S. The *core* of the game $\Gamma(N, c)$ is the set of vectors $a \in \mathbb{R}_+^{|N|}$ satisfying:

$$\begin{aligned} a(S) &\leq c(S) \quad \text{for all} \ \ S \subseteq N, \\ a(N) &= c(N). \end{aligned} \tag{1}$$

where $a(S) = \sum_{i \in S} a_i$. We say a vector $a \in \mathbb{R}_+^{|N|}$ satisfies the *core property* if $a(S) \leq c(S)$ for any $S \subseteq N$. Given $0 \leq \alpha \leq 1$, the *α-core* of the game $\Gamma(N, c)$ is the set of vectors $a \in \mathbb{R}_+^{|N|}$ satisfying:

$$a(S) \leq c(S) \quad \text{for all} \quad S \subseteq N,$$
$$a(N) \geq \alpha c(N). \tag{2}$$

A vector in the α-core guarantees that no coalition will cost more than the cost it makes on its own, and the total cost of that allocation is at least α times of the cost of all players. It is appealing to find the largest value α guaranteeing the α-core being non-empty. When the core is non-empty, the core is precisely the α-core for $\alpha = 1$.

Let $G = (V, E)$ be an undirected graph with vertices set V and edges set E. For any non-empty set $U \subseteq E$, the induced subgraph on U, denoted by $G[U]$, is a subgraph of G with edges in U. For any vertex subset $S \subseteq V$, $\delta(S)$ denotes the set of edges incident to exactly one vertex in S. If S contains a single vertex v, we use $\delta(\{v\})$ as an abbreviation for $\delta(v)$.

An *edge cover* of G is a set of edges $K \subseteq E$ such that $\delta(v) \cap K \neq \emptyset$ for any $v \in V$. Given a non-negative weight function w on E that assigns a cost to each edge, the *minimum weight edge cover problem* aims to find an edge cover such that the total weight of edges is minimized. The value of a minimum weight edge cover, denoted by $\gamma(G, w)$, is called the *weighted edge cover number*.

Edge cover games study how to allocate the total cost of the edge cover among all players. More precisely, $\Gamma_G(V, c)$ is the *edge cover game* defined on an edge-weighted graph $G = (V, E; w)$. The player set of $\Gamma_G(V, c)$ consists of the vertices in V. For any coalition $S \subseteq V$, the *cost function* $c : 2^V \to \mathbb{R}_+$ is defined by the minimum weight of edge set covering S. In other words, $c(S) = \gamma(G[E[S] \cup \delta(S)], w)$ where $E[S]$ denotes set of edges that both endpoints are contained in S.

3   The Approximate Core of Edge Cover Games

It is showed that the core of an edge cover game is non-empty if and only if there is no integrality gap for the underlying problem [6]. It turns out that the approximate core of edge cover games also admits a characterization with the integrality gap of the underlying problem. Moreover, the largest ratio guaranteeing the approximate core being non-empty is upper bounded by the reciprocal of integrality gap. Hence the problem of finding the largest ratio for the approximate core boils down to computing the integrality gap. This section is threefold. Subsection 3.1 shows how to compute an optimal half-integral edge cover. Subsection 3.2 utilizes the fractional edge cover computed in Subsect. 3.1 to prove the integrality gap of edge cover problems. Subsection 3.3 uses the integrality gap of edge cover problems to characterize the approximate core of edge cover games.

3.1   Computing an Optimal Half-Integral Edge Cover

To compute the integrality gap of edge cover problems, we resort to a class of special fractional edge covers, the optimal half-integral edge covers. We show that an optimal half-integral edge cover can always be found efficiently. Formally, a vector x is called *half-integral* if $2x$ is integral.

The following linear program (3) captures the minimum weight edge cover problem on $G = (V, E; w)$ by restricting variables to 0 and 1.

$$\text{minimize} \quad \sum_{e \in E} w_e x_e \tag{3a}$$

$$\text{subject to} \quad \sum_{e \in \delta(v)} x_e \geq 1 \quad v \in V, \tag{3b}$$

$$x_e \geq 0 \quad e \in E. \tag{3c}$$

A feasible solution to LP (3) is called a *fractional edge cover* in G. An optimal solution to LP (3) is called a *minimum fractional edge cover* in G. In the case of bipartite graphs, the weight of minimum fractional edge cover has the same value as weighted edge cover number $\gamma(G, w)$.

Lemma 1 (Schrijver [1]). *If G is a bipartite graph, then LP (3) has an integral optimal solution.*

In the following, we show how to obtain an optimal fractional solution to LP (3) that is half-integral. We employ the technique of edge doubling proposed by Nemhauser and Trotter [21], whereby we create two copies of the vertex set V, denoted as V' and V'', such that each vertex $v \in V$ corresponds to $v' \in V'$ and $v'' \in V''$. Next, we construct the graph $G' = (V' \cup V'', E')$, where $E' = \{u'v''|uv \in E\} \cup \{u''v'|uv \in E\}$, and assign weight w_{uv} to each of edges $u'v'' \in E'$ and $u''v' \in E'$. Since G' is bipartite, it is possible to efficiently compute a minimum edge cover of G', denoted as F. We set $\overline{x} = 1_F$, meaning that $\overline{x}_e = 1$ if $e \in F$ and $\overline{x}_e = 0$ otherwise.

Define x^* from $\overline{x}$ by

$$x^*_{uv} = \frac{1}{2} \left(\overline{x}_{u'v''} + \overline{x}_{u''v'} \right) \quad \text{for all} \quad uv \in E. \tag{4}$$

The following lemma shows that x^* defined in (4) is an optimal half-integral edge cover of G.

Lemma 2. x^* *is an optimal fractional edge cover of G.*

Proof. We first show that x^* is feasible for LP (3). For any vertex $v \in V$, we have

$$\sum_{uv \in \delta(v)} x^*_{uv} = \frac{1}{2} \sum_{u''v' \in \delta(v')} \overline{x}_{u''v'} + \frac{1}{2} \sum_{u'v'' \in \delta(v'')} \overline{x}_{u'v''} \geq 1, \tag{5}$$

which implies the feasibility of x^*. If x^* is not an optimal half-integral edge cover in G, there will be an optimal fractional edge cover z in G such that $\sum_{e \in E} w_e z_e < \sum_{e \in E} w_e x_e^*$. Then, we define a feasible edge cover $\bar{z}$ in G':

$$\bar{z}_{u'v''} = \bar{z}_{u''v'} = z_{vu} \quad \text{for all} \quad uv \in E. \tag{6}$$

This reduction get feasible of $\bar{z}$ and we have

$$2 \sum_{e \in E} w_e x_e^* = \sum_{u'v'' \in E'} w_{uv} \bar{x}_{u'v''} + \sum_{u''v' \in E'} w_{uv} \bar{x}_{u''v'}$$

$$\leq \sum_{u'v'' \in E'} w_{uv} \bar{z}_{u'v''} + \sum_{u''v' \in E'} w_{uv} \bar{z}_{u''v'} \tag{7}$$

$$= 2 \sum_{e \in E} w_e z_e$$

which contradicts with the assumption $\sum_{e \in E} w_e z_e < \sum_{e \in E} w_e x_e^*$.   □

We can adjust the half-integral edge cover in the Lemma 2 so that the subgraph induced by fractional components of x^* can be decomposed into vertex-disjoint odd cycles. This can be achieved by rounding x^* iteratively.

Lemma 3. *There exists an optimal half-integral edge cover $\tilde{x}$ such the subgraph induced by fractional components of $\tilde{x}$ can be decomposed into vertex-disjoint odd cycles.*

Proof. Initially, We set $\tilde{x}$ equal to x^* and describe a procedure that generates another optimal solution with strictly more integer coordinates than $\tilde{x}$. Let H be the subgraph of G induced by the set of edges $\{e \in E | \tilde{x}_e = \frac{1}{2}\}$. First, We round $\tilde{x}$ to eliminate all of paths and even cycles in H.

Let $P = v_1 v_2 \ldots v_k$ be the longest path in H. Note that if e is an edge incident to v_1 and different from $v_1 v_2$, then $x_e \neq \frac{1}{2}$; otherwise, H would contain either a cycle or a longer path. Therefore, the edge $v_1 v_2$ is the only edge connected to v_1 that has a half-integral value on $\tilde{x}$. As $\tilde{x}$ is feasible for LP (3), at least one edge incident to v_1 has a value of 1 in $\tilde{x}$. Thus we have $\tilde{x}(\delta(v_1)) \geq \frac{3}{2}$. Likewise, we can deduce that $\tilde{x}(\delta(v_k)) \geq \frac{3}{2}$. Define x' and x'' as follows:

$$x'_e = \begin{cases} \tilde{x}_e - \frac{1}{2}, & \text{if } e = v_i v_{i+1}, 1 \leq i \leq k-1 \text{ and } i \text{ is odd,} \\ \tilde{x}_e + \frac{1}{2}, & \text{if } e = v_i v_{i+1}, 1 \leq i \leq k-1 \text{ and } i \text{ is even,} \\ \tilde{x}_e, & \text{if } e \notin E(P), \end{cases}$$

and

$$x''_e = \begin{cases} \tilde{x}_e + \frac{1}{2}, & \text{if } e = v_i v_{i+1}, 1 \leq i \leq k-1 \text{ and } i \text{ is odd,} \\ \tilde{x}_e - \frac{1}{2}, & \text{if } e = v_i v_{i+1}, 1 \leq i \leq k-1 \text{ and } i \text{ is even,} \\ \tilde{x}_e, & \text{if } e \notin E(P). \end{cases}$$

There are two admissible solutions to LP (3). Moreover,

$$\sum_{e \in E} w_e \tilde{x}_e = \frac{1}{2} \left(\sum_{e \in E} w_e x'_e + \sum_{e \in E} w_e x''_e \right)$$

Vectors x' and x'' have integer coordinates in P, and share the same coordinates with $\tilde{x}$ on the other edges. As $\tilde{x}$ is an optimal fractional edge cover, x' and x'' are also optimal solutions.

For any even cycle $C = v_1 v_2 \ldots v_k$ in H with $v_1 = v_k$, we can use similar method to round $\tilde{x}$. Define x' and x'' as follows:

$$x'_e = \begin{cases} \tilde{x}_e - \frac{1}{2}, & \text{if } e = v_i v_{i+1}, 1 \leq i \leq k-1 \text{ and } i \text{ is odd}, \\ \tilde{x}_e + \frac{1}{2}, & \text{if } e = v_i v_{i+1}, 1 \leq i \leq k-1 \text{ and } i \text{ is even}, \\ \tilde{x}_e, & \text{if } e \notin E(C), \end{cases}$$

and

$$x''_e = \begin{cases} \tilde{x}_e + \frac{1}{2}, & \text{if } e = v_i v_{i+1}, 1 \leq i \leq k-1 \text{ and } i \text{ is odd}, \\ \tilde{x}_e - \frac{1}{2}, & \text{if } e = v_i v_{i+1}, 1 \leq i \leq k-1 \text{ and } i \text{ is even}, \\ \tilde{x}_e, & \text{if } e \notin E(C). \end{cases}$$

There are two admissible solutions to LP (3). Moreover,

$$\sum_{e \in E} w_e \tilde{x}_e = \frac{1}{2} \left(\sum_{e \in E} w_e x'_e + \sum_{e \in E} w_e x''_e \right)$$

Thus x' and x'' are also optimal solutions, have integer coordinates in C and share the same coordinates with $\tilde{x}$ on the other edges.

We continue this process until H does not contain any path or even cycle. Next, we proof that any two odd cycles in H are vertex-disjoint. If two odd cycles in H are vertex-disjoint but not edge-disjoint, we can combine them into an even cycle and then round it. Therefore, to prove that any two odd cycles are vertex-disjoint, it is sufficient to show that they are edge-disjoint. Suppose that H contains two cycles C_1 and C_2. Let $P = v_1 v_2 \ldots v_k$ be the longest path belong to $C_1 \cap C_2$. We use the same method to obtain two vectors x' and x'' which have integer coordinates on P. Then, we can replace $\tilde{x}$ by x' or x'' and continue this procedure until any two odd cycles in H are edge-disjoint.   □

3.2   Integrality Gap of Edge Cover Problems

This subsection studies the integrality gap of edge cover problems which will be used in characterizing the approximate core for edge cover games. The *edge cover polytope* of G, denoted by $\text{IP}(G)$, is the convex hull of incidence vectors of all edge covers of G. The *fractional edge cover polytope* of G, denoted by $\text{P}(G)$, is the convex hull of all fractional edge covers of G. It follows that $\text{P}(G)$ is precisely the polytope defined by constraints (3b) and (3c). According to Edmonds [7], $\text{IP}(G)$ can be described by $\text{P}(G)$ after imposing the following odd set constraints:

$$x(E[U] \cup \delta(U)) \geq \left\lceil \frac{1}{2}|U| \right\rceil \text{ for all } U \subseteq V, |U| \text{ odd}. \tag{8}$$

The *integrality gap* of the edge cover problem on G, denoted by $\rho(G)$, is defined by

$$\rho(G) = \max_{w:\mathbb{R}^{|E|} \to \mathbb{R}_+} \frac{\min\{wx : x \in \text{IP}(G)\}}{\min\{wx : x \in \text{P}(G)\}}. \tag{9}$$

We have the following result for $\rho(G)$.

Theorem 1. *Let $G = (V, E)$ be a graph. Then, $\rho(G) = 1 + \frac{1}{\ell(G)}$, where $\ell(G)$ is the length of the shortest odd cycle in G. Moreover, if G is bipartite, then $\rho(G) = 1$.*

Proof. We employ the result of Carr and Vempala [5]. A *dominant* $D(P)$ of a polyhedron $P \subseteq \mathbb{R}^n$ is the set of points $y \in \mathbb{R}^n$ which dominates some vector $x \in P$, i.e., $D(P) = \{y \in \mathbb{R}^n : \exists x \in P, y \geq x\}$.

Lemma 4 (Carr and Vempala [5]). *Given a polyhedron P and its convex hull of the integer points Z, the integrality gap of the linear programming on P is r if and only if $r \geq 1$ is the smallest real number such that for any point x^* of P, $rx^* \in D(Z)$.*

If G is bipartite, $\rho(G) = 1$ follows from Lemma 1 directly. Hence we assume that G is non-bipartite.

We first show that $\rho(G) \geq 1 + \frac{1}{\ell(G)}$. Let C^* be a shortest odd cycle in G. We obtain that

$$
\begin{aligned}
\rho(G) &= \max_{w:\mathbb{R}^{|E|} \to \mathbb{R}_+} \frac{\min\{wx : x \in \mathrm{IP}(G)\}}{\min\{wx : x \in \mathrm{P}(G)\}} \\
&\geq \frac{\min\{1_{C^*}x : x \in \mathrm{IP}(G)\}}{\min\{1_{C^*}x : x \in \mathrm{P}(G)\}} \\
&= \frac{(|C^*| + 1)/2}{|C^*|/2} \\
&= 1 + \frac{1}{\ell(G)}.
\end{aligned}
\tag{10}
$$

Here the second-to-last equality holds because the minimum edge cover of C^* can be attained by any matching in C^* that exposes exactly one vertex, while the minimum fractional edge cover of C^* corresponds to an half-integral edge cover.

Then we show that $\rho(G) \leq 1 + \frac{1}{\ell(G)}$. Let $\tilde{x}$ denote the optimal half-integral edge cover of G constructed in Lemma 3. Lemma 4 implies that the condition $\rho(G) \leq 1 + \frac{1}{\ell(G)}$ holds if and only if $(1 + \frac{1}{\ell(G)})\tilde{x}$ belongs to $\mathrm{IP}(G)$. Since that $(1 + \frac{1}{\ell(G)})\tilde{x}$ is a feasible fractional edge cover, we only need to show that it satisfies (8). Let H_1 and H_2 be the subgraph of G induced by the fractional and integral components of $\tilde{x}$ respectively. Then H_1 consists of vertex-disjoint cycles by Lemma 3. Moreover, by picking alternate edges in each path with a length greater than 3, we can assume that H_2 is composed of vertex-disjoint stars. Let U be any odd set of vertices in G. For any components K of $G[E[U] \cup \delta(U)]$, there are four possible cases:

1. K be a star in $H_2[E[U]]$, then $\tilde{x}(E[K]) = |E[K]|$,
2. K be a star in $H_2[\delta(U)]$, then $\tilde{x}(E[K]) = |E[K]|$,
3. K be a path in $H_1[E[U] \cup \delta(U)]$, then $\tilde{x}(K) = \frac{1}{2}|E[K]|$, and
4. K be an odd cycle in $H_1[E[U] \cup \delta(U)]$, then $\tilde{x}(K) = \frac{1}{2}|E[K]|$.

If there is no odd cycle in $G[E[U] \cup \delta(U)]$, then we have

$$\left(1 + \tfrac{1}{\ell(G)}\right) \tilde{x}(E[U] \cup \delta(U)) \geq \tilde{x}(E[U] \cup \delta(U))$$
$$= |E[U]| + |\delta(U)|$$
$$\geq \lceil \tfrac{1}{2}|U| \rceil.$$

The last inequality bases on the observation that $\tilde{x}(K) \geq \lceil \tfrac{1}{2}|V[K]| \rceil$ when K falls in the first three cases, where $V[K]$ denotes the set of vertices of K.

Otherwise, H_1 contains an odd cycle, thus $\ell(G) \leq |U|$. It follows that

$$\left(1 + \tfrac{1}{\ell(G)}\right) \tilde{x}(E[U] \cup \delta(U)) \geq \tfrac{|U|+1}{|U|} \tilde{x}(E[U] \cup \delta(U))$$
$$\geq \tfrac{|U|+1}{|U|} \cdot \tfrac{|U|}{2}$$
$$= \lceil \tfrac{|U|}{2} \rceil.$$

Therefore, we conclude that $(1 + \tfrac{1}{\ell(G)})\tilde{x}$ belongs to IP(G). $\qquad\square$

3.3   Characterizing Approximate Core with Integrality Gap

In this subsection, we introduce a characterization for the approximate core of edge cover games. For any vertex v, $N(v)$ denotes the set of vertices adjacent to v. If a vertex subset $S \subseteq N(v)$, $\delta(v, S)$ denotes the set of crossing edges between v and S. In additionally, the induced subgraph $G[\delta(v, S)]$ is called a *star* or a v-*star* with v being the *center*.

Liu and Fang [20] showed that the core of edge cover games is closely related to the stars in the underlying graph. Based on the observation that any minimum edge cover can be partitioned into some vertex-disjoint stars, we introduce the following characterizations for the core property.

Lemma 5. *A vector* $a \in \mathbb{R}_+^{|V|}$ *satisfies the core property of* $\Gamma_G(V, c)$ *if and only if for any vertex* $v \in V$ *and vertex subset* $T \subseteq N(v)$ *the inequality* $a(T \cup \{v\}) \leq \sum_{e \in \delta(v,T)} w_e$ *holds.*

Proof. Let $a \in \mathbb{R}_+^{|V|}$ be a vector satisfying the core property in $\Gamma_G(V, c)$, i.e., $a \geq 0$ and $a(S) \leq c(S)$ for any subset S of V. Since $T \subseteq N(v)$, $\delta(v, T)$ is an edge cover for $T \cup \{v\}$. It follows that $a(T \cup \{v\}) \leq c(T \cup \{v\}) \leq \sum_{e \in \delta(v,T)} w_e$.

To prove the converse, it suffices to show that $a(S) \leq c(S)$ for any $S \subseteq V$. Let K denote the set of edges which covers S with minimum weight. It is evident that K admits a star decomposition represented by $K_1, K_2, \ldots, K_l$. For each $i = 1, 2, \ldots, l$, we define u_i as the center of the star K_i, and T_i as the set of vertices in the star K_i except u_i. Hence, we have $a(S) \leq \sum_{i=1}^l a(T_i \cup \{u_i\}) \leq \sum_{i=1}^l \sum_{e \in \delta(u_i, S_i)} w_e = \sum_{e \in K} w_e = c(S)$. $\qquad\square$

Based on the linear programming formula of the α-core, we show that dual solutions of this game characterize the core property.

Lemma 6. *A vector $a \in \mathbb{R}_+^{|V|}$ satisfies the core property of $\Gamma_G(V,c)$ if and only if it is a dual feasible solution to the LP (3).*

Proof. Consider the dual of LP (3).

$$\text{maximize} \quad \sum_{v \in V} y_v \tag{11a}$$

$$\text{subject to} \quad y_u + y_v \leq w_{uv} \quad uv \in E, \tag{11b}$$

$$y_v \geq 0 \quad v \in V. \tag{11c}$$

On the one hand, let $a \in \mathbb{R}_+^{|V|}$ be a vector satisfying core property of edge cover game $\Gamma_G(V,c)$. By Lemma 5, it is easy to verify that $a_u + a_v \leq w_{uv}$ for any edge $uv \in E$. This implies that a is a feasible solution to LP (11).

On the other hand, let y be a feasible solution of LP (11), we show that y satisfies the core property. Let $v \in V$ and $T \subseteq N(v)$. Since $y_u + y_v \leq w_{uv}$, it follows that $y(T \cup \{v\}) \leq y(T) + |T|y_v \leq \sum_{e \in \delta(v,T)} w_{uv}$. By Lemma 5, a satisfies the core property.   □

Now we are ready to characterize the approximate ratio in terms of the integrality gap.

Theorem 2. *Let $\Gamma_G(V,c)$ be the edge cover game defined on graph $G = (V, E; w)$. Then the $\frac{1}{\ell(G)}$-core of $\Gamma_G(V,c)$ is always non-empty and can be computed efficiently. Moreover, $\frac{1}{\ell(G)}$ is the largest ratio guaranteeing the non-emptiness of the approximate-core of $\Gamma_G(V,c)$.*

Proof. An optimal solution a^* to the dual of LP (3) can be computed in polynomial time using standard linear programming techniques. By Lemma 6, a^* satisfies the core property, i.e., $a^*(S) \leq c(S)$ for any coalition $S \subseteq V$. Besides, we have

$$\rho(G)a^*(V) = \rho(G) \cdot \min\{wx : x \in P(G)\}$$
$$\geq \min\{wx : x \in IP(G)\}$$
$$= c(V).$$

According to the definition of the approximate core, a^* is a $\frac{1}{\ell(G)}$-core for $\Gamma_G(V,c)$. Algorithms for finding a shortest odd cycle of a graph can be finished in time $O(|V||E|)$ by using breadth-first search in [15]. Thus we can calculate $\rho(G)$ in polynomial time of $|V|$ and $|E|$.

Now we show that $\frac{1}{\ell(G)}$ is the largest ratio guaranteeing the non-emptiness of the approximate-core. Suppose a is a vector in the α-core of $\Gamma_G(V,c)$. By the definition, the a satisfies the core property. Thus a is a feasible dual solution to LP (3) by Lemma 6. We have

$$\alpha\gamma(G,w) = \alpha c(V) \leq a(V) \leq \min\{wx : x \in P(G)\},$$

where the last inequality follows by the weak duality theorem of linear programming. It follows that

$$\alpha \leq \frac{\min\{wx : x \in \mathrm{P}(G)\}}{\gamma(G, w)}.$$

By the definition of integrality gap, we have $\alpha \leq \frac{1}{\ell(G)}$.   □

Since the length of the shortest odd cycle is at least 3, the ratio will degenerate to $\frac{3}{4}$ if there is any triangle in the graph.

Corollary 1. *Let $\Gamma_G(V, c)$ be the edge cover game defined on graph $G = (V, E; w)$. Then the $\frac{3}{4}$-core of $\Gamma_G(V, c)$ is always non-empty. Moreover, an allocation in the $\frac{3}{4}$-core of $\Gamma_G(V, c)$ can be computed efficiently.*

4   Conclusion

In this paper, we considered a cost allocation problem for the edge cover game. We characterize approximate core by using the dual solution of the nature linear programming problem. Therefore, the best approximate factor depends on the integrality gap between the integer linear programming problem and its relaxation. To estimate this factor, we employ linear programming rounding techniques and prove that it is $1 + \frac{1}{\ell(G)}$. This result ensures that the $\frac{1}{\ell(G)}$-core of the edge cover game is always non-empty. Additionally, when the shortest odd cycle in the underlying graph is equal to 3, our proposed solution boils down to a factor of $\frac{3}{4}$.

One possible working direction for edge cover games is to study the nucleon. Faigle et al. [11] studied the nucleon of matching games and Kern and Paulusma [16] studied the nucleon of simple flow games. Our result might be helpful in computing the nucleon since the nucleon locals in the allocation of the largest satisfaction ratio. Besides, variants of edge cover games introduced by Liu and Fang [20] are also worth studying.

Acknowledgements. We would like to extend our sincere thanks to the anonymous reviewers for their thorough and constructive feedback, which helped us significantly improve the quality of this manuscript.

References

1. Schrijver, A.: Combinatorial Optimization: Polyhedra and Efficiency. AC, vol. 24. Springer, Berlin (2003)
2. Balinski, M.L.: Integer programming: methods, uses, computations. Manag. Sci. **12**(3), 253–313 (1965)
3. Chalkiadakis, G., Elkind, E., Wooldridge, M.: Computational aspects of cooperative game theory. Synth. Lect. Artif. Intell. Mach. Learn. **5**(6), 1–168 (2011)
4. Cornuéjols, G.: Combinatorial Optimization: Packing and Covering. SIAM (2001)
5. Carr, R., Vempala, S.: Towards a $\frac{4}{3}$ approximation for the asymmetric traveling salesman problem. In: SODA, pp. 116–125 (2000)

6. Deng, X., Ibaraki, T., Nagamochi, H.: Algorithmic aspects of the core of combinatorial optimization games. Math. Oper. Res. **24**(3), 751–766 (1999)
7. Edmonds, J.: Matching: a well-solved class of integer programs. Combinatorial Structures and their Applications, pp. 89–92 (1970)
8. Faigle, U., Fekete, S.P., Hochstättler, W., Kern, W.: On approximately fair cost allocation in Euclidean TSP games. Oper. Res. Spektrum **20**(1), 29–37 (1998)
9. Faigle, U., Kern, W.: On some approximately balanced combinatorial cooperative games. Z. Oper. Res. **38**(2), 141–152 (1993)
10. Faigle, U., Kern, W.: Approximate core allocation for binpacking games. SIAM J. Discret. Math. **11**(3), 387–399 (1998)
11. Faigle, U., Kern, W., Fekete, S.P., Hochstättler, W.: The nucleon of cooperative games and an algorithm for matching games. Math. Program. **83**(1–3), 195–211 (1998)
12. Gillies, D.B.: Solutions to general non-zero-sum games. Contrib. Theory Games **4**(40), 47–85 (1959)
13. Goemans, M.X., Skutella, M.: Cooperative facility location games. J. Algorithms **50**(2), 194–214 (2004)
14. Huang, D., Pettie, S.: Approximate generalized matching: f-factors and f-edge covers. Algorithmca **84**, 1952–1992 (2022)
15. Itai, A., Rodeh, M.: Finding a minimum circuit in a graph. SIAM J. Comput. **7**(4), 413–423 (1978)
16. Kern, W., Paulusma, D.: On the core and f-nucleolus of flow games. Math. Oper. Res. 12 (2009)
17. Kern, W., Qiu, X.: Integrality gap analysis for bin packing games. Oper. Res. Lett. **40**(5), 360–363 (2012)
18. Kim, H.K.: Note on fractional edge covering games. Glob. J. Pure Appl. Math. **12**(6), 4661–4675 (2016)
19. Kolen, A.: Solving covering problems and the uncapacitated plant location problem on trees. Eur. J. Oper. Res. **12**(3), 266–278 (1983)
20. Liu, Y., Fang, Q.: Balancedness of edge covering games. Appl. Math. Lett. **20**(10), 1064–1069 (2007)
21. Nemhauser, G., Trotter, L.E.: Vertex packings: structural properties and algorithms. Math. Program. **8**(1), 232–248 (1975)
22. Osborne, M.J., Rubinstein, A.: A Course in Game Theory. MIT Press, Cambridge (1994)
23. Park, B., Kim, S.R., Kim, H.K.: On the cores of games arising from integer edge covering functions of graphs. J. Comb. Optim. **26**(4), 786–798 (2013)
24. Peleg, B., Sudhölter, P.: Introduction to the Theory of Cooperative Games, vol. 34. Springer, Heidelberg (2007)
25. Potters, J.A., Curiel, I.J., Tijs, S.H.: Traveling salesman games. Math. Program. **53**(1), 199–211 (1992)
26. Qiu, X., Kern, W.: Approximate core allocations and integrality gap for the bin packing game. Theor. Comput. Sci. **627**, 26–35 (2016)
27. Sun, L., Karwan, M.H.: On the core of traveling salesman games. Oper. Res. Lett. **43**(4), 365–369 (2015)
28. Vazirani, V.V.: The general graph matching game: approximate core. Games Econom. Behav. **132**, 478–486 (2022)
29. van Velzen, B.: Dominating set games. Oper. Res. Lett. **32**(6), 565–573 (2004)
30. Xiao, H., Lu, T., Fang, Q.: Approximate core allocations for multiple partners matching games. arXiv:2107.01442 (2021)

Random Approximation Algorithms for Monotone k-Submodular Function Maximization with Size Constraints

YuYing Li, Min Li, Yang Zhou, and Qian Liu[✉]

School of Mathematics and Statistics, Shandong Normal University, Jinan, China
{liminEmily,zhouyang}@sdnu.edu.cn, lq_qsh@163.com

Abstract. A k-submodular function is an extension of the submodular function, which has received extensive attention due to its own value. In this paper, we design two random algorithms to improve the approximation ratio for maximizing the monotone k-submodular function with size constraints. With the total size constraint, we get an approximate ratio of $\frac{nk}{2nk-1}$, under which the total size of the k disjoint subsets is bounded by $B \in \mathbb{Z}_+$. With the individual size constraint, under which the individual size of the k disjoint subsets are bounded by $B_1, B_2, \ldots, B_k \in \mathbb{Z}_+$ respectively, satisfying $B = \sum_{i=1}^{k} B_i$, we get an approximate ratio of $\frac{nk}{3nk-2}$.

Keywords: k-submodular function · Approximation algorithms · Size constraints

1 Introduction

E is defined as a ground set with n elements. In this paper, we note that V is a set of all element position pairs (e, i), where e represents an arbitrary element in the finite set E, and i represents an arbitrary position in the k positions, represented by symbols $V = \{(e, i) | e \in E, i \in [k]\}$. The total number of element position pairs contained in V is nk. In this paper, we are interested in some subsets of V, whose element position pairs satisfy the following property: for any element $e \in E$, position $i \in [k]$, if (e, i) belongs to this set, then (e, j) must not belong to this set, where j is not equal to i. We denote by T the set of all subsets of V satisfying this property. Let $T = \{X \subseteq V | (e, i) \in X \Rightarrow (e, j) \notin X, \forall j \neq i, i, j \in [k]\}$. A set function $f : T \to R^+$ is called a k-submodular function, if it satisfies

$$f(X) + f(Y) \geq f(X \sqcup Y) + f(X \sqcap Y),$$

for any $X, Y \in T$, where

$$X \sqcap Y = X \cap Y,$$
$$X \sqcup Y = X \cup Y \backslash Z_{X \cup Y},$$
$$Z_{X \cup Y} = \{(e, i) \in X \cup Y | \exists j \neq i, (e, j) \in X \cup Y\}.$$

M. Li et al. (Eds.): IJTCS-FAW 2023, LNCS 13933, pp. 116–128, 2023.
https://doi.org/10.1007/978-3-031-39344-0_9

Obviously, when $k = 1$, the k-submodular function is equivalent to the submodular function. A k-submodular function is defined as monotone, if $f(X) \leq f(Y)$ for any $X, Y \in T$ with $X \subseteq Y$.

A k-submodular function is derived from submodular function [1–3] and bisubmodular function [4] and plays a significant role in many practical problems. For unconstrained maximization of the monotone k-submodular function, Ward and Živný [5] firstly proposed an approximate ratio of $\frac{1}{2}$ through a deterministic algorithm. Later, Iwata et al. [6] designed a random algorithm increasing the approximate ratio to $\frac{k}{2k-1}$. For unconstrained nonmonotone k-submodular maximization, Ward and Živný [5] obtained an approximate ratio of $\max\{\frac{1}{3}, \frac{1}{1+a}\}$, where $a = \max\{1, \sqrt{\frac{k-1}{4}}\}$. This approximate ratio had been increased to $\frac{1}{2}$ by Iwata et al. [6] and $\frac{k^2+1}{2k^2+1}$ by Oshima [7].

In recent years, constrained maximization of k-submodular functions have received increasing attention. For maximizing a monotone k-submodular function with the total size constraint, Ohsaka and Yoshida [8] obtained an approximate ratio of $\frac{1}{2}$ through a deterministic greedy algorithm. Meanwhile, for maximizing a monotone k-submodular function with the individual size constraint, Ohsaka and Yoshida [8] obtained an approximate ratio of $\frac{1}{3}$ through a greedy algorithm. For maximizing a monotone k-submodular function with the matroid constraint, Sakaue [9] obtained an approximate ratio of $\frac{1}{2}$ through the greedy algorithm. For nonmonotone cases, Li and Sun [10] obtained an approximate ratio of $\frac{1}{3}$. For dealing with large-scale problems, it is necessary to process elements in a flow manner, and streaming algorithm play an important role in it. For streaming k-submodular maximization under noise subject to size constraint, Nguyen and Thai [11] provided $\frac{2+2\epsilon B}{1-\epsilon}$ approximate ratio through greedy algorithm when f is monotone and $\frac{(2+2\epsilon+4\epsilon B)(1+\epsilon)}{(1-\epsilon)^2} + 1$ in case of non-monotonicity. Later for monotone k-submodular maximization with size constraints, Ene and Nguyen [12] also designed some new streaming algorithms. In addition, there are many good research results on maximizing and minimizing k-submodular functions with other constraints [13–22]. To the best of our knowledge, the approximation ratio of $\frac{1}{2}$ with the total size constraint is currently the best result for maximizing monotone k-submodular function in polynomial time, while the approximation ratio of $\frac{1}{3}$ with the individual size constraint is currently the best result.

In this paper, we study the maximization of monotone k-submodular function under different constraints, and we design two different random algorithms to improve the approximation ratio. Our main contributions are as follows:

- With the total size constraint, we design a random algorithm to obtain the approximate ratio of $\frac{nk}{2nk-1}$. The algorithm runs in $O(knB)$ time. Through the random algorithm, we select element position pairs with probability in each iteration, ensuring that the algorithm solution is closer to the optimal solution to a certain extent, thereby improving the approximation ratio.
- With the individual size constraint, we also design a random algorithm. Since the constraint is different with total size, we use a new method to analyze the

approximate ratio and get a result of $\frac{nk}{3nk-2}$. The algorithm runs in $O(knB)$ time.

The rest of the paper is organized as follows. In Sect. 2, we mainly define some symbols and describe some properties of k-submodular function. In Sect. 3, we design a random algorithm with the total size constraint to improve the approximation ratio of the monotone k-submodular function maximization. In Sect. 4, we propose a random algorithm for the problem of monotone k-submodular maximization with individual size constraint. In Sect. 5, we summarize the main work of this paper.

2   Preliminaries

In this section, we mainly introduce the representation of notations, and related properties of the k-submodular function. Let $[k] := \{1, 2, ..., k\}$, we define the support of X as $supp(X) = \{e \in E \mid (e, i) \in X, \forall i \in [k]\}$ and $supp_i(X) = \{e \in E \mid (e, i) \in X\}$ for any $i \in [k]$. Let $\varnothing$ represent a set without any element position pairs. For X and Y in T, we define a partial order $\preceq$, that is, $X \preceq Y$ if $X \subseteq Y$. We also define $\triangle_{e,i} f(X) := f(X \cup \{(e, i)\}) - f(X)$, for $X \in T, e \notin supp(X), i \in [k]$, which is a marginal gain when adding (e, i) to the $X \in T$. In addition, for any $X, Y \in T$, orthant submodularity is defined as:

$$\triangle_{e,i} f(X) \geq \triangle_{e,i} f(Y)$$

for $X \preceq Y, e \notin supp(Y)$. And the pairwise monotonicity is defined as:

$$\triangle_{e,i} f(X) + \triangle_{e,j} f(X) \geq 0$$

for any $X \in T, e \notin supp(X), i, j \in [k], i \neq j$.

Theorem 1. *(Ward and Živný [5]) A function f is k-submodular if and only if f is orthant submodular and pairwise monotone.*

In this paper, we consider the problem of maximizing a monotone k-submodular function with two different constraints, namely, total size constraint and individual size constraint. Then, we will briefly introduce the basic knowledge of the two constraints. Suppose X is a set composed of element position pairs, and $X \in T$. With the total size constraint, under which the total size of the k disjoint subsets is bounded by $B \in \mathbb{Z}_+$, we consider

$$\max \ f(X) \qquad \text{subject to } |supp(X)| \leq B,$$

where $f : T \to R^+$ is monotone k-submodular and $X \in T$. Since f is monotone, there exists an optimal solution $O \in T$, which satisfies $|supp(O)| = B$. With the individual size constraint, under which the individual size of the k disjoint subsets are bounded by $B_1, B_2, \ldots, B_k \in \mathbb{Z}_+$, that is,

$$\max \ f(X) \qquad \text{subject to } |supp_i(X)| \leq B_i, \ \forall i \in [k], X \in T.$$

Since f is monotone, there exists an optimal solution $O \in T$, which satisfies $|supp_i(O)| = B_i, \ \forall i \in [k]$ and $|supp(O)| = \sum_{i=1}^{k} B_i$.

3   Maximizing a Monotone k-Submodular Function with a Total Size Constraint

In the following, we introduce our random algorithm for maximizing a monotone k-submodular function with a total size constraint. We will select the element position pairs as probabilities, which are proportional to their marginal benefits.

Algorithm 1. A random algorithm for monotone k-submodular maximization with a total size constraint

Input: A nonnegative monotone k-submodular function f
Output: S satisfying $|supp(S)| = B$

1: $S \leftarrow \emptyset, t = nk - 1$
2: **for** $j = 1$ to B **do**
3:     $p \leftarrow 0_{n \times k}$
4:     $y \leftarrow 0_{n \times k}$
5:     **for** each $e \in E \backslash supp(S)$ and position $i \in [k]$ **do**
6:         $y_{e,i} \leftarrow \Delta_{e,i} f(S)$ for $(e, i) \in [n] \times [k]$
7:         $p_{e,i} \leftarrow y_{e,i}^t$
8:     **end for**
9:     $\beta \leftarrow \displaystyle\sum_{(e,i) \in [n] \times [k]} p_{e,i}$
10:     **if** $\beta \neq 0$ **then**
11:         $p \leftarrow \frac{p}{\beta}$
12:     **else**
13:         Arbitrarily choose an $e \in E \backslash supp(S)$ and $i \in [k]$, $p_{e,i} \leftarrow 1$
14:     **end if**
15:     Randomly choose $(e, i) \in [n] \times [k]$, with $Pr[(e, i)] = p_{e,i}$
16:     $S = S \cup \{(e, i)\}$
17: **end for**
18: **return** S

We consider the jth iteration of the algorithm, (e^j, i^j) is the element location pair selected in the jth iteration of the algorithm, $p_{e,i}^j$ represents the probability of selecting (e, i) in the jth iteration, and S^j is the output solution of the jth iteration algorithm, where S^0 is defined as an empty set. We iteratively construct $O^0 = O, O^1, ..., O^B$ as follows. For each $j \in [B]$, if $e^j \in supp(O^{j-1}) \backslash supp(S^{j-1})$, we set $o^j = e^j$. If $e^j \notin supp(O^{j-1}) \backslash supp(S^{j-1})$, we set o^j to be an arbitrary element in $supp(O^{j-1}) \backslash supp(S^{j-1})$. That is to say, in each iteration $j \in [B]$, we choose an element position pair (e^j, i^j), which corresponds to an element position pair $(o^j, O^{j-1}(o^j))$ in O^{j-1}. Firstly, we construct $O^{j-1/2}$: delete the element position pair $(o^j, O^{j-1}(o^j))$ in O^{j-1}, and the other elements and corresponding positions are exactly the same as those in O^{j-1}, that is, $O^{j-1/2} = O^{j-1} \backslash \{(o^j, O^{j-1}(o^j))\}$. Then we construct O^j: $O^j = O^{j-1/2} \cup \{(e^j, i^j)\}$. Note that for any $j \in \{0, 1, 2, ..., B\}$, there is $|supp(O^j)| = B$ and $O^B = S^B = S$. From this structure, we have $S^{j-1} \preceq O^{j-\frac{1}{2}}$. Next, for any $(e, i) \in V$, we define

$y^j_{e,i} = \triangle_{e,i} f(S^{j-1})$, $a^j_{e,i} = \triangle_{e,i} f(O^{j-\frac{1}{2}})$. The orthant submodularity implies $y^j_{e,i} \geq a^j_{e,i}$. Specifically, for each iteration j, we have the probability to select each element position pairs $(e,i) \in V$. For those element position pairs (e',i') where $e' \in supp(S^{j-1})$, the algorithm shows that, $y^j_{e',i'} = 0$, $p^j_{e',i'} = 0$.

Lemma 1. *For any $j \in [B]$, according to Algorithm 1, the following inequality holds.*

$$\sum_{S^{j-1}} \sum_{(e,i)\in[n]\times[k]} (a^j_{o^j,O^{j-1}(o^j)} - a^j_{e,i}) \cdot p^j_{e,i} \cdot p_{S^{j-1}}$$

$$\leq (1 - \frac{1}{nk}) \sum_{S^{j-1}} \sum_{(e,i)\in[n]\times[k]} y^j_{e,i} \cdot p^j_{e,i} \cdot p_{S^{j-1}}. \qquad (1)$$

Proof. To prove inequality (1), we only need to prove the following inequality holds,

$$\sum_{(e,i)\in[n]\times[k]} (a^j_{o^j,O^{j-1}(o^j)} - a^j_{e,i}) \cdot p^j_{e,i} \leq (1 - \frac{1}{nk}) \sum_{(e,i)\in[n]\times[k]} y^j_{e,i} \cdot p^j_{e,i}. \qquad (2)$$

In the following proof process, we omit the superscript j in the inequality in order to describe clearly. Firstly, let us consider the case of $\beta = 0$. At this time, $y_{e,i} = 0$, for any $(e,i) \in V$. Since f is monotone, we have $a_{e,i} \geq 0$. So

$$\begin{aligned}
a_{o^j,O^{j-1}(o^j)} - a_{e,i} &\leq a_{o^j,O^{j-1}(o^j)} \\
&= f(O^{j-\frac{1}{2}} \cup \{(o^j, O^{j-1}(o^j))\}) - f(O^{j-\frac{1}{2}}) \\
&\leq f(S^{j-1} \cup \{(o^j, O^{j-1}(o^j))\}) - f(S^{j-1}) \\
&= 0.
\end{aligned} \qquad (3)$$

Therefore, inequality (2) holds. Secondly, we need to prove that when $\beta \neq 0$, so our goal isto prove

$$\sum_{(e,i)\in[n]\times[k]} (a_{o^j,O^{j-1}(o^j)} - a_{e,i}) \cdot y^t_{e,i} \leq (1 - \frac{1}{nk}) \sum_{(e,i)\in[n]\times[k]} y^{t+1}_{e,i}. \qquad (4)$$

If $nk = 1$, it means $n = 1, k = 1$, so both sides of the inequality are equal to 0. If $nk \geq 2$, let $r = (nk-1)^{\frac{1}{t}} = t^{\frac{1}{t}}$. For each iteration $j \in [B]$, an element position pair (e,i) is added to S^{j-1}. According to the previous description, there is a corresponding $(o^j, O^{j-1}(o^j))$. We denote $I = O^{j-1} \setminus S^{j-1}$ and $y_{max} = max_{(e,i)\in I}(f(S^{j-1} \cup \{(e,i)\}) - f(S^{j-1}))$. When $(e,i) \in I$, we have $(e,i) = (o^j, O^{j-1}(o^j))$, then

$$\sum_{(e,i)\in[n]\times[k]} (a_{o^j,O^{j-1}(o^j)} - a_{e,i}) \cdot y_{e,i}^t$$

$$= \sum_{(e,i)\notin I} (a_{o^j,O^{j-1}(o^j)} - a_{e,i}) \cdot y_{e,i}^t$$

$$\leq \sum_{(e,i)\notin I} a_{o^j,O^{j-1}(o^j)} \cdot y_{e,i}^t$$

$$\leq \sum_{(e,i)\notin I} y_{o^j,O^{j-1}(o^j)} \cdot y_{e,i}^t \qquad (5)$$

$$\leq y_{max} \sum_{(e,i)\notin I} y_{e,i}^t$$

$$= \frac{1}{r}(r \cdot y_{max} \sum_{(e,i)\notin I} y_{e,i}^t).$$

Assume $a_1,...,a_n$ is a positive real number. If the sum of n non-negative real numbers is 1, $x_1 + x_2 + \cdots + x_n = 1$, the weighted AM-GM inequality holds, $a_1 x_1 + a_2 x_2 + \cdots + a_n x_n \geq a_1^{x_1} a_2^{x_2} \cdots a_n^{x_n}$. Let $a = (r \cdot y_{max})^{t+1}$, $b = (\sum_{(e,i)\neq(o^j,O^{j-1}(o^j))} y_{e,i}^t)^{\frac{t+1}{t}}$. Because $\beta > 0$, a and b cannot be 0 at the same time, when a=0 or b=0, inequality (4) clearly holds. Next, Let us discuss the case where a and b are both greater than 0. From the AM-GM inequality, $a^{\frac{1}{t+1}} b^{\frac{t}{t+1}} \leq \frac{1}{t+1}a + \frac{t}{t+1}b$, we have

$$\frac{1}{r}(r y_{max} \sum_{(e,i)\notin I} y_{e,i}^t)$$

$$\leq \frac{1}{r}[\frac{1}{t+1}(r y_{max})^{t+1} + \frac{t}{t+1}(\sum_{(e,i)\notin I} y_{e,i}^t)^{\frac{t+1}{t}}]. \qquad (6)$$

Holder's inequality, $a_1,...a_n$ and $b_1,...,b_n$ are non-negative real numbers, $\sum_{i=1}^{n} a_i b_i \leq (\sum_{i=1}^{n} a_i^p)^{\frac{1}{p}}(\sum_{i=1}^{n} b_i^q)^{\frac{1}{q}}$, let $a_i = y_{e,i}^t$, $b_i = 1$, we have

$$\frac{1}{r}(ry_{max}\sum_{(e,i)\notin I}y_{e,i}^t)$$

$$\le \frac{1}{r}[\frac{1}{t+1}(ry_{max})^{t+1}+\frac{t}{t+1}(\sum_{(e,i)\notin I}y_{e,i}^{t\cdot\frac{t+1}{t}})(\sum_{(e,i)\notin I}1^{t+1})^{\frac{1}{t}}]$$

$$\le \frac{1}{r}[\frac{1}{t+1}(ry_{max})^{t+1}+\frac{t}{t+1}(nk-1)^{\frac{1}{t}}\sum_{(e,i)\notin I}y_{e,i}^{t+1}]$$

$$=\frac{r^t}{t+1}y_{max}^{t+1}+\frac{t}{t+1}\sum_{(e,i)\notin I}y_{e,i}^{t+1} \tag{7}$$

$$=\frac{t}{t+1}y_{max}^{t+1}+\frac{t}{t+1}\sum_{(e,i)\notin I}y_{e,i}^{t+1}$$

$$\le \frac{t}{t+1}\sum_{(e,i)\in[n]\times[k]}y_{e,i}^{t+1}$$

$$=(1-\frac{1}{nk})\sum_{(e,i)\in[n]\times[k]}y_{e,i}^{t+1}.$$

To sum up, inequality (4) holds, thus inequality (1) holds. We obtained this lemma.

Theorem 2. *With the total size constraint, a $\frac{nk}{2nk-1}$-approximate solution can be obtained by Algorithm 1.*

Proof. We find the relationship between $f(O)$ and $\mathbb{E}[f(S)]$. Noting that

$$\mathbb{E}[f(O^{j-1})-f(O^j)]=\sum_{S^{j-1}}\sum_{(e,i)\in[n]\times[k]}(a_{o^j,O^{j-1}(o^j)}^j-a_{e,i}^j)\cdot p_{e,i}^j\cdot p_{S^{j-1}},$$

and

$$\mathbb{E}[f(S^j)-f(S^{j-1})]=\sum_{S^{j-1}}\sum_{(e,i)\in[n]\times[k]}y_{e,i}^j\cdot p_{e,i}^j\cdot p_{S^{j-1}}.$$

It can be known from Lemma 1, we have $\mathbb{E}[f(O^{j-1})-f(O^j)]\le (1-\frac{1}{nk})\mathbb{E}[f(S^j)-f(S^{j-1})]$. Hence

$$f(O)-\mathbb{E}[f(S)]=\sum_{j=1}^{B}\mathbb{E}[f(O^{j-1})-f(O^j)]$$

$$\le (1-\frac{1}{nk})\sum_{j=1}^{B}\mathbb{E}[f(S^j)-f(S^{j-1})] \tag{8}$$

$$=(1-\frac{1}{nk})(\mathbb{E}[f(S)]-f(\emptyset))$$

$$\le (1-\frac{1}{nk})\mathbb{E}[f(S)].$$

Then $\mathbb{E}[f(S)]\ge \frac{nk}{2nk-1}f(O)$, the result can be obtained.

4 Maximizing a Monotone k-Submodular Function with the Individual Size Constraint

In this section, we mainly introduce the problem of maximizing monotone k-submodular function with the individual size constraint.

Algorithm 2 A random algorithm for monotone k-submodular maximization with the individual size constraint

---

Input: A nonnegative monotone k-submodular function f
Output: S satisfying $|supp_i(S)| = B_i$ for each $i \in [k]$
1: $S \leftarrow \emptyset, t = nk - 1, B \leftarrow \sum\limits_{i \in [k]} B_i$
2: **for** $j = 1$ to B **do**
3:    $p \leftarrow 0_{n \times k}$
4:    $y \leftarrow 0_{n \times k}$
5:    **for** each $e \in E \backslash supp(S)$ and position $i \in [k]$ such that $|supp_i(S)| < B_i$ **do**
6:        $y_{e,i} \leftarrow \Delta_{e,i} f(S)$ for $(e,i) \in [n] \times [k]$
7:        $p_{e,i} \leftarrow y_{e,i}^t$
8:    **end for**
9:    $\beta \leftarrow \sum\limits_{(e,i) \in [n] \times [k]} p_{e,i}$
10:    **if** $\beta \neq 0$ **then**
11:        $p \leftarrow \frac{p}{\beta}$
12:    **else**
13:        Arbitrarily choose an $e \in E \backslash supp(S)$ and $i \in [k]$ such that $|supp_i(S)| < B_i$, $p_{e,i} \leftarrow 1$
14:    **end if**
15:    Randomly choose $(e,i) \in [n] \times [k]$, with $Pr[(e,i)] = p_{e,i}$
16:    $S = S \cup \{(e,i)\}$
17: **end for**
18: **return** S

The symbols defined in this section are mostly the same as those in Sect. 3. Below, we will provide a detailed description of the differences from those in Sect. 3. Firstly, we mainly describe how the O^j sequence is constructed. We iteratively construct $O^0 = O, O^1, ..., O^B$ as follows. Next, we will describe how to construct from O^{j-1} to O^j. For each $j \in [B]$, if $e^j \in supp_{ij}(O^{j-1}) \backslash supp(S^{j-1})$, let us set $o^j = e^j$. If $e^j \notin supp_{ij}(O^{j-1}) \backslash supp(S^{j-1})$, let o^j to be an arbitrary element in $supp_{ij}(O^{j-1}) \backslash supp(S^{j-1})$. For any $X \subseteq V$, if element $e \in supp(X)$, we denote $X(e)$ as the position of element e in X, then $(e, X(e)) \in X$. If element $e \notin supp(X)$, we remember that the position of element e in X is 0, which means $X(e) = 0$, then $(e, X(e))$ satisfying $f(X \cup \{(e, X(e))\}) - f(X) = 0$, such element position pairs have no meaning, but for a clearer description in the subsequent proof process, we still define a class of such element position pairs. Firstly we construct $O^{j-\frac{1}{2}}$: If $e^j \in supp(O^{j-1})$, $O^{j-\frac{1}{2}} = O^{j-1} \backslash \{(e^j, O^{j-1}(e^j))\}$, otherwise $O^{j-\frac{1}{2}} = O^{j-1}$. Construct $O^{j-\frac{1}{4}}$: $O^{j-\frac{1}{4}} = O^{j-\frac{1}{2}} \backslash \{(o^j, O^{j-1}(o^j))\}$.

Construct $O^{j-\frac{1}{8}}$: If $e^j \in supp_{i'}(O^{j-1})\backslash supp(S^{j-1})$, $i' \neq i^j$, $O^{j-\frac{1}{8}} = O^{j-\frac{1}{4}} \cup \{(o^j, O^{j-1}(e^j))\}$, otherwise $O^{j-\frac{1}{8}} = O^{j-\frac{1}{4}}$. Construct O^j: $O^j = O^{j-\frac{1}{8}} \cup \{(e^j, i^j)\}$, where $i^j = O^{j-1}(o^j)$. Note that $|supp_i(O^j)| = B_i$ holds for every $j \in \{0, 1, 2, \ldots, B\}$, and $O^B = S^B = S$. Next, for $\forall (e, i) \in V$, we define $y^j_{e,i} = \triangle_{e,i} f(S^{j-1})$, $a'^j_{e,i} = \triangle_{e,i} f(O^{j-\frac{1}{2}})$, $a''^j_{e,i} = \triangle_{e,i} f(O^{j-\frac{1}{4}})$, $a'''^j_{e,i} = \triangle_{e,i} f(O^{j-\frac{1}{8}})$. By constructing in this way, we have $S^{j-1} \preceq O^{j-\frac{1}{2}}$, $S^{j-1} \preceq O^{j-\frac{1}{4}}$, $S^{j-1} \preceq O^{j-\frac{1}{8}}$, $S^{j-1} \preceq O^j$, the orthant submodularity $y^j_{e,i} \geq a'^j_{e,i}$, $y^j_{e,i} \geq a''^j_{e,i}$, $y^j_{e,i} \geq a'''^j_{e,i}$. For each $j \in [B]$, let us note that L is the set of all these element position pairs (e, i) that make $|supp_i(S \cup (e, i))| \leq B_i$. These element position pairs (e, i) in L can be divided into three cases, we record them as L_1, L_2 and L_3, where $L = L_1 \cup L_2 \cup L_3$. L_1 represents the set of selected element position pairs (e, i) that are exactly in O^{j-1}, that is, $e \in supp_i(O^{j-1})\backslash supp(S^{j-1})$. If we choose the pair $(e, i) \in L_1$, $O^{j-\frac{1}{2}} = O^{j-1}\backslash\{(e, i)\}$, $O^{j-\frac{1}{2}} = O^{j-\frac{1}{4}} = O^{j-\frac{1}{8}}$, $O^j = O^{j-\frac{1}{8}} \cup \{(e, i)\}$. L_2 represents the set of selected element position pairs (e, i) whose elements are not in the support set of O^{j-1}, that is, $e \notin supp(O^{j-1})\backslash supp(S^{j-1})$. In this case, $O^{j-\frac{1}{2}} = O^{j-1}$, $O^{j-\frac{1}{4}} = O^{j-\frac{1}{2}}\backslash\{(o^j, O^{j-1}(o^j))\}$, $O^{j-\frac{1}{4}} = O^{j-\frac{1}{8}}$, $O^j = O^{j-\frac{1}{8}} \cup \{(e, i)\}$. L_3 represents set of selected element position pairs (e, i) that are not in O^{j-1}, but the elements in these element position pairs are in the support set of O^{j-1}, that is, $e \in supp_{i'}(O^{j-1})\backslash supp(S^{j-1})$, $i \neq i'$. At this point, $O^{j-1}, O^{j-\frac{1}{2}}, O^{j-\frac{1}{4}}, O^{j-\frac{1}{8}}, O^j$ are all different.

Lemma 2. *For any $j \in [B]$, according to Algorithm 2, the following inequality holds.*

$$\sum_{S^{j-1}} \sum_{(e,i)\in[n]\times[k]} [(a'^j_{e,O^{j-1}(e)} + a''^j_{o^j,O^{j-1}(o^j)}) - (a''^j_{o^j,O^{j-1}(e)} + a'''^j_{e,O^{j-1}(o^j)})]p^j_{e,i} \cdot p_{S^{j-1}}$$
$$\leq (2 - \frac{2}{nk}) \sum_{S^{j-1}} \sum_{(e,i)\in[n]\times[k]} y^j_{e,i} \cdot p^j_{e,i} \cdot p_{S^{j-1}}.$$
$$(9)$$

Proof. In the following proof process, we omit the superscript j in the inequality in order to describe clearly. From the above analysis, inequality (9) is equivalent to

$$\sum_{L_2} (\triangle_{o^j,O^{j-1}(o^j)} f(O^{j-\frac{1}{4}}) - \triangle_{e,i} f(O^{j-\frac{1}{8}})) p_{L_2} + \sum_{L_3} [(\triangle_{e,O^{j-1}(e)} f(O^{j-\frac{1}{2}})$$
$$+ \triangle_{o^j,O^{j-1}(o^j)} f(O^{j-\frac{1}{4}})) - (\triangle_{o^j,O^{j-1}(e)} f(O^{j-\frac{1}{4}}) + \triangle_{e,i} f(O^{j-\frac{1}{8}}))] p_{L_3} \quad (10)$$
$$\leq (2 - \frac{2}{nk}) \sum_L p_L y_L.$$

First, let us analyze the situation of $\beta = 0$, here $y_L = 0$,

$$\triangle_{o^j,O^{j-1}(o^j)} f(O^{j-\frac{1}{4}}) - \triangle_{e,i} f(O^{j-\frac{1}{8}})$$
$$\leq \triangle_{o^j,O^{j-1}(o^j)} f(O^{j-\frac{1}{4}})$$
$$\leq \triangle_{o^j,O^{j-1}(o^j)} f(S^{j-1})$$
$$= 0,$$
$$(11)$$

and

$$\triangle_{e,O^{j-1}(e)} f(O^{j-\frac{1}{2}}) + \triangle_{o^j,O^{j-1}(o^j)} f(O^{j-\frac{1}{4}})$$
$$- (\triangle_{o^j,O^{j-1}(e)} f(O^{j-\frac{1}{4}}) + \triangle_{e,i} f(O^{j-\frac{1}{8}}))$$
$$\leq \triangle_{e,O^{j-1}(e)} f(O^{j-\frac{1}{2}}) + \triangle_{o^j,O^{j-1}(o^j)} f(O^{j-\frac{1}{4}}) \qquad (12)$$
$$\leq \triangle_{e,O^{j-1}(e)} f(S^{j-1}) + \triangle_{o^j,O^{j-1}(o^j)} f(S^{j-1})$$
$$= 0,$$

so the left side of inequality (10) is equal or lesser than 0, and the right side is equal to 0, so inequality (10) holds.

Next, let us discuss the situation of $\beta \neq 0$, our goal is to show the following inequality.

$$\sum_{L_2} [(\triangle_{o^j,O^{j-1}(o^j)} f(O^{j-\frac{1}{4}}) - \triangle_{e,i} f(O^{j-\frac{1}{8}}))] y_{L_2}^t + \sum_{L_3} [(\triangle_{e,O^{j-1}(e)} f(O^{j-\frac{1}{2}})$$
$$+ \triangle_{o^j,O^{j-1}(o^j)} f(O^{j-\frac{1}{4}})) - (\triangle_{o^j,O^{j-1}(e)} f(O^{j-\frac{1}{4}}) + \triangle_{e,i} f(O^{j-\frac{1}{8}}))] y_{L_3}^t \qquad (13)$$
$$\leq (2 - \frac{2}{nk}) \sum_L y_L^{t+1}.$$

If $nk = 1$, it means $n = 1, k = 1$, so both sides of the inequality are 0. If $nk \geq 2$, let $r = (nk - 1)^{\frac{1}{t}} = t^{\frac{1}{t}}$. Any (e, i) belongs to L_1, we define $y_{L_1 max} = max_{(e,i) \in L_1} (f(S^{j-1} \cup \{(e, i)\}) - f(S^{j-1}))$, we know that if $e \in supp(O^{j-1})$, then $(o^j, O^{j-1}(o^j)) \in L_1$ and $(e, O^{j-1}(e)) \in L_1$, otherwise $\triangle_{e,O^{j-1}(e)} f(O^{j-1}) = 0$. Since f is monotone, so we have:

$$\sum_{L_2} [\triangle_{o^j,O^{j-1}(o^j)} f(O^{j-\frac{1}{4}}) - \triangle_{e,i} f(O^{j-\frac{1}{8}})] y_{L_2}^t + \sum_{L_3} [(\triangle_{e,O^{j-1}(e)} f(O^{j-\frac{1}{2}})$$
$$+ \triangle_{o^j,O^{j-1}(o^j)} f(O^{j-\frac{1}{4}})) - (\triangle_{o^j,O^{j-1}(e)} f(O^{j-\frac{1}{4}}) + \triangle_{e,i} f(O^{j-\frac{1}{8}}))] y_{L_3}^t$$
$$\leq \sum_{L_2} \triangle_{o^j,O^{j-1}(o^j)} f(O^{j-\frac{1}{4}}) y_{L_2}^t$$
$$+ \sum_{L_3} [\triangle_{e,O^{j-1}(e)} f(O^{j-\frac{1}{2}}) + \triangle_{o^j,O^{j-1}(o^j)} f(O^{j-\frac{1}{4}})] y_{L_3}^t$$
$$\leq \sum_{L_2} \triangle_{o^j,O^{j-1}(o^j)} f(S^{j-1}) y_{L_2}^t \qquad (14)$$
$$+ \sum_{L_3} [\triangle_{e,O^{j-1}(e)} f(S^{j-1}) + \triangle_{o^j,O^{j-1}(o^j)} f(S^{j-1})] y_{L_3}^t$$
$$\leq 2y_{L_1 max} \sum_{L_2 \cup L_3} y_L^t$$
$$= \frac{2}{r} (r y_{L_1 max} \sum_{L_2 \cup L_3} y_L^t).$$

According to the weighted AM-GM inequality and Holder's inequality, we have

$$\frac{2}{r}(ry_{L_1max}\sum_{L_2\cup L_3} y_L^t).$$

$$\leq \frac{2}{r}\{\frac{1}{t+1}(ry_{L_1max})^{t+1} + \frac{t}{t+1}[(\sum_{L_2\cup L_3} y_L^{t\cdot\frac{t+1}{t}})(\sum_{L_2\cup L_3} 1^{t+1})^{\frac{1}{t}}]\}$$

$$\leq \frac{2}{r}[\frac{1}{t+1}(ry_{L_1max})^{t+1} + \frac{t}{t+1}(nk-1)^{\frac{1}{t}}\sum_{L_2\cup L_3} y_L^{t+1}]$$

$$= \frac{2r^t}{t+1}y_{L_1max}^{t+1} + \frac{2t}{t+1}\sum_{L_2\cup L_3} y_L^{t+1} \tag{15}$$

$$= \frac{2t}{t+1}y_{L_1max}^{t+1} + \frac{2t}{t+1}\sum_{L_2\cup L_3} y_L^{t+1}$$

$$\leq \frac{2t}{t+1}\sum_L y_L^{t+1}$$

$$= (2 - \frac{2}{nk})\sum_L y_L^{t+1}.$$

To sum up, inequality (13) holds, thus inequality (9) holds. We obtained this lemma.

Theorem 3. *With the individual size constraint, a $\frac{nk}{3nk-2}$-approximate solution can be obtained by Algorithm 2.*

Proof. We use the random algorithm to find the relationship between $f(O)$ and $\mathbb{E}[f(S)]$. Here $\mathbb{E}[f(O^{j-1})] - f(O^j)] = \sum_{S^{j-1}}\sum_{(e,i)\in[n]\times[k]}[(a'^j_{e,O^{j-1}(e)} + a''^j_{o^j,O^{j-1}(o^j)}) - (a''^j_{o^j,O^{j-1}(e)} + a'''^j_{e,O^{j-1}(o^j)})] \cdot p^j_{e,i} \cdot p_{S^{j-1}},$

$$\mathbb{E}[f(S^j) - f(S^{j-1})] = \sum_{S^{j-1}}\sum_{(e,i)\in[n]\times[k]} y^j_{e,i} \cdot p^j_{e,i} \cdot p_{S^{j-1}}.$$

Therefore, according to Lemma 2, it can be inferred that $\mathbb{E}[f(O^{j-1})] - f(O^j)] \leq (2 - \frac{2}{nk})\mathbb{E}[f(S^j) - f(S^{j-1})]$. According to the proof process of Theorem 2, we obtain $\mathbb{E}[f(S)] \geq \frac{nk}{3nk-2}f(O)$. The result can be obtained.

5   Discusstion

In this paper, we first discuss the problem of maximizing monotone k-submodular function with the total size constraint, we obtain the approximate ratio of $\frac{nk}{2nk-1}$ by using the random algorithm. Next we discuss the problem of maximizing monotone k-submodular function with the individual size constraint and obtain the approximate ratio of $\frac{nk}{3nk-2}$ by using the random algorithm.

Acknowledgements. This paper was supported by the Natural Science Foundation of Shandong Province of China (Nos. ZR2020MA029 and ZR2021MA100) and the National Natural Science Foundation of China (No. 12001335).

References

1. Calinescu, G., Chekuri, C., Pál, M., Vondrák, J.: Maximizing a monotone submodular function subject to a matroid constraint. SIAM J. Comput. **40**(6), 1740–1766 (2011). https://doi.org/10.1137/080733991
2. Buchbinder, N., Feldman, M., Naor, J., Schwartz, R.: A tight linear time (1/2)-approximation for unconstrainted submodular maximization. SIAM J. Comput. **44**(5), 1384–1402 (2015). https://doi.org/10.1137/130929205
3. Sviridenko, M.: A note on maximizing a submodular set function to a knapsack constraint. Oper. Res. Lett. **32**(1), 41–43 (2004). https://doi.org/10.1016/S0167-6377(03)00062-2
4. Singh, A., Guillory, A., Bilmes, J.: On bisubmodular maximization. In: AISTATS, vol. 22, pp. 1055–1063 (2012). https://proceedings.mlr.press/v22/singh12.html
5. Ward, J., Živný,.S.: Maximizing k-submodular functions and beyond. ACM Trans. Algorithms **12**(4), 1–26 (2016). https://doi.org/10.1145/2850419
6. Iwata, S., Tanigawa, S., Yoshida, Y.: Improved approximation algorithms for k-submodular function maximization. In: Proceedings of the 27th Annual ACM-SIAM Symposium on Discrete Algorithms(SODA), pp. 404–413 (2016). https://doi.org/10.1137/1.9781611974331.ch30
7. Oshima, H.: Improved randomized algorithm for k-submodular function maximization. SIAM J. Discrete Math. **35**(1), 1–22 (2021). https://doi.org/10.1137/19M1277692
8. Ohsaka, N., Yoshida, Y.: Monotone k-submodular function maximization with size constraints. In: Proceedings of the 28th International Conference on Neural Information Processing Systems(NeurIPS), pp. 694–702 (2015). https://dl.acm.org/doi/abs/10.5555/2969239.2969317
9. Sakaue, S.: On maximizing a monotone k-submodular function subject to a matroid constraint. Discrete Optim. **23**, 105–113 (2017). https://doi.org/10.1016/j.disopt.2017.01.003
10. Sun, Y., Liu, Y., Li, M.: Maximization of k-submodular function with a matroid constraint. In: Proceedings of Theory and Applications of Models of Computation (TAMC), pp. 1–10 (2022). https://doi.org/10.1007/978-3-031-20350-3_1
11. Nguyen, L., Thai, M.: Streaming k-submodular maximization under noise subject to size constraint. In: Proceedings of the 37th International Conference on Machine Learning (ICML), pp. 7338–7347(2020). https://dlnext.acm.org/doi/10.5555/3524938.3525618
12. Ene, A., Nguyen, H.: Streaming algorithm for monotone k-Submodular maximization with cardinality constraints. In: Proceedings of the 39th International Conference on Machine Learning (ICML), pp. 5944–5967 (2022)
13. Huber, A., Kolmogorov, V.: Towards minimizing k-submodular functions. In: Proceedings of International Symposium on Combinatorial Optimization (ISCO), pp. 451–462 (2012). https://doi.org/10.1007/978-3-642-32147-4_40
14. Gridchyn, I., Kolmogorov, V.: Potts model, parametric maxflow and k-submodular functions. 2013 IEEE International Conference on Computer Vision, pp. 2320–2327 (2013). https://doi.org/10.1109/ICCV.2013.288

15. Pham, C.V., Vu, Q.C., Ha, D.K.T., Nguyen, T.T.: Streaming algorithms for budgeted k-submodular maximization problem. In: Proceedings of Computational Data and Social Networks (CSoNet), pp. 27–38 (2021). https://doi.org/10.1007/978-3-030-91434-9_3

16. Pham, C., Vu, Q., Ha, D., Nguyen, T., Le, N.: Maximizing k-submodular functions under budget constraint: applications and streaming algorithms. J. Comb. Optim. **44**, 723–751 (2022). https://doi.org/10.1007/s10878-022-00858-x

17. Qian, C., Shi, J., Tang, K., Zhou, Z.: Constrained monotone k-submodular function maximization using multiobjective evolutionary algorithms with theoretical guarantee. IEEE Trans. Evol. Comput. **22**(4), 595–608 (2018). https://doi.org/10.1109/TEVC.2017.2749263

18. Rafiey, A., Yoshida, Y.: Fast and private submodular and k-submodular functions maximization with matroid constraints. In: Proceedings of the 37th International Conference on Machine Learning (ICML), pp. 7887–7897 (2020). https://dl.acm.org/doi/abs/10.5555/3524938.3525669

19. Zheng, L., Chan, H., Loukides, G., Li, M.: Maximizing approximately k-submodular functions. In: Proceeding of the 2021 SIAM International Conference on Data Mining (SDM), pp. 414–422 (2021). https://doi.org/10.1137/1.9781611976700.47

20. Hiroshi, H., Yuni, I.: On k-submodular relaxation. SIAM J. Discrete Math. **30**(3), 1726–1736 (2016). https://doi.org/10.1137/15M101926X

21. Yu, K., Li, M., Zhou, Y., Liu, Q.: Guarantees for maximization of k-submodular functions with a knapsack and a matroid constraint. In: Proceeding of Algorithmic Applications in Management (AAIM), pp. 156–167 (2022). https://doi.org/10.1007/978-3-031-16081-3_14

22. Liu, Q., Yu, K., Li, M., Zhou, Y.: k-submodular maximization with a knapsack constraint and p matroid constraints. Tsinghua Sci. Technol. **28**(5), 896–905 (2023). https://doi.org/10.26599/TST.2022.9010039

Additive Approximation Algorithms for Sliding Puzzle

Zhixian Zhong$^{(\boxtimes)}$ (ID)

Beijing, China
zhongzx18@tsinghua.org.cn

Abstract. With the development of sport sliding puzzle, it is of great significance to study better algorithms to solve sliding puzzles. Since solving the puzzle optimally is hard, we hope to find an additive approximation algorithm, that is, the length of solution output by this algorithm is at most a low-order term more than optimal solution. $n \times 2$ rectangular puzzle, as a variation of $(n^2 - 1)$-puzzle, can be solved by an algorithm in divide-and-conquer scheme. We proved that it's an additive approximation algorithm within $O(n \log n)$ additive constant. For $(n^2 - 1)$-puzzle, finding a good approximation algorithm is more difficult. We designed a new poly-time algorithm to solve $(n^2 - 1)$-puzzle, which consists of several phases: First build the board into a "clear state", then transport the tiles in this clear state, after arranging some tiles the puzzle is divided into smaller parts, finally solve each of parts. And we proved that it is an additive approximation algorithm within $O(n^{2.75})$ additive constant. Also, using these approximation algorithms, we analyzed the optimal solution length asymptotically in the average case and the worst case (God's number). For $n \times 2$ rectangular puzzle, the average optimal solution length is $n^2 + O(n \log n)$ and the God's number is $2n^2 + O(n \log n)$. For $(n^2 - 1)$-puzzle, the average optimal solution length is $\frac{2}{3}n^3 + O(n^{2.75})$ and the God's number is $n^3 + O(n^{2.75})$.

Keywords: sliding puzzle · approximation algorithm · potential function · asymptotic analysis · God's number

1 Introduction

Sliding puzzle is a classical game that has a history of over 100 years. In $(n^2 - 1)$-puzzle, there is an $n \times n$ grid board with $n^2 - 1$ tiles, distinct integers in $[1, n^2 - 1]$ are written on all tiles, and there is exactly one empty cell. Initially tiles are scrambled (arranged out of order). Player can move any tile to its adjacent cell if it's empty, for any times. The goal is to arrange the tiles in increasing order from top to bottom, from left to right.

In recent years, sliding puzzle becomes a mind sport. Players are required to solve sliding puzzle boards as fast as possible (speedsolving event), or using as least moves as possible (fewest-move event).

© Springer Nature Switzerland AG 2023
M. Li et al. (Eds.): IJTCS-FAW 2023, LNCS 13933, pp. 129–146, 2023.
https://doi.org/10.1007/978-3-031-39344-0_10

With the development of sport sliding puzzle, studying better algorithms for sliding puzzle is useful for training and competing. Since solving it optimally is hard, it's important to find better approximation algorithms.

In 1990, Danial Ratner and Manfred Warmuth proved that computing the optimal solution of $(n^2 - 1)$-puzzle is NP-hard, even if $O(1)$ additive constant is allowed. Also, they designed an approximation algorithm within a large constant multiplicative factor [1]. In 1995, Ian Parberry analyzed the performance of naive tile-by-tile algorithm for $(n^2 - 1)$-puzzle in the worst case, and provided upper and lower bound for "God's number" (optimal solution length in the worst case) [2] However, the bound is not tight: the upper bound of God's number $5n^3 + o(n^2)$ is up to 5 times the lower bound $n^3 + O(n^2)$. Parberry's subsequent work also analyzed the bound of its average optimal solution length, which is also tight: The upper bound $\frac{8}{3}n^3 + O(n^2)$ is up to 4 times the lower bound $\frac{2}{3}n^3 + O(n^2)$ [4].

Some work focuses on small boards. In 2005, Richard E. Korf and Peter Schultze used large-scaled parallel breadth-first search to compute the optimal solution for all states, and find the God's number of 15-puzzle is 80 [3]. The God's number of $(n^2 - 1)$-puzzle for $n \geq 5$ is still open. With the development of artificial intelligence, in 2019, Forest Agostinelli et al. developed an approximation algorithm to solve Rubik's cube and $(n^2 - 1)$-puzzle for $n \leq 7$ based on reinforcement learning, named DeepCubeA [5]. AI method can find near-optimal solution in practice, but the bound of difference between solution length and optimal solution length is not guaranteed.

In this work:

- We design approximation algorithms in polynomial time for two types of sliding puzzle: $n \times 2$ rectangular puzzle and $(n^2 - 1)$-puzzle. These are additive approximation algorithms, that is, the difference of solution length and optimal solution length is bounded by a lower-order term, which is better than common multiplicative approximation in most cases.
- Then, we use them to analyze optimal solution length in average/worst case asymptotically. The upper bounds of average optimal solution length and the God's number meet the lower bounds in the asymptotic sense.

2   Preliminaries

2.1   Basic Definitions

Definition 1 (Board). *An $n \times m$ **board** B is a rectangular array of $n \geq 2$ rows and $m \geq 2$ columns, and element 0 appears exactly once in it. For integers $0 \leq x < n, 0 \leq y < m$, (x, y) represents the cell on the $(x + 1)$-th row, the $(y + 1)$-th column, and $B(x, y)$ is the element at cell (x, y) in B.*

*If $B(x, y) \neq 0$, then cell (x, y) contains a **tile** with number $B(x, y)$ on it; if $B(x, y) = 0$, then (x, y) is an **empty cell**, denoted as $(x_0(B), y_0(B))$.*

Definition 2 (Operation sequence). *An **operation sequence** σ is a sequence of form $\delta_1\delta_2\cdots\delta_l$, where each $\delta_i \in \{U, D, L, R\}$ is a single-tile movement[1], meaning that move a tile up, down, left or right to the empty cell. The board after acting σ on B, denoted by $B\sigma$, is the board obtained by successively acting $\delta_1, \delta_2, \cdots, \delta_l$ on B. σ is applicable to B if all operations are applicable when acting σ on B.*

Formally, let $x_0 = x_0(B), y_0 = y_0(B)$, the result of acting move δ on B is to swap $B(x_0, y_0)$ and $B(x_0 - \Delta x_\delta, y_0 - \Delta y_\delta)$, where movement vector $(\Delta x_\delta, \Delta y_\delta)$ is given in Table 1. δ is applicable to B if cell $(x_0 - \Delta x_\delta, y_0 - \Delta y_\delta)$ exists.

Table 1. Movement vector

δ	U	D	L	R
$(\Delta x_\delta, \Delta y_\delta)$	$(-1,0)$	$(1,0)$	$(0,-1)$	$(0,1)$

In this work, we use single-tile movement metric, that is, the **length** of solution $\sigma = \delta_1\delta_2\cdots\delta_l$, denoted by $|\sigma|$, is the number of moves in σ, i.e. $|\sigma| = l$. Now we can give the formal definition of sliding puzzle.

Definition 3 (Sliding puzzle). *A sliding puzzle is a scrambled board B_S along with its target board B_T, and a **solution** to this puzzle is an operation sequence σ such that $B_T = B_S\sigma$. The goal is to minimize the solution length $|\sigma|$.*

Standard $n \times m$ **rectangular puzzle** has target board

$$B_T = \begin{pmatrix} 1 & 2 & \cdots & m-1 & m \\ m+1 & m+2 & \cdots & 2m-1 & 2m \\ \vdots & \vdots & \ddots & \vdots & \vdots \\ nm-m+1 & nm-m+2 & \cdots & nm-1 & 0 \end{pmatrix}$$

here, $B_T(x,y) = (mx + y + 1) \bmod mn$. In particular, if $n = m$, the puzzle is called $(n^2 - 1)$-puzzle.

In fact, we can use any target board B_T containing distinct elements and $B_T(n-1, m-1) = 0$ in $n \times m$ rectangular puzzle, which is equivalent to standard one by relabelling the tiles to $1, 2, \cdots, nm - 1$.

2.2 Some Properties

Here are some properties of boards and operations, they will be used in further discussion.

Definition 4 (Subarray). *For $n \times m$ board B and integers $0 \le u < d \le n, 0 \le l < r \le m$, define $B(u:d, l:r)$ as the subarray of B in rectangle $(u:d, l:r)$: from the $(u+1)$-th to the d-th row and from the $(l+1)$-th to the r-th column. $B(u:d, l:r)$ is called a **sub-board** of B if it is a board.*

---

[1] The following text calls it a **move** for short.

Proposition 1. *For $n \times m$ board B and its sub-board $B(u : d, l : r)$, if operation sequence σ is applicable to $B(u : d, l : r)$, then σ is applicable to B, $(B\sigma)(u : d, l : r) = B(u : d, l : r)\sigma$ and $(B\sigma)(x, y) = B(x, y)$ for all cells (x, y) not in $(u : d, l : r)$. In other words, $B\sigma$ is the board by replacing subarray $B(u : d, l : r)$ from B with $B(u : d, l : r)\sigma$.*

Definition 5 (Composition). *The **composition** of operation sequences σ_1, σ_2 is the operation sequence by concatenating σ_1 and σ_2, denoted as $\sigma_1\sigma_2$. For $n \in \mathbb{N}$, σ^n is the composition of n identical operation sequences σ.*

Definition 6 (Mapping). *For array B and map π defined on all elements in B, define πB as the array obtained by replacing each element $B(x, y)$ in B with $\pi(B(x, y))$.*

Proposition 2. *For board B:*

- $B(\sigma_1\sigma_2) = (B\sigma_1)\sigma_2$, *for any operation sequences σ_1, σ_2 if applicable;*
- $(\pi_2\pi_1)B = \pi_2(\pi_1 B)$, *for any mapping π_1, π_2;[2]*
- $(\pi B)\sigma = \pi(B\sigma)$, *for any operation sequence σ applicable to B and permutation π s.t. $\pi(0) = 0$.*

Definition 7 (Orbit). *For board B, set $\{B\sigma : \sigma \text{ is applicable to } B\}$ is called the **orbit** of B.*

In the area of sliding puzzle, there is a well-known result:

Proposition 3. *Let B_1, B_2 be two $n \times m$ boards that has the same set of elements X ($|X| = nm$, $0 \in X$), then B_1, B_2 are in the same orbit if and only if integer $x_0(B_1) + y_0(B_1) + x_0(B_2) + y_0(B_2)$ and permutation $\pi : X \to X$ s.t. $B_2 = \pi B_1$ are both odd or both even.*

2.3 Algorithms

An algorithm for sliding puzzle requires a puzzle B_S along with its target board B_T, outputs a solution σ if exists. We will introduce some definitions and notations for sliding puzzle algorithms.

Online Algorithm. In sliding puzzle speedsolving area, because of the limitation of time, player can only operate according to current state to solve the puzzle. Similarly, we mainly use **online algorithm** to solve the puzzle in polynomial time, that is, the algorithm directly acts moves on current board, and terminates when the board is solved. Formally, an online algorithm implies the following structure:

- Initially, let current board $B \leftarrow B_S$, and current operation sequence $\sigma \leftarrow \epsilon$;[3]

---

[2] $\pi_2\pi_1$ is short for $\pi_2 \circ \pi_1$.

[3] ϵ represents empty sequence.

- In statement "**act** τ", update $B \leftarrow B\tau$, $\sigma \leftarrow \sigma\tau$;
- When the algorithm terminates, $B = B_T$, return σ.

Sometimes, instead of using "**act** τ" directly, we use an alternative way to describe the actions in our description of algorithms:

- **move-x**(x): means **act** $D^{x_0(B)-x}$ (if $x < x_0(B)$) or $U^{x-x_0(B)}$ (otherwise), after that $x_0(B) = x$;
- **move-y**(y): means **act** $R^{y_0(B)-y}$ (if $y < y_0(B)$) or $L^{y-y_0(B)}$ (otherwise), after that $y_0(B) = y$.

Example 1. In 1995, Ian Parberry analyzed a simple tile-by-tile algorithm for (n^2-1)-puzzle, as Algorithm 1 [2]. It's an online algorithm: each τ only depends on current position of tile $B_T(x,y)$ and empty cell in B.

Algorithm 1. Parberry's algorithm for $(n^2 - 1)$-puzzle

---

for $k = 0$ to $n - 2$ **do**
    for (x,y) in $(k, k..n-1)$ and $(k+1..n-1, k)$ **do**
        let τ be an operation sequence such that, when acting on $B(k:n, k:n)$, it places tile $B_T(x,y)$ into $(x-k, y-k)$ in $B(k:n, k:n)$ preserving tiles that are previously placed
        act τ ▷ now $B(x,y) = B_T(x,y)$
    end for
end for

Proposition 4 (Parberry). *It costs at most $O(n)$ moves to put any tile into a particular cell in the first row/column, i.e. $|\tau| \leq O(n)$ in Algorithm 1. Hence, it costs at most $O(n^2)$ moves to put the first row/column into place, and at most $O(n^3)$ moves to solve (n^2-1)-puzzle.*

Multi-phase Algorithm. In practice, we usually design **multi-phase algorithm**, consisting of several phases, while each phase reduce current board B to a simpler one. By implementing each phase, we will obtain the whole algorithm. Algorithm 2 gives the paradigm of a 2-phase algorithm as follows:

$$\{^{\text{initial}}_{\text{board}}\} \xrightarrow{\text{phase I}} \{^{\text{board after}}_{\text{Phase I}}\} \xrightarrow{\text{phase II}} \{B_T\}$$

To implement it, we only need to implement phase I and II respectively.

Algorithm 2. 2-phase algorithm

---

act SOLVE-PHASE-I(B, B_T)
act SOLVE-PHASE-II(B, B_T)

Example 2. In (n^2-1)-puzzle speedsolving area, there are two multi-phase methods that are widely used: reduction-of-order and divide-and-conquer.

The scheme of reduction-of-order is as follows, in fact Algorithm 1 uses this scheme:

$$\{{}^{\text{initial}}_{\text{board}}\} \xrightarrow[\text{1st row}]{\text{solve}} \left\{ {}^{\text{board s.t.}}_{{}^{B(0,0:n)}_{=B_T(0,0:n)}} \right\} \xrightarrow[\text{column}]{\text{solve 1st}} \left\{ {}^{\text{board s.t.}}_{{}^{B(0,0:n)=B_T(0,0:n)}_{B(0:n,0)=B_T(0:n,0)}} \right\} \xrightarrow[\text{by recursion}]{{}^{\text{solve}}_{{}^{(n-1)\times(n-1)}_{\text{block}}}} \{B_T\}$$

The scheme of divide-and-conquer is as follows[4], later we'll use it in $n \times 2$ rectangular puzzle:

$$\{{}^{\text{initial}}_{\text{board}}\} \xrightarrow[m=\lfloor\frac{n}{2}\rfloor]{\text{partition}} \left\{ {}^{\text{board s.t. } B(0:n,0:m)}_{{}^{\text{contains all elements}}_{\text{in } B_T(0:n,0:m)}} \right\} \xrightarrow[{}^{\text{by}}_{\text{recursion}}]{{}^{\text{solve}}_{\text{left part}}} \left\{ {}^{\text{board s.t.}}_{{}^{B(0:n,0:m)}_{=B_T(0:n,0:m)}} \right\} \xrightarrow[{}^{\text{by}}_{\text{recursion}}]{{}^{\text{solve}}_{\text{right part}}} \{B_T\}$$

Approximation. For $(n^2 - 1)$-puzzle, when n is large, the orbit of B_T is too large to use brute force (e.g. BFS) to find the optimal solution. So we turn to consider to design approximation algorithms in polynomial time.

For an algorithm and an instance B_S in the orbit $\mathcal{O}$ of B_T, denote

– OPT: the length of optimal solution, i.e. $\min\{|\sigma| : B_T = B_S\sigma\}$;
– SOL: the length of solution found by this algorithm.

Definition 8 (Approximation). *For an algorithm for sliding puzzle, if* SOL $\leq$ OPT $+\alpha$ *holds, where α is a function of the size of puzzle, then it is an **additive approximation** algorithm within an **additive constant** of α.*

A non-trivial additive approximation should satisfy $\alpha = o(\overline{\text{OPT}})$,[5] where $\overline{\text{OPT}} = \mathbb{E}[\text{OPT}]$ when B_S is uniformly distributed in $\mathcal{O}$.

To analyze such algorithms, we introduce a concept:

Definition 9 (Potential function). *Let $\mathcal{O}$ be an orbit of boards, then a map $\varphi : \mathcal{O} \to \mathbb{R}$ is called a potential function if for all $B \in \mathcal{O}$ and move $\delta \in \{U, D, L, R\}$ that is applicable to B, $\varphi(\delta B) \geq \varphi(B) - 1$. Move δ acting on B is called an **efficient move** if $\varphi(B\delta) = \varphi(B) - 1$, and it's called an **inefficient move** otherwise.*

Proposition 5. *For $B \in \mathcal{O}$, if $B' = B\sigma$ for some operation sequence σ, then $\varphi(B) - \varphi(B') \leq |\sigma| \leq \varphi(B) - \varphi(B') + 2k$, where k is the number of inefficient moves when acting σ on B.*

If an algorithm always solves within α inefficient moves, then SOL $\leq \varphi(B_S) - \varphi(B_T) + 2\alpha \leq$ OPT $+ 2\alpha$, that is, it is also within additive constant 2α.

---

[4] In fact, the algorithm needs to solve rectangular puzzle for recursion.
[5] In this article, all asymptotic notations assume that the size of puzzle tends to infinity.

3 Approximation Algorithm for $n \times 2$ Rectangular Puzzle

In this section, we discuss a type of puzzle: $n \times 2$ rectangular puzzle. It is closely related to $(n^2 - 1)$-puzzle solving. The target board is specified as

$$B_T = \begin{pmatrix} 1 & 2 \\ 3 & 4 \\ \vdots & \vdots \\ 2n - 1 & 0 \end{pmatrix}$$

Here we give a potential function for $n \times 2$ rectangular puzzle:

Definition 10. *For board B, its **inversion number** $I(B)$ is the inversion number of non-zero elements in sequence $\{b_i\}_{0 \leq i < 2n}$, where $b_i = B(\lfloor \frac{i}{2} \rfloor, i \bmod 2)$, that is:*

$$I(B) = \sum_{0 \leq i < j < 2n} [b_i > b_j > 0] \tag{1}$$

Proposition 6. *I defined by (1) is a potential function.*

Proof. Let $i = 2x_0(B) + y_0(B)$, then $b_i = 0$. When acting move δ on B:

- If $\delta \in \{L, R\}$, then b_i and $b_{i \pm 1}$ are swapped, $I(B\delta) = I(B)$;
- If $\delta = U$, then b_i and b_{i+2} swap, $I(B\delta) = I(B) - [b_{i+1} > b_{i+2}] + [b_{i+2} > b_{i+1}]$;
- If $\delta = D$, then b_i and b_{i-2} swap, $I(B\delta) = I(B) - [b_{i-2} > b_{i-1}] + [b_{i-1} > b_{i-2}]$.

Since $I(B\delta) \geq I(B) - 1$ holds, I is a potential function. When $\delta = U, b_{i+1} > b_{i+2}$ or $\delta = D, b_{i-1} < b_{i-2}$, δ is an efficient move.

In this section, we will give an approximation algorithm for $n \times 2$ rectangular puzzle to show that:

Theorem 1. *There is a poly-time additive approximation algorithm to solve $n \times 2$ rectangular puzzle within $O(n \log n)$ additive constant.*

This algorithm use divide-and-conquer scheme (mentioned in Example 2) for $n \times 2$ rectangular puzzle:

$$\left\{ \begin{matrix} \text{initial} \\ \text{board} \end{matrix} \right\} \xrightarrow[m = \lceil \frac{n}{2} \rceil]{\text{partition}} \left\{ \begin{matrix} \text{board s.t. } B(0{:}m,0{:}2) \\ \text{contains all elements} \\ \text{in } B_T(0{:}m,0{:}2) \end{matrix} \right\} \xrightarrow[\text{recursion}]{\substack{\text{upper} \\ \text{part} \\ \text{by}}} \left\{ \begin{matrix} \text{board s.t.} \\ B(0{:}m{-}1,0{:}2) \\ = B_T(0{:}m{-}1,0{:}2) \end{matrix} \right\} \xrightarrow[\text{recursion}]{\substack{\text{lower} \\ \text{part} \\ \text{by}}} \{B_T\}$$

3.1 Partition

The implementation of partition phase is shown in Algorithm 3.

Algorithm 3. Partition

$v \leftarrow B_T(\lceil \frac{n}{2} \rceil, 0); \ u \leftarrow \min\{u : \exists y, B(u,y) \geq v\}$

move-y(1); **move-x**(u)   ▷ empty cell is adjacent to a tile $\geq v$

if $B(u-1,1) \geq v$ **then** $u \leftarrow u - 1$ ▷ all tiles in $B(0:u,0:2)$ are $< v$

for $x = x_0(B) + 1$ to $n - 1$ **do**

   if $B(x,1) \geq v$ **then act** U

   else

      if $B(u:x,0)$ are all $< v$ **then** $x' \leftarrow u, \ u \leftarrow u + 1$

      else $x' \leftarrow \max\{x' : x' < x, B(x',0) \geq v\}$

      act $D^{x-x'-1}RU^{x-x'}L$ ▷ (a)

   end if   ▷ $B(x,1) = 0$, all tiles in $B(u:x,1)$ are $\geq v$

end for

$d \leftarrow \max\{d : B(d,0) < v\}$ ▷ all tiles in $B(d+1:n,0)$ are $\geq v$

move-x(d)

for $x = d - 1$ to u decreasing **do**

   act D

   if $B(x,0) \geq v$ **then act** $RU^{d-x}LD^{d-x}, \ d \leftarrow d - 1$ ▷ (b)

end for   ▷ $B(x,1) = 0$, all tiles in $B(x:d+1,0)$ are $< v$

for $i = 0$ to $(d-u-1)/2$ **do**

   act $RU^{d-u-2i}LD^{d-u-2i-1}$ ▷ (c)

end for

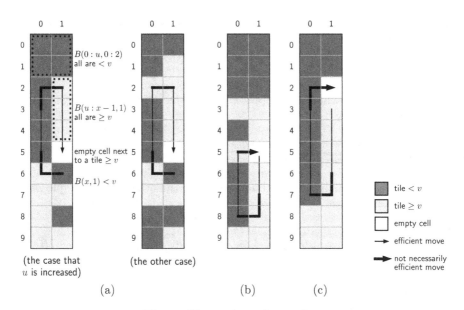

Fig. 1. Illustration of partition

Figure 1 illustrates the actions (a)(b)(c) in Algorithm 3. Take (a) for example, denote $x_0 = x_0(B), y_0 = y_0(B)$:

- Before each D move, $B(x_0 - 1, 1) \geq v > B(x_0, 0)$, by Proposition 6, this D move is an efficient move (according to potential function (1), same below);
- Before each U move (except the first and last one), $B(x_0 + 1, 0) < v \leq B(x_0, 1)$, this U move is also efficient;
- So at most $O(1)$ moves are inefficient in (a).

It's the same for each (b) and (c). Thus, partition costs at most $O(n)$ inefficient moves during up to n loops containing (a), (b) and (c).

3.2   Solve Two Parts

After partition, we can use recursion to solve upper part $((m-1) \times 2)$ and lower part $((n-m+1) \times 2)$, until the size of the puzzle is small (typically $n \leq 5$) so that a brute force can be used. The steps are as follows:

(i) Put $v - 4 = B_T(m - 2, 0)$ into $B(m - 1, 0)$ within $O(n)$ moves (Proposition 4), keeping that $B(0 : m, 0 : 2)$ contains all elements in $B_T(0 : m, 0 : 2)$;

(ii) Do **move-y**(0), **move-x**$(m - 1)$, now $B(m - 1, 0) = 0, B(m, 0) = v - 4$, then recursively solve $B' = (v - 1\ v - 2\ v - 3\ v - 4)B(0 : m, 0 : 2)$ while the target board is $B'_T = (v - 1\ 0)B_T(0 : m, 0 : 2)$,[6] after acting the solution, $B(0 : m, 0 : 2) = (v - 4\ v - 3\ v - 2\ v - 1\ 0)B_T(0 : m, 0 : 2)$;
   - Notice that B' may not be in the orbit of B'_T, if so, B' is in the orbit of $(v - 3\ v - 2)B'_T$ (Proposition 3); So we allow that $v - 3$ and $v - 2$ in B' are swapped; In this case, $B(0 : m, 0 : 2) = (v - 4\ v - 3\ v - 1\ 0)B_T(0 : m, 0 : 2)$ after action;

   Then, **act** $\mathrm{RULD^2RU}$, now $B(0 : m - 1, 0 : 2) = B_T(0 : m - 1, 0 : 2)$;

(iii) Finally, recursively solve $B(m - 1 : n, 0 : 2)$ while the target board is $B_T(m - 1 : n, 0 : 2)$, it must be solvable if B is solvable (Proposition 3).

Let's analyze the bound of inefficient moves in this algorithm. Denote $T(n)$ as the upper bound of inefficient moves of this algorithm in $n \times 2$ board, note that in step (ii), any efficient move on B' is also efficient when applying on B (see Proposition 6), so there are at most $T\left(\lceil \frac{n}{2} \rceil\right)$ inefficient moves in step (ii), and it's similar in step (iii). Then

$$T(n) = T\left(\left\lceil \frac{n}{2} \right\rceil\right) + T\left(\left\lfloor \frac{n}{2} \right\rfloor + 1\right) + O(n)$$

solved as $T(n) = O(n \log n)$. So far, we have proved that this is an additive approximation algorithm within $O(n \log n)$ additive constant.

---

[6] Elements in B'_T (except the last row) are also increasing from top to bottom, so the correctness of Algorithm 3 is guaranteed. Recall that $(a_1\ a_2\ \cdots\ a_k)$ represents a cyclic permutation such that $a_1 \mapsto a_2, a_2 \mapsto a_3, \cdots, a_k \mapsto a_1$.

4 Approximation Algorithm for $(n^2 - 1)$-Puzzle

Now let's turn to general $(n^2 - 1)$-puzzle. The target board is specified as

$$
B_T = \begin{pmatrix}
1 & 2 & \cdots & n-1 & n \\
n+1 & n+2 & \cdots & 2n-1 & 2n \\
\vdots & \vdots & \ddots & \vdots & \vdots \\
n^2-n+1 & n^2-n+2 & \cdots & n^2-1 & 0
\end{pmatrix}
$$

Also, there is an obvious potential function for $n \times 2$ rectangular puzzle:

Definition 11. *For $n \times n$ board B, let (x_i, y_i) and (x_i', y_i') be the position of element i in B and B_T respectively, define its **Manhattan distance** to B_T as*

$$
D(B) = \sum_{i=1}^{n^2-1} (|x_i - x_i'| + |y_i - y_i'|) \tag{2}
$$

Proposition 7. *D defined by (2) is a potential function.*

Unlike $n \times 2$ rectangular puzzle, existing algorithm schemes, such as Example 2, can't achieve an additive approximation for (n^2-1)-puzzle, which is what we'll improve. In this section, we will show this result by designing a new algorithm:

Theorem 2. *There is a poly-time additive approximation algorithm to solve $(n^2 - 1)$-puzzle within $O(n^{2.75})$ additive constant.*

4.1 Algorithm when $n = k^4$

Suppose $n = k^4$ for some integer $k \geq 2$, let's split target board B_T into k^2 squares of size $k^3 \times k^3$. For $0 \leq i < k^2$, denote S_i as the set of tiles (non-zero elements) in square $B_T(a_i k^3 : (a_i + 1)k^3, b_i k^3 : (b_i + 1)k^3)$, where $a_i = \lfloor \frac{i}{k} \rfloor$, $b_i = i \bmod k$, as Fig. 2(a). Clearly $|S_i| = k^6 - [i = k^2 - 1]$. Then, let's split current board B into these parts, as Fig. 2(b):

- $H_i(B)$: Set of elements in $B(a_i k^3 + b_i, 0 : n)$;
- $V_{i,j}(B)$ $(0 \leq j < k^2)$: Set of elements in $B(a_i k^3 + k : (a_i + 1)k^3, b_i k^3 + j)$;
- $R_i(B)$: Set of elements in $B(a_i k^3 + k : (a_i + 1)k^3, b_i k^3 + k^2 : (b_i + 1)k^3)$.

We will omit "(B)" in $H_i(B)$, $V_{i,j}(B)$ and $R_i(B)$ in the following text. Remember that H_i, $V_{i,j}$ and R_i change as current board B changes.

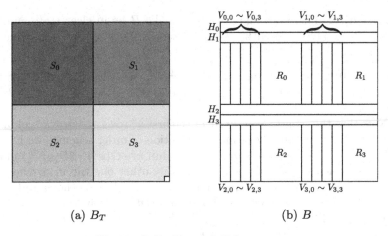

(a) B_T (b) B

Fig. 2. Split B_T and B into parts

Clearly $|H_i| = n$, $|V_{i,j}| = k^3 - k$, $|R_i| = (k^3 - k)(k^3 - k^2)$, and these parts exactly cover B. We call that board B is in **clear state** if

- $\forall 0 \leq i < k^2$, $H_i \subseteq S_i$;
- $\forall 0 \leq i,j < k^2$, $V_{i,j} \subseteq S_j$.

Now, we can design the phases of our algorithm:

I. **Build**: To make B be in clear state and $\forall 0 \leq i < k^2 - 1$, $R_{k^2-1} \cap S_i \neq \varnothing$;
II. **Transport**: To make B be in clear state and $\forall 0 \leq i < k^2$, $R_i \subseteq S_i$;
III. **Arrange**: To make each square $B(a_i k^3 : (a_i + 1)k^3, b_i k^3 : (b_i + 1)k^3)$ be filled with tiles in S_i;
IV. **Finish**: To make $B = B_T$.

Figure 3 illustrates its first three phases: (a) is the initial board, (b) is the board after phase I (**Build**), (c) is the board after phase II (**Transport**), and (d) is the board after phase III (**Arrange**). For better understanding, tiles are colored corresponding to Fig. 2(a) (tiles of the same color belong to the same S_i) and the board is split according to Fig. 2(b).

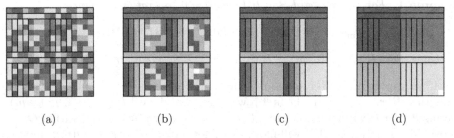

(a)   (b)   (c)   (d)

Fig. 3. Illustration of the algorithm

Phase I: Build. There are many ways to make B be in clear state and $\forall 0 \leq i < k^2 - 1, R_{k^2-1} \cap S_i \neq \varnothing$ within $O(n^{2.75})$ moves. Here we give one, which has following steps:

(i) For each $0 \leq i < k^2$, choose any $n + k^2(k^3 - k)$ tiles in S_i and put them into $B(a_i k^3 + b_i, 0 : k^3)$, $B(i, k^3 : n)$ and $B(a_j k^3 + k : (a_j + 1)k^3, b_j k^2 + i)$ for every $0 \leq j < k^2$;

  - Note that these parts exactly cover the first k^2 rows and first k^3 columns of B, so we can use an algorithm in reduction-of-order scheme (see Example 2 and Proposition 4) to solve them within $k^3 \cdot O(n^2) = O(n^{2.75})$ moves. Then, for each $0 \leq i < k^2 - 1$, choose any other one tile in S_i and put it into $B(n - k^2 : n, n - k^2 : n)$, so that $R_{k^2-1} \cap S_i \neq \varnothing$, as shown in Fig. 4(a) (note that only tiles which are chosen and put are drawn);

(ii) For $i = k^2 - 1, k^2 - 2, \cdots, 0$, move $B(i, k^3 : n)$ down to $B(a_i k^3 + b_i, k^3 : n)$, so that $H_i \subseteq S_i$, as shown in Fig. 4(b);

(iii) For $i = k^2 - 1, k^2 - 2, \cdots, 0$, move $B(a_i k^3 + k : (a_i + 1)k^3, b_i k^2 : b_i k^2 + k^2)$ right to $B(a_i k^3 + k : (a_i + 1)k^3, b_i k^3 : b_i k^3 + k^2)$, so that $V_{i,j} \subseteq S_j$ for all j, as shown in Fig. 4(c).

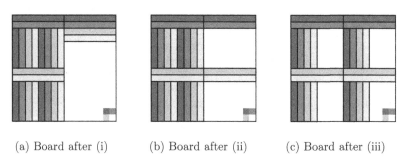

(a) Board after (i)   (b) Board after (ii)   (c) Board after (iii)

Fig. 4. Illustration of phase I

In (ii) and (iii), we need to do translational movement for a $1 \times m$ or $m \times 1$ subarray. Here we describe a way to move a $1 \times m$ subarray down by 1 cell within $O(m)$ moves, while other cases of translational movement are similar:

Lemma 1. *For $3 \times m$ board B' s.t. $B'(2,0) = 0$, operation sequence*

$$\theta_m = \text{LDRDL}^{m-1}(\text{RULDR})^{m-2}\text{U}^2\text{RD}^2\text{LURU} \tag{3}$$

satisfies that $(B'\theta_m)(1,0:m) = B'(0,0:m)$, $(B'\theta_m)(2,0:m) = B'(2,0:m)$.

If we need to move $B(x, y : y+m)$ down, first move the empty cell to $(x+2, y)$, note that $B(x, y : y+m)$ is the first row of $3 \times m$ sub-board $B(x : x+3, y : y+m)$, acting θ_m on B will move tiles in $B(x, y : y + m)$ down by one cell within $O(m)$ moves. Then acting $(\theta_m \text{U})^t$ will move them down by t cells within $O(nm)$ moves. Therefore, it costs $k^2 \cdot O(n^2) + k^2 \cdot k^2 \cdot O(nk^3) = O(n^{2.75})$ moves in (ii)(iii).

Phase II: Transport. After B is built to such a state, the transport phase begins. Before implementation, let's introduce a kind of operation:

Lemma 2. *For $m \times 2$ or $2 \times m$ board B' and any cell (x,y) s.t. $x + y + x_0 + y_0$ is odd ($x_0 = x_0(B'), y_0 = y_0(B')$), there is an operation sequence τ s.t. $B'\tau = (B'(x,y)\ 0)B'$ within $O(m)$ moves.*

Proof. Here only show one case. Other cases are similar and we omit them.

For $2 \times m$ board s.t. $x_0 = 0$, $x = 1$, $y = y_0 + 2l$, denote $b_i = B'(1, y_0 + 2l - i)$, $\tau_i = \mathrm{L}^{2(l-i)}\mathrm{UL}^2\mathrm{DR}$, $\rho = \mathrm{URDL}$, then $B'\tau_i \rho = (b_{2i-2}\ b_{2i-1}\ b_{2i})B'\tau_i$, that is, $B'\tau_i \rho \tau_i^{-1} = (b_{2i-2}\ b_{2i-1}\ b_{2i})B'.$[7] Let $\tau = \tau_1(\rho \tau_{l-1}^{-1}\tau_l)^{l-1}\rho \tau_l^{-1}\mathrm{UL}^{2l}$, then

$$\begin{aligned}
B'\tau &= B'\tau_1 \rho \tau_{l-1}^{-1}\tau_l \rho \tau_{l-1}^{-1}\tau_l \cdots \rho \tau_{l-1}^{-1}\tau_l \rho \tau_l^{-1}\mathrm{UL}^{2l} \\
&= B'\tau_1 \rho \tau_1^{-1}\tau_2 \rho \tau_2^{-1}\tau_3 \cdots \rho \tau_{l-1}^{-1}\tau_l \rho \tau_l^{-1}\mathrm{UL}^{2l} \\
&= (b_0\ b_1\ b_2)(b_2\ b_3\ b_4)\cdots(b_{2l-2}\ b_{2l-1}\ b_{2l})B'\mathrm{UL}^{2l} \\
&= (b_0\ b_1\ \cdots\ b_{2l})B'\mathrm{UL}^{2l} = (b_0\ 0)B'
\end{aligned}$$

For current board B, let $x_0 = x_0(B), y_0 = y_0(B)$, if cell (x,y) is in an $m \times 2$ or $2 \times m$ sub-board and $x + y + x_0 + y_0$ is odd, then we can act an operation sequence within $O(m)$ moves to swap the empty cell and $B(x,y)$. We use term "jump (x,y)" to describe such operation.

The implementation is shown in Algorithm 4.

Algorithm 4. Transport

$c_{i,j} \leftarrow |R_i \cap S_j|$ for all $0 \le i, j < k^2$
$i \leftarrow$ integer s.t. $0 \in R_i$
while $\exists j \neq i, c_{j,i} > 0$ **do**
    $j \leftarrow \min\{j : j \neq i : c_{j,i} > 0\}$ ▷ (*)
    choose any $B(x,y) \in R_j \cap S_i$
    jump $(a_i k^3 + b_i, y')$ for $y' = y_0(B)$ or $y_0(B) - 1$ ▷ (i)
    move-y$(b_j k^3 + i)$ ▷ (ii)
    for $x' = x_0(B)$ to x increasing or decreasing **do**
        if $x' \bmod k^3 \ge k$ **then**
            if $x' = x_0(B) \pm 1$ **then move-x**(x') ▷ (iii)
            else
                jump $(x', b_j k^3 + i)$ or $(x' \pm 1, b_j k^3 + i)$ in S_i ▷ (iv)
            end if   ▷ $H_i \subseteq S_i$ and $V_{*,i} \subseteq S_i$ holds
        end if
    end for   ▷ Now $0 \in V_{j,i}$
    jump (x,y), or **act** U or D (keeping $0 \in V_{j,i}$) then jump   ▷ (v)
    $c_{j,i} \leftarrow c_{j,i} - 1, c_{i,i} \leftarrow c_{i,i} + 1$ ▷ $c_{i,j} = |R_i \cap S_j|$ holds
    $i \leftarrow j$ ▷ $0 \in R_i$ holds
end while   ▷ B is still in clear state

---

[7] σ^{-1} is the sequence by reversing σ and replacing $\mathrm{U} \mapsto \mathrm{D}, \mathrm{D} \mapsto \mathrm{U}, \mathrm{L} \mapsto \mathrm{R}, \mathrm{R} \mapsto \mathrm{L}$, clearly $B\sigma\sigma^{-1} = B$.

Figure 5 shows the operation in each loop of Algorithm 4.

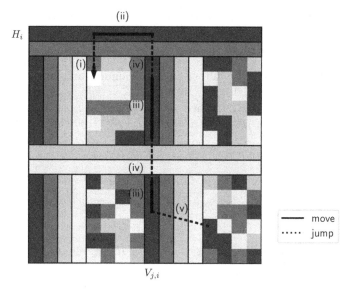

Fig. 5. Operations in each loop of phase II

Here are some facts of Algorithm 4 to show the correctness of this algorithm:

- **Fact 1:** The process terminates in at most n^2 loops.
  Because $\sum_i \sum_{j \neq i} c_{i,j} \leq n^2$ initially, and $\sum_i \sum_{j \neq i} c_{i,j}$ decreases by 1 after each loop.
- **Fact 2:** When the process terminates, $i = k^2 - 1$, and $R_j \subseteq S_j$ for all j.
  Note that when $0 \in R_i$, $c_{i,i} \leq |R_i| - 1$, since $H_i, V_{j,i} \subseteq S_i$, $|S_i| = |H_i| + \sum_j |V_{j,i}| + \sum_j c_{j,i}$, so

$$\sum_{j \neq i} c_{j,i} = |S_i| - |H_i| - \sum_j |V_{j,i}| - c_{i,i}$$

$$= |S_i| - n - k^2(k^3 - k) + (|R_i| - c_{i,i}) - (k^3 - k)(k^3 - k^2)$$

$$= |S_i| - k^6 + (|R_i| - c_{i,i}) \geq |S_i| - k^6 + 1$$

When the process terminates, $\sum_{j \neq i} c_{j,i} = 0$, so $|S_i| = k^6 - 1$, $i = k^2 - 1$. Now $c_{i,i} = |R_i| - 1$, that means $R_i - \{0\} \subseteq S_i$, hence $c_{i,j} = 0$ for all $j \neq i = k^2 - 1$. Also note that for all $i' < k^2 - 1$, $c_{k^2-1,i'}$ is positive at the beginning (because $R_{k^2-1} \cap S_{i'} \neq \varnothing$) and zero at the end, so there is some loop that changes $c_{k^2-1,i'}$ from 1 to 0, in that loop, $i = i'$, $j = k^2 - 1$, by (*), $\forall j' < k^2 - 1 \wedge j' \neq i', c_{j',i'} = 0$, that means after this loop, $\forall j' \neq i', c_{j',i'} = 0$ holds. Finally, $c_{j,i} = 0$ holds for all j, i s.t. $i \neq j$, that is, $R_j \subseteq S_j$.

– **Fact 3:** This phase costs at most $O(n^{2.75})$ inefficient moves according to potential function D defined by (2).
By Lemma 2, each (i) and (v) costs $O(k^3)$ moves while they are executed for at most n^2 times, each (iv) costs $O(k)$ moves while it is executed for at most $n^2 k$ times, so they cost at most $n^2 \cdot O(k^3) + n^2 k \cdot O(k) = O(n^{2.75})$ moves. Since (ii) and (iii) only move tiles in S_i towards rectangle area $(a_i k^3 : (a_i + 1)k^3, b_i k^3 : (b_i + 1)k^3)$ of B, when $y_0(B) \notin [b_i k^3, (b_i + 1)k^3)$, L/R moves in (ii) are efficient, and similarly, when $x_0(B) \notin [a_i k^3, (a_i + 1)k^3)$, U/D moves in (iii) are efficient. Therefore, in each loop, (ii) and (iii) cost at most $O(k^3)$ inefficient moves, and at most $O(n^{2.75})$ inefficient moves in total.

Phase III: Arrange. In this phase, we only need to:

– Swap $V_{i,j}$ and $V_{j,i}$ for all $0 \leq i, j \leq k^2, i \neq j$;
– Swap elements in $B(ak^3 + b, ck^3 : (c + 1)k^3)$ and elements in $B(ak^3 + c, bk^3 : (b + 1)k^3)$ for all $0 \leq a, b, c < k, b \neq c$;
Now tiles in H_i are moved into $B(a_i k^3 : a_i k^3 + k, b_i k^3 : (b_i + 1)k^3)$.

To swap the sets of tiles P, Q of two $1 \times k^3$ or $k^3 \times 1$ subarrays in B, we can use any way to move them into a $2 \times k^3$ or $k^3 \times 2$ rectangle within at most $O(k^3 n)$ moves (such as tile-by-tile, or using the way in Lemma 1 to move the subarray). Assume that operation sequence σ_0 places tiles in P, Q into $(x : x+3, y : y+k^3)$, while the first row of $B' = (B\sigma_0)(x : x + 3, y : y + k^3)$ is filled with tiles in P, the second row of B' is filled with tiles in Q, $B'(2,0) = 0$, then, by Lemma 1, $B\sigma_0\theta_{k^3} = \pi B\sigma_0$ for permutation π which maps P to Q and maps Q to P (θ_{k^3} is given by (3)), so $B\sigma_0\theta_{k^3}\sigma_0^{-1} = \pi B$. Act $\sigma_0\theta_{k^3}\sigma_0^{-1}$ on B, then sets P, Q are swapped on B, preserving other tiles.

Use this method, swapping the sets of tiles of two $1 \times k^3$ or $k^3 \times 1$ subarrays costs $O(k^3 n)$ moves, so there are at most $k^4 \cdot O(k^3 n) = O(n^{2.75})$ moves in this phase.

Phase IV: Finish. Finally, we need to solve k^2 parts $B(ak^3 : (a + 1)k^3, bk^3 : (b+1)k^3)$. The method is similar to the phase to solve upper and lower parts of $n \times 2$ rectangular puzzle, see Sect. 3.2. However, it's not necessary to do recursion, instead, we can use any poly-time algorithm within $O(k^9)$ time to solve each part, such as Algorithm 1. So there are at most $k^2 \cdot O(k^9) = O(n^{2.75})$ moves in this phase.

4.2 Algorithm in General

For general $(n^2 - 1)$-puzzle, we can easily reduce it to the case which $n = k^4$:

– Let $k = \lfloor \sqrt[4]{n} \rfloor$, then $k^4 \leq n < (k + 1)^4 = k^4 + O(k^3)$;
– Use algorithm in reduction-of-order scheme to solve first $n - k^4$ rows and first $n - k^4$ columns of B, it costs $(n - k^4) \cdot O(n^2) = O(n^{2.75})$ moves (Proposition 4;

- Solve $k^4 \times k^4$ block $B(n - k^4 : n, n - k^4 : n)$ to finish.
  Obviously $D_{B_T}(B) = D_{B_T(n-k^4:n,n-k^4:n)}(B(n - k^4 : n, n - k^4 : n))$,[8] so it costs at most $O(n^{2.75})$ inefficient moves for B too.

So far, we have proved that this is an additive approximation algorithm for $(n^2 - 1)$-puzzle within $O(n^{2.75})$ additive constant.

5 Application: Average/Worst Case Analysis

For a type of sliding puzzle, how many moves are needed to solve the puzzle in average case, and in worst case?

Let $\mathcal{O}$ be the orbit of the target board B_T, and the scrambled board $B_S \in \mathcal{O}$.

- In average case, the **average optimal solution length** $\overline{\text{OPT}}$ is the average value of OPT of all $B_S \in \mathcal{O}$;
- In worst case, the **maximum optimal solution length** $\max \text{OPT}$, also known as the **God's number**, is the maximum value of OPT for all $B_S \in \mathcal{O}$.

For $n \times 2$ rectangular puzzle or $(n^2 - 1)$-puzzle, computing $\overline{\text{OPT}}$ or $\max \text{OPT}$ is hard. However, using additive approximation algorithms, we can get asymptotic results for these puzzles. For an additive approximation algorithm within α inefficient moves based on potential function φ, we have

$$\varphi(B_S) - \varphi(B_T) \leq \text{OPT} \leq \text{SOL} \leq \varphi(B_S) - \varphi(B_T) + \alpha$$

when B_S is uniformly distributed in $\mathcal{O}$:

$$\mathbb{E}[\varphi(B_S)] - \varphi(B_T) \leq \overline{\text{OPT}} \leq \mathbb{E}[\varphi(B_S)] - \varphi(B_T) + \alpha$$

$$\max_{B_S \in \mathcal{O}} \varphi(B_S) - \varphi(B_T) \leq \max \text{OPT} \leq \max_{B_S \in \mathcal{O}} \varphi(B_S) - \varphi(B_T) + \alpha$$

so $\overline{\text{OPT}} = \mathbb{E}[\varphi(B_S)] - \varphi(B_T) + O(\alpha)$, $\max \text{OPT} = \max_{B_S \in \mathcal{O}} \varphi(B_S) - \varphi(B_T) + O(\alpha)$. The remaining task is to find $\mathbb{E}[\varphi(B_S)]$ and $\max_{B_S \in \mathcal{O}} \varphi(B_S)$.

5.1 Analysis for $n \times 2$ Rectangular Puzzle

Proposition 8. *For $n \times 2$ rectangular puzzle, the average optimal solution length is $n^2 + O(n \log n)$, and the God's number is $2n^2 + O(n \log n)$.*

Proof. Assume $n \geq 3$. For uniformly distributed $B_S \in \mathcal{O}$, let $b_i = B_S(\lfloor \frac{i}{2} \rfloor, i \bmod 2)$, for $i = 1, 2, \cdots, 2n - 1$, denote p_i as the integer s.t. $b_{p_i} = i$, then

$$\mathbb{E}[I(B_S)] = \mathbb{E}\left[\sum_{1 \leq i < j < 2n} [p_i > p_j] \right] = \sum_{1 \leq i < j < 2n} \Pr\{p_i > p_j\}$$

---

[8] Here the subscript represents the corresponding target board.

For any $1 \leq i < j < 2n$, choose $1 \leq i', j' < 2n$ s.t. i, j, i', j' are distinct, consider a map $f : \mathcal{O} \to \mathcal{O}$ defined by $B \mapsto (i\ j)(i'\ j')B$, then f maps $\{B_S \in \mathcal{O} : p_i < p_j\}$ to $\{B_S \in \mathcal{O} : p_i > p_j\}$ and vice versa, so these two sets have the same size, $\Pr\{p_i > p_j\} = \frac{1}{2}$, thus $\mathbb{E}[I(B_S)] = \frac{1}{4}(2n-1)(2n-2)$. So $\overline{\mathrm{OPT}} = \mathbb{E}[I(B_S)] + O(n \log n) = n^2 + O(n \log n)$.

Note that $I(B_S) \leq \sum_{1 \leq i < j < 2n} 1 = \frac{1}{2}(2n-1)(2n-2)$ and the equality holds when B_S is obtained by rotating B_T half a turn:

$$
B_S = \begin{pmatrix} 0 & 2n-1 \\ 2n-2 & 2n-3 \\ \vdots & \vdots \\ 2 & 1 \end{pmatrix}
$$

so $\max \mathrm{OPT} = \max_{B_S \in \mathcal{O}} I(B_S) + O(n \log n) = 2n^2 + O(n \log n)$.

5.2   Analysis for $(n^2 - 1)$-Puzzle

Proposition 9. *For (n^2-1)-puzzle, the average optimal solution length is $\frac{2}{3}n^3 + O(n^{2.75})$, and the God's number is $n^3 + O(n^{2.75})$.*

Proof. For uniformly distributed $B_S \in \mathcal{O}$, here directly use the conclusion in [2]:

$$
\mathbb{E}[D(B_S)] = \frac{2}{3}n^3 + O(n^2), \quad \max_{B_S \in \mathcal{O}} D(B_S) \geq n^3 - O(n^2)
$$

so $\overline{\mathrm{OPT}} = \mathbb{E}[D(B_S)] + O(n^{2.75}) = \frac{2}{3}n^3 + O(n^{2.75})$. Let $(x_i, y_i), (x_i', y_i')$ be current and target position of tile i in B_S, then

$$
D(B_S) \leq \sum_{i=1}^{n^2-1} \left(\left| x_i - \frac{n-1}{2} \right| + \left| x_i' - \frac{n-1}{2} \right| + \left| y_i - \frac{n-1}{2} \right| + \left| y_i' - \frac{n-1}{2} \right| \right)
$$

$$
\leq 4n \sum_{x=0}^{n-1} \left| x - \frac{n-1}{2} \right| = n^3 + O(n^2)
$$

so $\max \mathrm{OPT} = \max_{B_S \in \mathcal{O}} D(B_S) + O(n^{2.75}) = n^3 + O(n^{2.75})$.

6   Conclusion

This research focus on the design of poly-time additive approximation algorithm for two types of sliding puzzle: $n \times 2$ rectangular puzzle and $(n^2 - 1)$-puzzle.

Some methods in sport sliding puzzle area are useful in approximation algorithm design. For $n \times 2$ rectangular puzzle, which is a variation of $(n^2 - 1)$-puzzle, directly using an algorithm in divide-and-conquer scheme has a good performance. Using inversion number as potential function, we have proved that it achieves an additive approximation within $O(n \log n)$ additive constant.

However, to achieve additive approximation for $(n^2 - 1)$-puzzle is more difficult. Therefore, we designed a new poly-time algorithm for $(n^2 - 1)$-puzzle to solve this problem successfully. Using Manhattan distance as potential function, we have proved that it achieves an additive approximation within $O(n^{2.75})$ additive constant.

An important application of these approximation algorithms is average/worst case analysis. Using the potential functions and approximation algorithms we have designed, we can analyze optimal solution length in average case and worst case (God's number) for $n \times 2$ rectangular puzzle and $(n^2 - 1)$-puzzle asymptotically. The result is shown in Table 2.

Table 2. Average/worst case analysis

Type	Average optimal solution length	God's number
$n \times 2$ rectangular puzzle	$n^2 + O(n \log n)$	$2n^2 + O(n \log n)$
$(n^2 - 1)$-puzzle	$\frac{2}{3}n^3 + O(n^{2.75})$	$n^3 + O(n^{2.75})$

Acknowledgements. I would like to thank my supervisor, Ran Duan, for his guidance during this research.

I would also like to thank professional sliding puzzle players: jrj5423, Samarra, dphdmn, for bring me some data and idea.

References

1. Ratner, D., Warmuth, M.: The $(n^2 - 1)$-puzzle and related relocation problems. J. Symb. Comput. **10**(2), 111–137 (1990)
2. Parberry, I.: A real-time algorithm for the $(n^2 - 1)$-puzzle. Inf. Process. Lett. **56**(1), 23–28 (1995)
3. Korf, R., Schultze, P.: Large-scale parallel breadth-first search. In: Proceedings of the 20th National Conference on Artificial Intelligence (AAAI-2005), pp. 1380–1385 (2005)
4. Parberry, I.: Solving the $(n^2 - 1)$-puzzle with $\frac{8}{3}n^3$ expected moves. Algorithms **8**(3), 459–465 (2015)
5. Agostinelli, F., McAleer, S., Shmakov, A., Baldi, P.: Solving the Rubik's cube with deep reinforcement learning and search. Nat. Mach. Intell. **1**, 356–363 (2019)

Differential Game Analysis for Cooperation Models in Automotive Supply Chain Under Low-Carbon Emission Reduction Policies

Yukun Cheng[1,3]([✉])([iD]), Zhanghao Yao[2], and Xinxin Wang[2]

[1] School of Business, Jiangnan University, Wuxi 214122, China
ykcheng@amss.ac.cn
[2] School of Business, Suzhou University of Science and Technology,
Suzhou 215009, China
zhyao@post.usts.edu.cn
[3] Think Tank for Urban Development, Suzhou University of Science and Technology,
Suzhou 215009, China

Abstract. In the context of reducing carbon emissions in the automotive supply chain, collaboration between vehicle manufacturers and retailers has proven to be an effective measure for enhancing carbon emission reduction within the enterprise. This study aims to evaluate the effectiveness of such collaboration by constructing a differential game model that incorporates carbon trading and consumer preferences for low-carbon products. The model examines the decision-making process of an automotive supply chain comprising a vehicle manufacturer and multiple retailers. By utilizing the Hamilton-Jacobi-Bellman equation, we analyze the equilibrium strategies of the participants under both a decentralized model and a Stackelberg leader-follower game model. In the decentralized model, the vehicle manufacturer optimizes its carbon emission reduction efforts, while each retailer independently determines its low-carbon promotion efforts and vehicle retail price. In the Stackelberg leader-follower game model, the vehicle manufacturer cooperates with the retailers by offering them a subsidy. Consequently, the manufacturer plays as the leader, making decisions on carbon emission reduction efforts and the subsidy rate, while the retailers, as followers, compute their promotion efforts and retail prices accordingly. Through theoretical analysis and numerical experiments considering the manufacturer's and retailers' efforts, the low-carbon reputation of vehicles, and the overall system profits under both models, we conclude that compared to the decentralized model, where each party pursues individual profits, the collaboration in the Stackelberg game yields greater benefits for both parties. Furthermore, this collaborative approach promotes the long-term development of the automotive supply chain.

This work is supported by the National Nature Science Foundation of China (No. 11871366) and the Research Innovation Program for College Graduate Students of Jiangsu Province (No. KYCX22 3249).

M. Li et al. (Eds.): IJTCS-FAW 2023, LNCS 13933, pp. 147–159, 2023.
https://doi.org/10.1007/978-3-031-39344-0_11

Keywords: Differential game · Carbon trading · Automotive · Supply chain · Carbon emission reduction efforts

1   Introduction

The automotive industry, in particular, has been identified as one of the major contributor to global carbon emissions, and as a result, has been the focus of regulatory and societal pressures to reduce its carbon emissions. In response to these pressures, many manufactures have set ambitious goals to reduce their carbon emissions, including improving energy efficiency and adopting electric vehicles [1]. However, achieving these goals is not always straightforward, as manufacturers operate in a complex ecosystem that includes not only themselves but also retailers, consumers, and other participants.

One important part in this ecosystem is the interaction between the manufacturers and their retailers. The manufacturer is responsible for carbon emission reduction efforts by conducting research and development of low-carbon technologies and producing low-carbon vehicles. Retailers play a critical role in the distribution of vehicles, using low-carbon promotions to attract consumers to purchase low-carbon products. Besides, the manufacturer and retailers may also cooperate in their emission reduction activities. However, retailers and manufacturer are primarily driven by their own interests to maximize profits, which may lead to suboptimal resource allocation and market efficiency. Therefore, designing appropriate cooperation models between manufacturers and retailers is crucial. Generally, decentralized decision-making and the Stackelberg leader-follower game model are widely adopted [2–4]. In decentralized decision-making, manufacturers and retailers can make independent decisions based on their own information and considerations. This decision design encourages them to make optimal decisions based on their own interests. At the same time, decentralized decision-making can reduce the costs of information exchange and coordination. Through the Stackelberg leader-follower decision-making, manufacturer can first formulate the carbon emission reduction effort and subsidy rate, while retailers react based on the manufacturer's decisions. This arrangement of decision sequences can fully leverage the manufacturer's advantage and achieve market efficiency to some extent.

When it comes to government efforts to reduce carbon emissions, various administrative measures are implemented, including carbon taxes, building carbon trading markets, increasing green investments, and more. Among these measures, carbon trading is the most widely used, which has been implemented in many countries, including China and the European Union [5]. Under a carbon trading system, the government initially allocates a certain amount of carbon quotas to each enterprise based on specific allocation rules. If an enterprise's actual carbon emissions are lower than the initial quota, they can sell the excess quotas in the carbon trading market to generate profit. On the other hand, enterprises that exceed their carbon quotas need to buy additional quotas to comply with government regulations [6]. It is evident that the carbon trading policy influences the production and operation decisions of enterprises [7]. Taking the impact of carbon trading into consideration, Yang et al. [8] designed pricing and emission reduction models for two competitive supply chains under a carbon trading scheme. Sun et al. [3] studied multi-period continuous production subject to dynamically changing characteristic conditions within the framework of carbon trading, and proposed different differential game models to explore cooperation models between the government and enterprises. Unlike existing works that typically involve two participants, such as one manufacturer and one supplier or one manufacturer and one retailer, this paper focuses on studying the interactions among one manufacturer and multiple retailers.

Many studies have investigated the influence of consumers' low-carbon preferences on the supply chain. Xia et al. [5] incorporated reciprocal preferences and consumers' low-carbon awareness into a supply chain model consisting of a single manufacturer and a single retailer, and examined how these preferences affect the decisions, performance, and efficiency of the supply chain members. Wang et al. [2] studied the carbon reduction decisions of automotive supply chain members under total control and transaction rules using differential game theory, where consumers' low-carbon preferences were also a significant factor affecting companies' emission reduction decisions. Chen et al. [9] developed a closed-loop supply chain model for recycling and remanufacturing based on Stackelberg leader-follower game theory, considering consumers' low-carbon preferences and government subsidies, and established profit models for each stakeholder under centralized and decentralized decision-making models. They also examined the impact of consumers' low-carbon preferences on the profits and decisions of supply chain members. Xu et al. [4] developed a cost-sharing model for the automotive supply chain, considering the dynamic changes in consumers' low-carbon preferences. They argued that consumer demand would be affected by the low-carbon reputation of products, which in turn would impact product sales. Similar insights can also be found in the studies of [3,10,11], which suggest that market demand depends on a company's green reputation and the level of environmental friendliness of its products, and that the formation of a green reputation requires coordination among supply chain members. These studies highlight that consumers' low-carbon preferences have become an essential factor in supply chain decision-making. Additionally, we noted that [2,4] considered dynamic consumer low-carbon preferences in their research. Dynamic preferences are more reflective of their impact on the supply chain compared to static preferences. Therefore, drawing inspiration from [3,4], our paper aims to introduce dynamic consumer low-carbon preferences based on green low-carbon reputation.

In reality, the process of reducing carbon emissions within the supply chain is a dynamic and long-term endeavor. Throughout this process, the level of emission reduction and the green low-carbon reputation of vehicles, as well as consumers' low-carbon preferences, are continuously changing. This prompts us to adopt a dynamic perspective to study the interaction between the manufacturer and n retailers. Differential game theory, as an important dynamic game model, can provide solutions for dynamic equilibrium outcomes in continuous time [12]. By utilizing the Hamilton-Jacobi-Bellman equation, we analyze the equilibrium strategies of participants in two distinct models: a decentralized model and a Stackelberg leader-follower game model. In the decentralized model, each participant makes individual decisions, whereas in the Stackelberg game model, the manufacturer engages in collaboration with the retailers by offering a subsidy. Unlike previous studies that mainly focus on the manufacturer's strategy of carbon emission reduction efforts and the retailers' strategies for low-carbon promotion, our study goes further by incorporating the retailers' decision-making process regarding retail pricing. Furthermore, to enhance the realism of our model, we introduce carbon trading as an additional element.

The remaining parts of this article are organized as follows. Section 2 provides the introduction of the problem and the necessary assumptions. In Sect. 3, two differential cooperation models are established and the corresponding equilibrium solutions are analyzed. Section 4 conducts the experiments and the numerical analysis to verify our theoretical results. Finally, Sect. 5 presents the conclusion and discussion on further directions. The detailed proofs of the main results are placed in the full version.

2   Problem Assumptions and Notations

This paper focuses on a two-tier automotive supply chain consisting of a manufacturer and n retailers. The manufacturer is responsible for carbon emission reduction efforts to develop low-carbon technology and producing vehicles under a brand, and its investment in low-carbon technology can enhance the low-carbon reputation of its products and the emissions reduction level. The retailers strive to improve the low-carbon reputation of the brand of vehicles through carbon promotion efforts, such as publicity, subsidies, and other efforts to attract more consumers. Both the manufacturer and the retailers' efforts can improve the low-carbon reputation of this brand of vehicle. In addition, due to the existence of cap-and-trade regulation, the manufacturer needs to consider the impact of carbon trading when making decisions.

Considering the interaction between the manufacturer and the retailers, we propose two cooperation models: the decentralized model and the Stackelberg leader-follower game model, to explore the strategy selection and interaction among the participants, respectively. Each model accounts for the interdependence between the manufacturer and the retailers, as well as the impact of carbon trading and consumer green preferences on their decision-making process.

Table 1 provides the notations used in this paper.

2.1   Model Assumptions

Assumption 1. Both the carbon emission reduction effort E_m of the manufacturer and the low-carbon promotion effort E_{r_i} of retailer i affect the low-carbon reputation G of the brand of vehicles. Similar to [4], this dynamic process of $G(t)$ is described by the following differential equation:

$$\dot{G}(t) = \mu_m E_m(t) + \sum_{i=1}^{n} \mu_{r_i} E_{r_i}(t) - \delta G(t),$$

Table 1. Notations and description.

Notation	Descriptions
t	Time period
$G(t), G(0)$	Low carbon reputation of the vehicle at time t, and initial value of the low carbon reputation, $G(0) \geq 0$.
E_m, E_{r_i}	Manufacturer's carbon emission reduction effort, retailer i's low-carbon promotion effort.
λ_m, λ_{r_i}	Manufacturer's cost coefficient related to carbon emission reduction, retailer cost coefficient related to the promotion of low-carbon vehicle, $\lambda_m, \lambda_{r_i} > 0$.
μ_m, μ_{r_i}	Influence coefficient of manufacturer emission reduction efforts on the reputation, influence coefficient of retailer's low-carbon promotion efforts on reputation, $\mu_m, \mu_{r_i} > 0$.
ω	Influence coefficient of manufacturer emission reduction efforts on the emission reduction level, $\omega > 0$.
$p, \, p_i, \, p_c$	Manufacturer's wholesale price, retail price, price per unit of carbon emission credit, $p, p_i, p_c > 0$.
$Q_i(t)$	Demand function for retailer i at time t and the total demand is $Q(t) = \sum_{i=1}^{n} Q_i(t)$.
θ	Low-carbon preference of consumers, $\theta > 0$.
a_i	retailer i's potential sales, $a_i > 0$.
b_i, c	Influence coefficient of price on sales, influence coefficient of other retailers' prices on sales, $b_i > 0, 0 \leq c \leq 1$.
F_0	Carbon emission quota.
$F(t)$	Total quantity of carbon quota trading at time t.
δ	Decay coefficient of low-carbon reputation, $\delta > 0$.
ρ	Discount rate of profit, $\rho > 0$

where μ_m and μ_{r_i}, $i = 1, \cdots, n$, are the influence coefficients of manufacturer's and retailers' efforts on the reputation, $\delta > 0$ is the decay coefficient of low-carbon reputation.

Assumption 2. By assumptions for demand in [13], the demand Q_i of the vehicles sold by retailer i is decreasing with the retail price p_i and increasing with the price p_j, $j \neq i$ set by others. In addition, the higher lower-carbon reputation $G(t)$ can positively influence demand $Q_i(t)$. Thus

$$Q_i(t) = (a_i - b_i p_i + c \sum_{k=1, k \neq i}^{n} \frac{b_k p_k}{n-1}) \theta G(t),$$

where $b_i > 0$ is the coefficient of the effect from the retail price p_i on demand Q_i. Since the effect from other retail price $p_k \neq p_i$ on Q_i is no more than the effect from p_i, we let the coefficient $0 \leq c \leq 1$.

Assumption 3. The carbon emissions level of the vehicle is determined by the manufacturer's efforts in emission reduction (through low-carbon technology

investment and development), with the emission reduction level being proportional to the manufacturer's efforts, denoted by a coefficient ω. Thus, the vehicle's
emission reduction level is equal to $\omega \cdot E_m(t)$. Similar to the assumption in [2,5],
we assume that the initial carbon emissions per unit product is 1, indicating
the emission reduction per unit product is $1 \cdot \omega \cdot E_m(t)$, the per unit product
carbon emission is $1 - \omega \cdot E_m(t)$, and the total emission of the manufacturer is
$(1 - \omega E_m(t)) \sum_{i=1}^{n} Q_i(t)$. Let us denote the carbon quota of the manufacturer
at time t as $F(t)$, with F_0 being the initial carbon quota. Therefore, the carbon
quota of the manufacturer is expressed as:

$$F(t) = F_0 - (1 - \omega E_m(t)) \sum_{i=1}^{n} Q_i(t).$$

It is worth to note that when $F(t) > 0$, the manufacturer possesses excess carbon
quotas that can be traded for financial gain in the carbon market. Conversely,
if $F(t) < 0$, the manufacturer is obligated to procure carbon quotas from the
market to comply with the government's regulations.

Assumption 4. The costs of emission reduction efforts paid by the manufacturer and retailer are quadratic functions of $E_m(t)$ and $E_{r_i}(t)$, respectively, which
are popular and have been adopted by a lot of literatures [2,3,5]. Therefore, the
costs of efforts respectively paid by the manufacture and retailer i are

$$C_m(t) = \frac{1}{2}\lambda_m E_m^2(t), \quad C_{r_i}(t) = \frac{1}{2}\lambda_{r_i} E_{r_i}^2(t).$$

Assumption 5. Due to the fact that the n retailers are the retailers of the
same brand of automobile vehicles, which are provided by the manufacturer,
their business scale, target customers, and sales models are all similar. Therefore,
we suppose these n retailers to be homogeneous, and thus the cost coefficients,
the impact coefficients of effort levels on reputation, and the price sensitivity
coefficient are all set to be the same, that is, $\lambda_r = \lambda_{r_1} = \lambda_{r_2} = \ldots = \lambda_{r_i}$,
$\mu_r = \mu_{r_1} = \mu_{r_2} = \ldots = \mu_{r_i}$, $b = b_1 = b_2 = \ldots = b_i$.

3   Model Formulation

Based on the assumptions in Sect. 2, this section establishes two differential
game models to analyze the equilibrium strategies of the manufacture and the
retailers, by considering the long-term impact of carbon trading and low-carbon
reputation of vehicles. For the sake of convenience, we omit t in the following.

3.1   Decentralized Model

Before establishing the decentralized model, it is necessary for us to clarify that
the objectives of the manufacture and the retailers are all to maximize their own

profits individually over an infinite time, and the discount rate is denoted by $\rho > 0$. Therefore, their objective functions are formulated as follows.

$$\max J_m^D = \int_0^\infty e^{-\rho t} \left[p \sum_{i=1}^n Q_i + p_c \left(F_0 - (1 - \omega E_m) \sum_{i=1}^n Q_i \right) - \frac{1}{2} \lambda_m E_m^2 \right] dt,$$

$$\max J_{r_i}^D = \int_0^\infty e^{-\rho t} \left[p_i Q_i - \frac{1}{2} \lambda_r E_{r_i}^2 \right] dt,$$

where p is the wholesale price, p_c is the price of per carbon quota, which are given in advance; and p_i is the retail price set by retailer i. Let V_m^D, $V_{r_i}^D$ denote the profit function of manufacture and retailer i, respectively. Thus, the corresponding Hamiltonian-Jacobi- Bellman (HJB) equations are formulated as:

$$\rho V_m^D = \max \left[p \sum_{i=1}^n Q_i + p_c (F_0 - (1 - \omega E_m) \sum_{i=1}^n Q_i) - \frac{1}{2} \lambda_m E_m^2 + V_m^{D'} (\mu_m E_m \right.$$

$$\left. + \sum_{i=1}^n \mu_r E_{r_i} - \delta G) \right],$$

$$\rho V_{r_i}^D = \max \left[p_i Q_i - \frac{1}{2} \lambda_r E_{r_i}^2 + V_{r_i}^{D'} (\mu_m E_m + \sum_{i=1}^n \mu_r E_{r_i} - \delta G) \right].$$

Let us denote $q_i = a_i - bp_i + c \sum_{k=1, k \neq i}^n \frac{bp_k}{n-1}$, meaning that $Q_i(t) = \theta G(t) q_i$. So

$$\sum_{i=1}^n q_i = \sum_{i=1}^n (a_i - bp_i + c \sum_{k=1, k \neq i}^n \frac{bp_k}{n-1}) = \sum_{i=1}^n a_i - (1-c)b \sum_{i=1}^n p_i.$$

Proposition 1. *Under the decentralized model, the optimal carbon emission reduction effort of the manufacturer and the optimal low-carbon promotion efforts of retailers are*

$$E_m^{D*} = \frac{p_c \theta \omega G \sum_{i=1}^n q_i + (2A^D G + B^D) \mu_m}{\lambda_m}, \quad E_{r_i}^{D*} = \frac{\mu_r D_i^D}{\lambda r},$$

the optimal retailer price set by retailer i is

$$p_i^{D*} = \frac{(n-1)(2-c)a_i + c \sum_{k=1}^n a_k}{b(2-c)(2n-2+c)},$$

the optimal trajectory of low carbon reputation is:

$$G^{D*} = \frac{4B^D \lambda_r \mu_m^2 + 4 \lambda_m \mu_r^2 \sum_{i=1}^n D_i^D}{\lambda_r \sqrt{\Delta^D} - \lambda_m \lambda_r \rho},$$

and the optimal profits of manufacture and retailer i are:

$$V_m^D = A(G^{D*})^2 + BG^{D*} + C^D, \quad V_{r_i}^D = D_i^D G^{D*} + H_i^D,$$

where A^D, B^D, C^D and D_i^D are the coefficients in the value functions $V_m^D(G) = A^D G^2 + B^D G + C^D$ and $V_{r_i}^D = D_i^D G + H_i^D$,

$$A^D = \frac{4\lambda_m \delta + 2\lambda_m \rho - 4p_c \theta \omega \mu_m \sum_{i=1}^n q_i - \sqrt{\triangle^D}}{8\mu_m^2},$$

$$B^D = \frac{4\lambda_m \lambda_r \theta(p - p_c) \sum_{i=1}^n q_i + 8A\lambda_m \mu_r^2 \sum_{i=1}^n D_i}{2\lambda_m \lambda_r \rho + \sqrt{\triangle^D}},$$

$$C^D = \frac{p_c F_0}{\rho} + \frac{{B^D}^2 \mu_m^2}{2\lambda_m \rho} + \frac{B^D \sum_{i=1}^n \mu_r E_{r_i}}{\rho},$$

$$D_i^D = \frac{\lambda_m p_i q_i \theta}{\lambda_m(\rho + \delta) - p_c \theta \omega \mu_m \sum_{i=1}^n q_i - 2A\mu_m^2},$$

where $\triangle^D = (4p_c \theta \omega \mu_m \sum_{i=1}^n q_i - 4\lambda_m \delta - 2\lambda_m \rho)^2 - 16(\mu_m p_c \theta \omega \sum_{i=1}^n q_i)^2 > 0$.

The proof of Proposition 1 is provided in the full version.

3.2 Stackelberg Leader-Follower Game Model

In the Stackelberg leader-follower game model, the manufacturer supports the retailers by offering a subsidy. In this model, the manufacturer plays as a leader to disclose its carbon emission reduction effort E_m and subsidy rate x_i to retailer i in the first stage, and its objective function is

$$\max J_m^S = \int_0^\infty e^{-\rho t} \left[p \sum_{i=1}^n Q_i + p_c(F_0 - (1 - \omega E_m) \sum_{i=1}^n Q_i) - \frac{1}{2}\lambda_m E_m^2 \right.$$
$$\left. - \frac{1}{2} \sum_{i=1}^n \lambda_r x_i E_{r_i}^2 \right] dt.$$

Then, in the second stage retailer i determines its low-carbon promotion effort E_{r_i} after observing the manufacturer actions as a follower. Therefore, the objective functions of retailer i is given by:

$$\max J_{r_i}^S = \int_0^\infty e^{-\rho t} \left[p_i Q_i - \frac{1 - x_i}{2}\lambda_r E_{r_i}^2 \right] dt.$$

Let $V_m^S, V_{r_i}^S$ denote the value functions of manufacture and retailer i, respectively. We have the Hamiltonian-Jacobi- Bellman (HJB) equations as

$$\rho V_m^S = \max \left[p \sum_{i=1}^n Q_i + p_c(F_0 - (1 - \omega E_m) \sum_{i=1}^n Q_i) - \frac{1}{2}\lambda_m E_m^2 - \frac{1}{2} \sum_{i=1}^n \lambda_r x_i E_{r_i}^2 \right.$$
$$\left. + V_m^{S'}(\mu_m E_m + \sum_{i=1}^n \mu_r E_{r_i} - \delta G) \right],$$

$$\rho V_{r_i}^S = \max \left[p_i Q_i - \frac{1 - x_i}{2}\lambda_r E_{r_i}^2 + V_{r_i}^{S'}(\mu_m E_m + \sum_{i=1}^n \mu_r E_{r_i} - \delta G) \right].$$

Proposition 2. *Under the Stackelberg leader-follower game model, the equilibrium carbon emission reduction effort of the manufacturer and the low-carbon promotion effort of retailer i are presented as follows:*

$$E_m^{S*} = \frac{p_c\theta\omega G \sum_{i=1}^{n} q_i + (2A^S G + B^S)\mu_m}{\lambda_m}, \quad E_{r_i}^{S*} = \frac{\mu_r D_i^S}{\lambda_r(1-x_i)},$$

the equilibrium retailer price p_i^{S*} is

$$p_i^{S*} = \frac{(n-1)(2-c)a_i + c\sum_{k=1}^{n} a_k}{b(2-c)(2n-2+c)},$$

the equilibrium subsidy rate x_i^* is

$$x_i^* = \frac{2(A^S G_s^S + B^S) - D_i^S}{2(A^S G_s^S + B^S) + D_i^S},$$

the equilibrium trajectory of low carbon reputation is:

$$G^{S*} = \frac{4B^S \lambda_r \mu_m^2 + 4nB^S \lambda_m \mu_r^2 + 2\lambda_m \mu_r^2 \sum_{i=1}^{n} D_i^S}{\sqrt{\triangle^S} - \lambda_m \lambda_r \rho}. \tag{1}$$

and the equilibrium profits of manufacture and retailer i under the Stackelberg leader-follower game are

$$V_m^S = A^S G_s^{D^2} + B^S G_s^S + C^S, \quad V_{r_i}^S = D_i^S G_s^S + H_i^S,$$

where A^S, B^S, C^S and D_i^S are the coefficients of the value functions $V_m^S = A^S G^2 + B^S G + C^S$ and $V_{r_i}^S = D_i^S G + H_i^S$, with the formulations as:

$$A^S = \frac{(4\delta\lambda_m\lambda_r + 2\lambda_m\lambda_r\rho - 4\lambda_r\mu_m p_c\omega\theta \sum_{i=1}^{n} q_i) - \sqrt{\triangle^S}}{8(\lambda_r\mu_m^2 + n\lambda_m\mu_r^2)},$$

$$B^S = \frac{4\lambda_m\lambda_r\theta(p - p_c) \sum_{i=1}^{n} q_i + 4A\lambda_m\mu_r^2 \sum_{i=1}^{n} D_i}{2\lambda_m\lambda_r\rho + \sqrt{\triangle^S}},$$

$$C^S = \frac{p_c F_0}{\rho} + \frac{B^{S^2}\mu_m^2}{2\lambda_m\rho} + \frac{4nB^2\mu_r^2 - 4B^S\mu_r^2 \sum_{i=1}^{n} D_i^{S^2} + \mu_r^2 \sum_{i=1}^{n} D_i^S}{8\lambda_r\rho},$$

$$D_i^S = \frac{4\lambda_m\lambda_r\theta p_i q_i}{2\lambda_m\lambda_r\rho + A^S\lambda_m\mu_r^2 + \sqrt{\triangle^S}},$$

where

$$\triangle^S = (4\lambda_m\lambda_r\delta + 2\lambda_m\lambda_r\rho - 4\lambda_r\mu_m p_c\omega\theta \sum_{i=1}^{n} q_i)^2 - 16\lambda_r(\lambda_r\mu_m^2 + n\lambda_m\mu_r^2)(p_c\theta\omega \sum_{i=1}^{n} q_i)^2.$$

The proof of Proposition 2 is provided in the full version.

4   Numerical Analysis

This section performs numerical experiments to assess the models' validity, conduct the sensitivity analysis of key parameters, and provide managerial insights. Moreover, we simulate the impact of changes in low-carbon reputation and supply chain members' profits under the scenario without carbon trading, aiming to examine the decision-makings under various policy conditions.

Drawing upon the works of Xu et al. [4] and Wang et al. [2], we set the relevant parameters for our numerical experiments as: $G(0) = 0, \lambda_m = 500, \lambda_r = 100, \mu_m = 2, \mu_r = 0.5, \omega = 0.4, p = 15, \theta = 0.6, b = 0.9, c = 0.8, F_0 = 500, \delta = 0.8, \rho = 0.6, n = 6$.

We distinguish the existence of carbon trading policy by setting $p_c = 1 > 0$ and the scenario without carbon trading by setting $p_c = 0$.

The results of the identical decision-making model are depicted using the same color scheme. Specifically, the decentralized model is represented by the color cyan, while the Stackelberg game model is represented by the color blue. Furthermore, solid and dashed lines are employed to differentiate between cases with and without carbon trading, respectively. Finally, we use the superscript 'N' to denote the scenario where the carbon trading policy is not implemented.

4.1   Changes in Manufacturer's and Retailer's Profits over Time

This section discusses the variation of profits for the manufacturer and the retailer over time for the decentralized model and the Stackelberg game model respectively in Fig. 1-(a) and Fig. 1-(b).

From the figures, we can observe that conditional cooperation leads to improved profits for both manufacturers and retailers. This is achieved through manufacturers providing subsidies to retailers to incentivize the adoption of more low-carbon promotional measures. Additionally, in the absence of carbon trading, both decision-making models result in lower profitability compared to the scenarios with carbon trading. This indicates that manufacturers derive benefits from engaging in carbon trading.

It is worth noting that in the decentralized model with carbon trading, retailers experience higher profitability than in scenarios without carbon trading. This suggests that retailers benefit from the enhanced reputation resulting from manufacturers' carbon emission reduction efforts. Overall, if the vehicles produced by the manufacturer are environmentally friendly, they can generate more profits from the carbon trading market and allocate subsidies to support retailers' promotion efforts. By channeling a portion of the carbon trading revenue into the supply chain, both the manufacturer and retailers are incentivized to actively participate in low-carbon management, ultimately leading to the decarbonization of the entire supply chain.

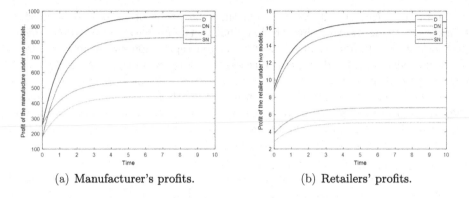

(a) Manufacturer's profits.   (b) Retailers' profits.

Fig. 1. Manufacturer's and retailers' profits under different models.

4.2 Changes of Low-Carbon Reputations and Supply Chain Profits over Time

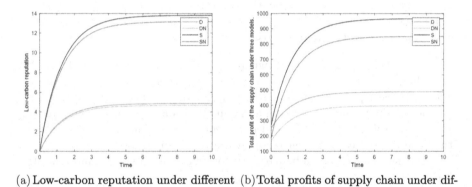

(a) Low-carbon reputation under different   (b) Total profits of supply chain under dif-
models.   ferent models.

Fig. 2. Low-carbon reputations and total profits of supply chain under different models.

Figure 2 illustrates the trajectories of low-carbon reputation under three scenarios, which increase with the increase of time t and stabilize as t approaches infinity. In Fig. 2-(a), we can observe that the low-carbon reputation under the Stackelberg game model is higher than that under the decentralized model, which indicating that cooperation can enhance low-carbon reputation even if it is one-way. Furthermore, reputations were improved across all models where carbon trading was implemented. Therefore, supply chains should collaborate rather than act independently in the search for low-carbon solutions and the government should consider carbon trading as an alternative policy after subsidy cancellation.

From the Fig. 2.-(b), it can be seen that the profit of supply chain in the Stackelberg game model is higher than that in the decentralized model, which demonstrates a similar trend as the low-carbon reputation shown in Fig. 2.-(a). These findings highlight the importance of prioritizing cooperation among supply chain members in the long run, as it can yield better profit performance and cost reduction. Moreover, the implementation of carbon trading enhances supply chain profits. Therefore, it is imperative for supply chain members to remain vigilant and adapt to policy changes accordingly.

5 Conclusion

This study employs a differential game framework to examine the dynamic decision-making process related to low-carbon strategies within the automotive supply chain. Departing from previous assumptions of fixed marginal profit, this study places emphasis on the joint decision-making regarding pricing and low-carbon operations. By considering the influence of carbon trading policies and the reputation of low-carbon vehicles, two interaction models between the manufacturer and n retailers are discussed: the decentralized model and the Stackelberg game model. For each model, the equilibrium decisions of all participants and the trajectory of low-carbon reputation are analyzed. The key conclusions derived from this study are as follows:

– In the Stackelberg game model, where the manufacturer plays as the leader, providing subsidies to retailers to incentivize their low-carbon promotion efforts, several positive outcomes are observed. Compared to decentralized decision-making, the Stackelberg game model leads to improvements in supply chain profit, effort levels, and low-carbon reputation. This demonstrates that conditional cooperation through the Stackelberg model can effectively motivate supply chain members to actively engage in emission reduction activities.
– The low-carbon reputation of vehicles has a significant impact on market demand. Both retailers and manufacturers can increase their respective efforts to enhance the reputation of their vehicles, expand market share, and boost revenue.
– The implementation of carbon trading policies contributes to carbon emission reduction within the automotive supply chain and enhances the industry's low-carbon reputation to a certain extent. However, it is important to note that the initial adoption of carbon trading may impose challenges on the supply chain. Therefore, it is crucial for the government to dynamically adjust relevant policies to ensure a smooth transition towards a low-carbon supply chain.

However, this study does have some limitations that should be acknowledged. Firstly, the model settings do not incorporate the participation of retailers in carbon trading, which could be a valuable aspect to explore in future research. Secondly, the decision-making process within the automotive supply chain can be influenced by various other government policies, such as subsidies and carbon taxes, which were not extensively considered in this study. Finally, the research primarily focuses on manufacturers' direct retail channels and does not delve into the analysis of dual-channel sales, which could provide additional insights and avenues for further investigation.

References

1. Al-Buenain, A., Al-Muhannadi, S., Falamarzi, M., Kutty, A.A., Kucukvar, M., Onat, N.C.: The adoption of electric vehicles in Qatar can contribute to net carbon emission reduction but requires strong government incentives. Vehicles **3**(3), 618–635 (2021)
2. Wang, Y., Xin, X., Zhu, Q.: Carbon emission reduction decisions of supply chain members under cap-and-trade regulations: a differential game analysis. Comput. Ind. Eng. **162**, 107711 (2021)
3. Sun, H., Gao, G., Li, Z.: Differential game model of government-enterprise cooperation on emission reduction under carbon emission trading policy. Pol. J. Environ. Stud. **31**(5), 4859–4871 (2022)
4. Xu, C., Jing, Y., Shen, B., Zhou, Y., Zhao, Q.Q.: Cost-sharing contract design between manufacturer and dealership considering the customer low-carbon preferences. Expert Syst. Appl. **213**, 118877 (2023)
5. Xia, L., Guo, T., Qin, J., Yue, X., Zhu, N.: Carbon emission reduction and pricing policies of a supply chain considering reciprocal preferences in cap-and-trade system. Ann. Oper. Res. **268**, 149–175 (2018). https://doi.org/10.1007/s10479-017-2657-2
6. Zang, J., Wan, L., Li, Z., Wang, C., Wang, S.: Does emission trading scheme have spillover effect on industrial structure upgrading? Evidence from the EU based on a PSM-DID approach. Environ. Sci. Pollut. Res. **27**, 12345–12357 (2020). https://doi.org/10.1007/s11356-020-07818-0
7. Benjaafar, S., Li, Y., Daskin, M.: Carbon footprint and the management of supply chains: insights from simple models. IEEE Trans. Autom. Sci. Eng. **10**(1), 99–116 (2012)
8. Yang, L., Zhang, Q., Ji, J.: Pricing and carbon emission reduction decisions in supply chains with vertical and horizontal cooperation. Int. J. Prod. Econ. **191**, 286–297 (2017)
9. Chen, Y., Wang, Z., Liu, Y., Mou, Z.: Coordination analysis of the recycling and remanufacturing closed-loop supply chain considering consumers' low carbon preference and government subsidy. Sustainability **15**(3), 2167 (2023)
10. Bian, J., Zhang, G., Zhou, G.: Manufacturer vs. consumer subsidy with green technology investment and environmental concern. Eur. J. Oper. Res. **287**(3), 832–843 (2020)
11. Zhang, Z., Liying, Yu.: Dynamic optimization and coordination of cooperative emission reduction in a dual-channel supply chain considering reference low-carbon effect and low-carbon goodwill. Int. J. Environ. Res. Public Health **18**(2), 539 (2021)
12. Li, Y.: Research on supply chain CSR management based on differential game. J. Clean. Prod. **268**, 122171 (2020)
13. Zhang, J., Chiang, W.K., Liang, L.: Strategic pricing with reference effects in a competitive supply chain. Omega **44**, 126–135 (2014)

Adaptivity Gap for Influence Maximization with Linear Threshold Model on Trees

Yichen Tao[1], Shuo Wang[2], and Kuan Yang[3(✉)]

[1] School of Electronic, Information and Electrical Engineering,
Shanghai Jiao Tong University, Shanghai 200240, China
taoyc0904@sjtu.edu.cn
[2] Zhiyuan College, Shanghai Jiao Tong University, Shanghai 200240, China
shuo_wang@sjtu.edu.cn
[3] John Hopcroft Center for Computer Science, Shanghai Jiao Tong University,
Shanghai 200240, China
kuan.yang@sjtu.edu.cn

Abstract. We address the problem of influence maximization within the framework of the linear threshold model, focusing on its comparison to the independent cascade model. Previous research has predominantly concentrated on the independent cascade model, providing various bounds on the adaptivity gap in influence maximization. For the case of a (directed) tree (in-arborescence and out-arborescence), [CP19] and [DPV23] have established constant upper and lower bounds for the independent cascade model.

However, the adaptivity gap of this problem on the linear threshold model is not so extensively studied as on the independent cascade model. In this study, we present constant upper bounds for the adaptivity gap of the linear threshold model on trees. Our approach builds upon the original findings within the independent cascade model and employs a reduction technique to deduce an upper bound of $\frac{4e^2}{e^2-1}$ for the in-arborescence scenario. For out-arborescence, the equivalence between the two models reveals that the adaptivity gap under the linear threshold model falls within the range of $[\frac{e}{e-1}, 2]$, as demonstrated in [CP19] under the independent cascade model.

1 Introduction

The *influence maximization problem*, initially introduced in [DR01, RD02], is a well-known problem that lies at the intersection of computer science and economics. It focuses on selecting a specific number of agents, referred to as seeds, in a social network to maximize the number of agents influenced by them. To analyze this problem mathematically and formally, social networks are represented as weighted graphs, where vertices correspond to agents and edges represent their connections, with each edge assigned a weight indicating the strength of the connection. The *independent cascade model* [KKT03] and the *linear threshold*

© Springer Nature Switzerland AG 2023
M. Li et al. (Eds.): IJTCS-FAW 2023, LNCS 13933, pp. 160–170, 2023.
https://doi.org/10.1007/978-3-031-39344-0_12

model [KKT03] are two prominent diffusion models that have received significant attention in previous studies. These models have been applied to various fields such as viral marketing, meme usage, and rumor control.

More recently, the *adaptive influence maximization* problem has gained considerable attention. Unlike the original setting where all seeds are selected at once, the adaptive version allows seeds to be selected based on observations of the propagation of previously chosen seeds. Of particular interest are two feedback models [GK11], namely *myopic feedback* and *full adoption feedback*. When considering myopic feedback, only the status of the seeds' neighbors can be observed. Conversely, the full adoption feedback allows the whole propagation process of previously selected seeds to be considered when selecting the next seed. While the introduction of adaptive seed selection might enhance the influence of the seed set, it also presents significant technical challenges. Therefore, it becomes imperative to evaluate the benefits of adaptivity, which is measured by the *adaptivity gap*. The adaptivity gap is informally defined as the supremum value of the ratio between the optimal influence spread of an adaptive policy and a non-adaptive one. It provides insights into the performance improvement achieved by the adaptive strategy and gives us a taste of whether it is worth the effort to develop the adaptive strategy for the problem.

Regarding the adaptivity gap, a number of previous works have explored this concept in the context of the independent cascade model [CP19, DPV23, PC19]. In [CP19], the adaptivity gap for the independent cascade model with full adoption feedback was studied for certain families of influence graphs. It was demonstrated that the adaptivity gap lies in the range of $[\frac{e}{e-1}, \frac{2e}{e-1}]$ for in-arborescence, $[\frac{e}{e-1}, 2]$ for out-arborescence, and exactly $\frac{e}{e-1}$ for bipartite graphs. Another recent work [DPV23] improved upon these results by providing a tighter upper bound of $\frac{2e^2}{(e^2-1)}$ for the adaptivity gap of in-arborescence. Furthermore, this work established an upper bound of $(\sqrt[3]{n} + 1)$ for general graphs, where n stands for the number of vertices in the graph. For the myopic feedback setting, it has been proved in [PC19] that the adaptivity gap for the independent cascade model with myopic feedback is at most 4 and at least $\frac{e}{e-1}$. However, despite the progress made in analyzing the adaptivity gap for the independent cascade model, to the best of our knowledge, no existing results are available on the adaptivity gap for the linear threshold model.

1.1 Our Results

In this work, we give an upper bound for the adaptivity gap for in-arborescence under the linear threshold model as follows.

Theorem 1. *The adaptivity gap AG_{LT} for in-arborescence under the linear threshold model is no more than $\frac{4e^2}{e^2-1}$.*

Also, for out-arborescence, the linear threshold model is equivalent to the independent cascade model since each vertex has at most in-degree 1. Thus, the results under the independent cascade model for out-arborescence given by [CP19] can be also used in the linear threshold model (Table 1).

Theorem 2. *The adaptivity gap AG_{LT} for out-arborescence under the linear threshold model satisfies that $AG_{LT} \in [\frac{e}{e-1}, 2]$.*

Table 1. The previous results and the results of this paper are summarized in the table. New results of this paper are in blue.

Diffusion Model	Feedback Model	Graph Family	Lower Bound of Adaptivity Gap	Upper Bound of Adaptivity Gap
Independent Cascade	Full adoption feedback	In-arborescence	$\dfrac{e}{e-1}$	$\dfrac{2e^2}{(e^2-1)}$
		Out-arborescence	$\dfrac{e}{e-1}$	2
		Bipartite graphs	$\dfrac{e}{e-1}$	$\dfrac{e}{e-1}$
		General graphs		$(\sqrt[3]{n}+1)$
	Myopic feedback	General graphs	$\dfrac{e}{e-1}$	4
Linear Threshold	Full adoption feedback	In-arborescence		$\dfrac{4e^2}{e^2-1}$
		Out-abborescence	$\dfrac{e}{e-1}$	2

1.2   Related Works

The influence maximization problem was initially proposed in [DR01] and [RD02]. Subsequently, the two most extensively studied diffusion models, namely the independent cascade model and the linear threshold model, were introduced in [KKT03], which also demonstrated their submodularity. For any submodular diffusion model, the greedy algorithm is shown to obtain a $(1 - 1/e)$-approximation to the optimal influence spread [NWF78, KKT03, KKT05, MR10]. A later work [STY20] shows that the approximation guarantee of the greedy algorithm for the influence maximization problem under the linear threshold model is asymptotically $(1 - 1/e)$.

The adaptive influence maximization problem is first considered in [GK11]. The results relevant to adaptivity gaps under the independent cascade model have been discussed before, and we discuss further related work here. Later, Asadpour and Nazerzadeh studied the adaptivity gap for the problem of maximizing stochastic monotone submodular functions [AN16].

The adaptivity gap compares the optimal adaptive solution to the optimal nonadaptive solution. Motivated by that the inapproximability of the influence maximization problem [KKT03, ST20] and the fact that most influence maximization algorithms are based on greedy, the concept of greedy adaptivity gap is introduced in [CPST22], which depicts how much adaptive greedy policy would outperform its non-adaptive counterpart. This work also showed that the greedy adaptivity gap is at least $(1 - 1/e)$ for an arbitrary combination of diffusion and feedback models.

2   Preliminaries

2.1   Linear Threshold Model

In the *linear threshold model* (LT), we have a weighted directed graph called the influence graph $G = (V = [n], E, \{p_{u,v} \mid (u, v) \in E\})$, satisfying $\sum_u p_{u,v} \leq 1$. Fix a seed set $S \subseteq V$, the diffusion process in the LT model is defined as follows. Define the current activated vertex set T, and initialize $T = S$. Before the diffusion process starts, every vertex first independently samples a value $a_i \in [0, 1]$ uniformly at random. In each iteration, if a non-activated vertex x satisfies that $\sum_{u \in T} p_{u,x} \geq a_x$, it will be activated and let $T = T \cup \{x\}$. The diffusion process terminates when there is no more activated vertex in an iteration.

It is mentioned in [KKT03] that the LT model has another equivalent interpretation (Fig. 1). Fix a seed set $S \subseteq V$.

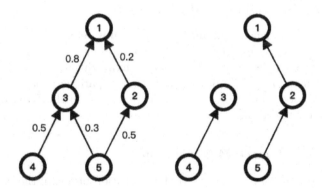

Fig. 1. The above pictures are an example of weighted directed influence graph and an instance of its live-edge graph. In the LT model, the right live-edge graph appears for a probability $0.5 \cdot 0.5 \cdot 0.2 = 0.05$. However, in the IC model, the appearing probability is $0.5 \cdot 0.7 \cdot 0.5 \cdot 0.2 \cdot 0.2 = 0.007$. Readers can find the differences between the two models in this example.

Then sample a *live-edge graph* $L = (V, L(E))$ of G, which is a random graph generated from the base graph G as follows. For each vertex i, sample at most one in-edge, where the edge (u, i) is selected with probability $p_{u,i}$, and add this edge (if exists) to $L(E)$. In this case, the diffusion process will activate all the vertices that can be reached from S. Given a live-edge graph L, use $R(S, L)$ to denote all the vertices activated at the end of this process. Given a seed set S, the expected influence spread of S is defined as $\sigma(S) := \mathbf{E}_L[\|R(S, L)\|]$.

2.2   Independent Cascade Model

The *independent cascade model* (IC) also involve a weighted directed influence graph $G = (V = [n], E, \{p_{u,v} \mid (u, v) \in E\})$. Before the beginning of the propagation, a *live-edge graph* $L = (V, L(E))$ is sampled. The sampling of the live-edge graph in the IC model is simpler than that of the LT model. Each edge $e \in E$ appears in $L(E)$ independently with probability p_e. When an edge is present in the live-edge graph, we say that it is *live*. Otherwise, we say that it is *blocked*. Denote the set of all possible live-edge graphs by $\mathcal{S}$, and the distribution over $\mathcal{L}$ by $\mathcal{P}$. Given a seed set $S \subseteq V$, the vertices affected, denoted by $\Gamma(S, L)$ is exactly the set of vertices reachable from S in the live-edge graph L. The influence reach of a certain seed set on a live-edge graph $f : \{0, 1\}^V \times \mathcal{L} \to \mathbf{R}^+$ is defined as the number of affected vertices, i.e., $f(S, L) = |\Gamma(S, L)|$. Then we define the influence spread of a seed set $\sigma(S)$ to be the expected number of affected vertices at the end of the diffusion process, i.e., $\sigma(S) = \mathbf{E}_{\mathcal{L} \sim \mathcal{P}}[f(S, L)]$.

2.3   Non-adaptive Influence Maximization

The *non-adaptive* influence maximization problem is defined as a computational problem that, given an influence graph G and an integer $k \geq 1$, we are asked to find a vertex set S satisfying that $|S| = k$ and maximizing $\sigma(S)$. Use $\mathrm{OPT}_N(G, k)$ to denote the maximal $\sigma(S)$ under graph G and parameter k. The subscript "N" stands for "non-adaptive", which is in contrast with the "adaptive" model defined in the next section.

2.4   Adaptive Influence Maximization

Compared with non-adaptive influence maximization problem, the *adaptive* setting allows to activate the seeds sequentially and adaptively in k iterations. One can first choose a vertex, activates it, and see how it goes. After observing the entire diffusion process of the first vertex, we can change their strategy optimally adaptive to the diffusion process. Similarly, the choices of the following vertices are based on the previous observation. We consider the *full-adoption* feedback model, which means the adaptive policy observes the entire influence spread from the previous chosen vertices.[1]

An *adaptive policy* can be formally defined as follows. Given a live-edge graph L, the realization $\phi_L : V \to 2^V$ denotes a function from a vertex to a vertex set. For a fix vertex v, define $\phi_L(v) := R(v, L)$, i.e., the set of vertices activated by v under the live-edge graph L. Given a subset S of V satisfying that $|S| = k$,

---

[1] Another commonly considered model is called the *myopic* feedback model, where only one iteration of the spread can be observed.

define the partial realization $\psi\colon S \to 2^V$ restricted to S to be the part of some realization, which can be used to represent the graph observed by the player at some point of the adaptive algorithm. For a fixed partial realization, let its domain (the chosen seed vertices) be $\mathrm{dom}(\psi) := S$, let $R(\psi) = \cup_{v \in S}\psi(v)$, and let $f(\psi) = |R(\psi)|$. A partial realization ψ' is called a sub-realization of another partial realization ψ if and only if that $\mathrm{dom}(\psi') \subseteq \mathrm{dom}(\psi)$ and $\psi'(v) = \psi(v)$ for any $v \in \mathrm{dom}(\psi')$.

2.5   Adaptivity Gap

The adaptivity gap for the LT model is defined as follows

$$AG_{\mathrm{LT}} = \sup_{G,k} \frac{\mathrm{OPT}_A^{\mathrm{LT}}(G,k)}{\mathrm{OPT}_N^{\mathrm{LT}}(G,k)},$$

where $\mathrm{OPT}_A^{\mathrm{LT}}(G,k)$ is the optimal influence spread with a k-vertex seed set on graph G in the adaptive setting, and $\mathrm{OPT}_N^{\mathrm{LT}}(G,k)$ is its counterpart in the non-adaptive setting.

Similarly for the IC model, the adaptivity gap can be defined as follows

$$AG_{\mathrm{IC}} = \sup_{G,k} \frac{\mathrm{OPT}_A^{\mathrm{IC}}(G,k)}{\mathrm{OPT}_N^{\mathrm{IC}}(G,k)}.$$

3   Adaptivity Gap for In-Arborescence

An *in-arborescence* is a directed graph $G = (V, E)$ that can be constructed by the following process: fix a rooted tree $T = (V, E')$, and add edge (u, v) if v is the parent of u in T. An upper bound for AG_{IC} for in-arborescence is given by [DPV23]. This bound also plays an essential role in our proof of the constant upper bound for AG_{LT} for in-arborescence.

We prove the following theorem:

Theorem 3. $AG_{LT} \leq \frac{4e^2}{e^2-1}$ *for in-arborescence.*

The key technique is to reduce the influence maximization problem in the LT model to the influence maximization problem in the IC model.

To find a relation between the LT model and IC model, we construct a new instance G' in the IC model, but the graph G' is the same as G both in structures and weights of edges. The following lemma is the technical lemma in our proof, which reveals the relation between the two models (Fig. 2).

Lemma 4. $OPT_A^{LT}(G,k) \leq OPT_A^{IC}(G',2k)$

Proof. The proof outline is to construct an algorithm for G' based on the optimal adaptive algorithm for G. There is an observation that after choosing the same

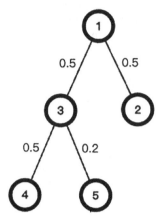

Fig. 2. This figure gives an example. In the first round, we choose vertex indexed 4, and the diffusion process stops at it self. In the second, round we choose vertex 5, while the process also stops at itself. According to our reduction, we have the probability $p_{5,3}(1/pt_3 - 1) = 0.2$ to add vertex 3 to our seed set in this case.

first seed vertex both in G and G', the diffusion process shares the same distribution on the in-arborescence. However, in the following process, the appearing probability of the edge would increase in the LT model, and we need a larger seed set in the IC model to compensate for the boosted probability.

To formalize the intuitions above, we design an k-round adaptive algorithm for G'. Let π be the optimal adaptive algorithm for G, and π' be the algorithm constructed for G'. First, we maintain a current partial realization ψ, which is an empty set at first. In the first round, we simulate the algorithm π in G' to get the first seed of π'. In π', we also add $\pi(\psi)$ ($\psi = \emptyset$ at first) to our seed set, and add $(\pi(\psi), R(\pi(\psi)))$ to ψ. Suppose that the diffusion process of $\pi(\psi)$ end at vertex u, we mark the parent of u (if exists) as a *critical point*. Also, we maintain a value for each vertex v called *remaining potential* defined as

$$pt_v := \sum_{u \in \text{Child}(v), u \text{ is not activated}} p_u,$$

where we use $\text{Child}(v)$ to denote the set of children of v on the tree.

In the next $k - 1$ rounds, we first choose $x = \pi(\psi)$ to be our new seed. However, the existing probabilities of some edges increased because of the previous rounds. Suppose the diffusion process of x stops at a vertex u, and let its parent be v. If v is a critical point, which means the existing probability of edges (u, v) has already increased, we flip a biased coin which appears heads with probability $p_{u,v}(1/pt_v - 1)$. If the coin appears heads, we choose v to be our next seed, remove v from critical vertex sets, and continue this round. On the contrary, we just go to the next round. Obviously, this process eventually ends at some vertex y. Before the end of this round, we update ψ with $\psi \cup (x, \{\text{the paths from } x \text{ to } y)\}$, and mark y's parent (if exists) as a new critical point.

First of all, it's easy to verify that, the ψ after i rounds shares the same distribution in the LT model after choosing the first i seeds. Thus, $\mathbf{E}_{\mathrm{LT}}(\pi) = \mathbf{E}_{\mathrm{IC}}(\pi')$. Also, there is an observation that there are at most k vertices marked as critical points. Thus, we have that

$$|\pi'| \leq k + k = 2k,$$

and

$$\mathrm{OPT}_A^{\mathrm{LT}}(G, k) = \mathbf{E}_{\mathrm{LT}}(\pi) = \mathbf{E}_{\mathrm{IC}}(\pi') \leq \mathrm{OPT}_A^{\mathrm{IC}}(G', 2k),$$

as desired.

Lemma 4 builds a connection between the two models. Further analysis is needed to give an upper bound of AG_{LT}.

Lemma 5. $OPT_N^{LT}(G, k) \geq OPT_N^{IC}(G', k)$

Proof. This can be proved by an easy reduction. We want to prove that for every fixed seed set S, it holds that,

$$\mathbf{E}_{L \sim \mathrm{LT}}(R(L, S)) \geq \mathbf{E}_{L \sim \mathrm{IC}}(R(L, S)).$$

First, by the linearity of the expectation, it holds that,

$$\mathbf{E}_{L \sim \mathrm{LT}}(R(L, S)) = \sum_{v \in V} \Pr_{L \sim \mathrm{LT}}[v \text{ is activated}].$$

Similarly, we have that,

$$\mathbf{E}_{L \sim \mathrm{IC}}(R(L, S)) = \sum_{v \in V} \Pr_{L \sim \mathrm{IC}}[v \text{ is activated}].$$

Then, we will prove by induction that

$$\Pr_{L \sim \mathrm{LT}}[v \text{ is activated}] \geq \Pr_{L \sim \mathrm{IC}}[v \text{ is activated}]$$

with a decreasing order of v's depth.

For a vertex v with the largest depth, if it is in the seed set, it holds that

$$\Pr_{L \sim \mathrm{LT}}[v \text{ is activated}] = \Pr_{L \sim \mathrm{IC}}[v \text{ is activated}] = 1.$$

Otherwise, it holds that

$$\Pr_{L \sim \mathrm{LT}}[v \text{ is activated}] = \Pr_{L \sim \mathrm{IC}}[v \text{ is activated}] = 0.$$

For a vertex w of another depth, assume that every child of w satisfies the induction hypothesis. We have that,

$$\Pr_{L\sim LT}[w \text{ is activated}] = \sum_{u\in\text{Child}(v)} p_{u,w} \Pr_{L\sim LT}[u \text{ is activated}]$$

$$\geq \sum_{u\in\text{Child}(v)} p_{u,w} \Pr_{L\sim IC}[u \text{ is activated}]$$

$$\overset{\text{Union bound}}{\geq} 1 - \prod_{u\in\text{Child}(v)} (1 - p_{u,w} \Pr_{L\sim IC}[u \text{ is activated}])$$

$$= \Pr_{L\sim IC}[w \text{ is activated}],$$

as desired.

The last step is to bound $\text{OPT}_A^{IC}(G', 2k)$ by $\text{OPT}_A^{IC}(G', k)$. This comes from the following submodularity lemma:

Lemma 6 (Adaptive Submodularity for the IC model, [GK11]).
  Let G be an arbitrary influence graph. For any partial realizations ψ, ψ' of G such that $\psi \subseteq \psi'$, and any node $u \notin R(\psi')$, we have that $\Delta(u \mid \psi') \leq \Delta(u \mid \psi)$, where $\Delta(u \mid \psi)$ represents the expected increasing influence to choose u under ψ.

This lemma gives a good property of the IC model, leading to the following submodularity lemma of the optimal adaptive algorithm:

Lemma 7. $OPT_A^{IC}(G', 2k) \leq 2OPT_A^{IC}(G', k)$

Proof. First, we divide the optimal adaptive algorithms π' for G' with a fixed seed set size $2k$. We want to argue that the expected influence of each part is less than $\text{OPT}_A^{IC}(G', k)$.

For the first part, it is an adaptive algorithm with seed set size equaling k. Thus, the total influence should be not more than $\text{OPT}_A^{IC}(G', k)$.

After the selection of the first k seeds, there exists a non-empty partial realization ψ. We want to prove that if we select k more seeds, the expected extra influence is no more than $\text{OPT}_A^{IC}(G', k)$. This is a natural corollary of adaptive submodularity.

Thus, we have $\text{OPT}_A^{IC}(G', 2k) \leq 2\text{OPT}_A^{IC}(G', k)$ as desired.

Lemma 8 ([DPV23]). $AG_{IC} \leq \frac{2e^2}{e^2-1}$.

This bound is given by [DPV23]. And can be used to give a bound of AG_{LT}:

Proof (Proof of Theorem 1) Putting together Lemma 4, 5, 7, 8, we have

$$
\begin{aligned}
AG_{\mathrm{LT}} = \sup_{G,k} \frac{\mathrm{OPT}_A^{\mathrm{LT}}(G,k)}{\mathrm{OPT}_N^{\mathrm{LT}}(G,k)} \\
\leq \sup_{G,k} \frac{\mathrm{OPT}_A^{\mathrm{LT}}(G,k)}{\mathrm{OPT}_N^{\mathrm{IC}}(G',k)} \\
\leq \sup_{G,k} \frac{\mathrm{OPT}_A^{\mathrm{IC}}(G',2k)}{\mathrm{OPT}_N^{\mathrm{IC}}(G',k)} \\
\leq \sup_{G,k} \frac{2\mathrm{OPT}_A^{\mathrm{IC}}(G',k)}{\mathrm{OPT}_N^{\mathrm{IC}}(G',k)} \\
\leq 2AG_{\mathrm{IC}} \\
\leq \frac{4e^2}{e^2-1},
\end{aligned}
$$

as desired.

4   Discussions and Open Questions

In [CP19] and [DPV23], they also give some constant upper bounds under the IC model for some other graphs such as one-directional bipartite graphs. Also, [DPV23] gives an upper bound for general graphs though it is not a constant bound. We have the following conjecture.

Conjecture 1 The adaptivity gap for general graphs under the LT model has a constant upper bound.

However, adaptivity algorithms gain more profits in the LT model than the IC model. Thus, we also have the following conjecture.

Conjecture 2 The adaptivity gap for general graphs under the IC model has a constant upper bound.

Also, we have a conjecture about the relation between the two models on the general graphs, and we believe that our approach can help to build a similar argument on general graphs.

Conjecture 3 There is a Lemma 4 like argument holds for general graphs.

Thus, as a corollary of the above conjecture, we are able to claim that the adaptivity gap on LT model is linearly upper bounded by the adaptivity gap on IC model.

Acknowledgements. The research of K. Yang is supported by the NSFC grant No. 62102253. Y. Tao, S. Wang and K. Yang sincerely thank the anonymous reviewers for their helpful feedback. Y. Tao, S. Wang and K. Yang also want to express the gratitude to Panfeng Liu and Biaoshuai Tao for their insightful suggestions during the composition of this paper.

References

[AN16]   Asadpour, A., Nazerzadeh, H.: Maximizing stochastic monotone submodular functions. Manage. Sci. **62**(8), 2374–2391 (2016)

[CP19]   Chen, W., Peng, B.: On adaptivity gaps of influence maximization under the independent cascade model with full adoption feedback. arXiv preprint arXiv:1907.01707 (2019)

[CPST22] Chen, W., Peng, B., Schoenebeck, G., Tao, B.: Adaptive greedy versus non-adaptive greedy for influence maximization. J. Artif. Intell. Res. **74**, 303–351 (2022)

[DPV23]   D'Angelo, G., Poddar, D., Vinci, C.: Better bounds on the adaptivity gap of influence maximization under full-adoption feedback. Artif. Intell. **318**, 103895 (2023)

[DR01]   Domingos, P., Richardson, M.: Mining the network value of customers. In: Proceedings of the Seventh ACM SIGKDD International Conference on Knowledge Discovery and Data Mining, pp. 57–66 (2001)

[GK11]   Golovin, D., Krause, A.: Adaptive submodularity: theory and applications in active learning and stochastic optimization. J. Artif. Intell. Res. **42**, 427–486 (2011)

[KKT03]   Kempe, D., Kleinberg, J., Tardos, É.: Maximizing the spread of influence through a social network. In: Proceedings of the Ninth ACM SIGKDD International Conference on Knowledge Discovery and Data Mining, pp. 137–146 (2003)

[KKT05]   Kempe, D., Kleinberg, J., Tardos, É.: Influential nodes in a diffusion model for social networks. In: Caires, L., Italiano, G.F., Monteiro, L., Palamidessi, C., Yung, M. (eds.) ICALP 2005. LNCS, vol. 3580, pp. 1127–1138. Springer, Heidelberg (2005). https://doi.org/10.1007/11523468_91

[MR10]   Mossel, E., Roch, S.: Submodularity of influence in social networks: from local to global. SIAM J. Comput. **39**(6), 2176–2188 (2010)

[NWF78]   Nemhauser, G.L., Wolsey, L.A., Fisher, M.L.: An analysis of approximations for maximizing submodular set functions-I. Math. Program. **14**, 265–294 (1978)

[PC19]   Peng, B., Chen, W.: Adaptive influence maximization with myopic feedback. In: Advances in Neural Information Processing Systems vol. 32 (2019)

[RD02]   Richardson, M., Domingos, P.: Mining knowledge-sharing sites for viral marketing. In: Proceedings of the Eighth ACM SIGKDD International Conference on Knowledge Discovery and Data Mining, pp. 61–70 (2002)

[ST20]   Schoenebeck, G., Tao, B.: Influence maximization on undirected graphs: toward closing the $(1-1/e)$ gap. ACM Trans. Econ. Comput. (TEAC) **8**(4), 1–36 (2020)

[STY20]   Schoenebeck, G., Tao, B., Yu, F.-Y.: Limitations of greed: influence maximization in undirected networks re-visited. In: Proceedings of the 19th International Conference on Autonomous Agents and MultiAgent Systems, pp. 1224–1232 (2020)

Physically Verifying the First Nonzero Term in a Sequence: Physical ZKPs for ABC End View and Goishi Hiroi

Suthee Ruangwises[✉] [iD]

Department of Informatics, The University of Electro-Communications, Tokyo, Japan
ruangwises@gmail.com

Abstract. In this paper, we develop a physical protocol to verify the first nonzero term of a sequence using a deck of cards. This protocol enables a prover to show a verifier the value of the first nonzero term in a given sequence without revealing which term it is. Our protocol uses $\Theta(1)$ shuffles, making it simpler and more practical than a similar protocol recently developed by Fukusawa and Manabe in 2022, which uses $\Theta(n)$ shuffles, where n is the length of the sequence. We also apply our protocol to construct zero-knowledge proof protocols for two famous logic puzzles: ABC End View and Goishi Hiroi. These zero-knowledge proof protocols allow a prover to physically show that he/she know solutions of the puzzles without revealing them.

Keywords: zero-knowledge proof · card-based cryptography · sequence · ABC End View · Goishi Hiroi

1 Introduction

A *zero-knowledge proof (ZKP)* is an interactive protocol between a prover P and a verifier V, which enables P to convince V that a statement is correct without revealing any other information. The concept of a ZKP was introduced by Goldwasser et al. [8] in 1989. A ZKP with perfect completeness and soundness must satisfy the following three properties.

1. **Perfect Completeness:** If the statement is correct, then V always accepts.
2. **Perfect Soundness:** If the statement is incorrect, then V always rejects.
3. **Zero-knowledge:** V obtains no information other than the correctness of the statement.

Goldreich et al. [7] proved that a computational ZKP exists for every NP problem. Therefore, one can construct a computational ZKP for any NP problem via a reduction to an NP-complete problem with a known ZKP. Such construction, however, requires cryptographic primitives and thus is not intuitive and looks unconvincing. As a result, many researchers instead aim to construct ZKPs using physical objects such as a deck of cards. These protocols have benefits that they do not require computers and allow external observers to check

© Springer Nature Switzerland AG 2023
M. Li et al. (Eds.): IJTCS-FAW 2023, LNCS 13933, pp. 171–183, 2023.
https://doi.org/10.1007/978-3-031-39344-0_13

that the prover truthfully executes the protocol (which is often a challenging task for digital protocols). They also have didactic values and can be used to teach the concept of a ZKP to non-experts.

1.1 Related Work

In the recent years, card-based ZKP protocols for many popular logic puzzles have been developed: ABC End View [6], Akari [2], Ball sort puzzle [20], Bridges [25], Heyawake [16], Hitori [16], Juosan [12], Kakuro [2,13], KenKen [2], Makaro [3,26], Masyu [11], Nonogram [4,19], Norinori [5], Numberlink [23], Nurikabe [16], Nurimisaki [17], Ripple Effect [24], Shikaku [22], Slitherlink [11], Sudoku [9,21,27], Suguru [15], Takuzu [2,12], and Usowan [18].

Many of these protocols employ methods to physically verify specific sequence-related functions, as shown in the following examples.

- A protocol in [27] verifies that a sequence is a permutation of all given numbers in some order without revealing their order.
- A protocol in [3] verifies that a term in a sequence is the largest one in that sequence without revealing any term in the sequence.
- A protocol in [23] counts the number of terms in a sequence that are equal to a given secret value without revealing that value, which terms in the sequence are equal to it, or any term in the sequence.
- A protocol in [24], given a secret number x and a sequence, verifies that x does not appear among the first x terms in the sequence without revealing x or any term in the sequence.

Protocol of Fukusawa and Manabe. One of interesting work is to develop a protocol that can verify the first nonzero term of a sequence. Given a sequence $S = (x_1, x_2, ..., x_n)$ of numbers, P wants to show V the value of the first nonzero term in S without revealing which term it is, i.e. show the term a_i with the lowest index i such that $a_i \neq 0$ without revealing i.

Very recently, Fukasawa and Manabe [6] developed a protocol to verify the first nonzero term of a sequence in the context of constructing a ZKP protocol for a logic puzzle called *ABC End View*. However, their protocol uses three cards (one numbered card and two binary cards: ♣, ♡) to encode each term in the sequence, which is not optimal. Also, their protocol involves computing multiple Boolean functions using the binary cards, making it very complicated. It also uses as many as $\Theta(n)$ shuffles, where n is the length of the sequence, making it very impractical to implement in real world.

1.2 Our Contribution

In this paper, we develop a much simpler protocol to verify the first nonzero term of a sequence. Our protocol uses only $\Theta(1)$ shuffles and uses only one card to encode each term in the sequence.

We also apply our protocol to construct ZKP protocols for two popular logic puzzles: ABC End View (the same as in [6] but with asymptotically better performance) and *Goishi Hiroi* (which is more complicated than that of ABC End View).

2   Preliminaries

2.1   Cards

Each card used in our protocol has a number written on the front side. All cards have indistinguishable back sides denoted by ?.

2.2   Pile-Shifting Shuffle

Given a matrix M of cards, a *pile-shifting shuffle* [28] shifts the columns of M by a uniformly random cyclic shift unknown to all parties (see Fig. 1). It can be implemented by putting all cards in each column into an envelope, and taking turns to apply *Hindu cuts* (taking several envelopes from the bottom of the pile and putting them on the top) to the pile of envelopes [29].

Note that each card in M can be replaced by a stack of cards, and the protocol still works in the same way as long as every stack in the same row consists of the same number of cards.

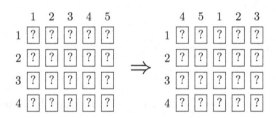

Fig. 1. An example of a pile-shifting shuffle on a 4×5 matrix

2.3   Chosen Cut Protocol

Given a sequence of n face-down cards $C = (c_1, c_2, ..., c_n)$, a *chosen cut protocol* [10] enables P to select a card c_i he/she wants (to use in other operations) without revealing i to V. This protocol also reverts C back to its original state after P finishes using c_i.

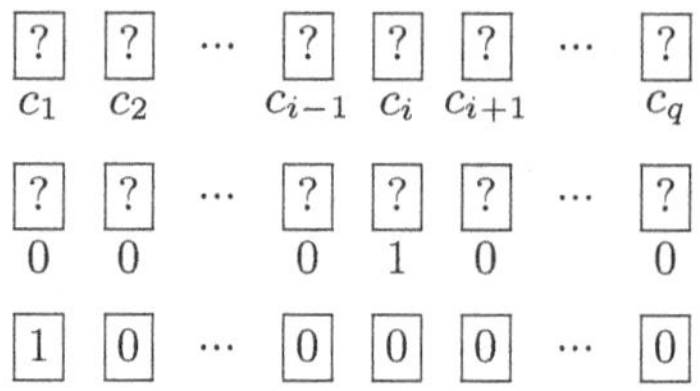

Fig. 2. A $3 \times n$ matrix M constructed in Step 1 of the chosen cut protocol

1. Construct the following $3 \times n$ matrix M (see Fig. 2).
   (a) In Row 1, place the sequence C.
   (b) In Row 2, secretly place a face-down $\boxed{1}$ at Column i and a face-down $\boxed{0}$ at each other column.
   (c) In Row 3, publicly place a $\boxed{1}$ at Column 1 and a $\boxed{0}$ at each other column.
2. Turn over all face-up cards and apply the pile-shifting shuffle to M.
3. Turn over all cards in Row 2. Locate the position of the only $\boxed{1}$. A card in Row 1 directly above this $\boxed{1}$ will be the card c_i as desired.
4. After finishing using c_i in other operations, place c_i back into M at the same position.
5. Turn over all face-up cards and apply the pile-shifting shuffle to M.
6. Turn over all cards in Row 3. Locate the position of the only $\boxed{1}$. Shift the columns of M cyclically such that this $\boxed{1}$ moves to Column 1. This reverts M back to its original state.

Note that each card in C can be replaced by a stack of cards, and the protocol still works in the same way as long as every stack consists of the same number of cards.

3   Verifying the First Nonzero Term in a Sequence

Given a sequence $S = (x_1, x_2, ..., x_n)$ of numbers, P wants to show V the value of the first nonzero term in S without revealing which term it is, i.e. show the term a_i with the lowest index i such that $a_i \neq 0$ without revealing i. Optionally, P may publicly edit the value of that term. The protocol also preserves the sequence S.

We propose the following protocol, which we call the *FirstNonZero* protocol.

Let x_k be the first nonzero term of S. P constructs a sequence $A = (a_1, a_2, ..., a_n)$ of face-down cards, with each a_i being a face-down $\boxed{x_i}$. Then, P performs the following steps.

1. Publicly append cards $b_1, b_2, ..., b_{n-1}$, all of them being $\boxed{0}$s, to the left of A. Call the new sequence $C = (c_1, c_2, ..., c_{2n-1}) = (b_1, b_2, ..., b_{n-1}, a_1, a_2, ..., a_n)$.
2. Turn over all face-up cards. Apply the chosen cut protocol to C to choose the card $c_{k+n-1} = a_k$.

3. As the chosen cut protocol preserves the cyclic order of C, turn over cards $c_k, c_{k+1}, ..., c_{k+n-2}$ to show that they are all $\boxed{0}$s (otherwise V rejects). Also, turn over card c_{k+n-1} to show that it is not a $\boxed{0}$ (otherwise V rejects).
4. Optionally, P may publicly replace the card c_{k+n-1} with a card he/she wants.
5. End the chosen cut protocol. Remove cards $b_1, b_2, ..., b_{n-1}$. The sequence S is now reverted back to its original order.

Note that in step 3, V is convinced that a_k is not zero while $a_1, a_2, ..., a_{k-1}$ are all zeroes without knowing k.

Our protocol uses $\Theta(n)$ cards and $\Theta(1)$ shuffles, in contrast to the similar protocol of Fukusawa and Manabe [6], which uses $\Theta(n)$ cards and $\Theta(n)$ shuffles. Moreover, our protocol uses one card to encode each number, while their protocol uses three. This will benefit the performance when the protocol is applied several times.

4   ZKP Protocol for ABC End View

ABC End View is a logic puzzle consisting of an $n \times n$ empty grid, with some letters written outside the edge of the grid. The player has to fill letters from a given range (e.g. A, B, and C) into some cells in the grid according to the following rules.

1. Each row and column must contain every letter exactly once.
2. A letter outside the edge of the grid indicates the first letter in the corresponding row or column from that direction (see Fig. 3).

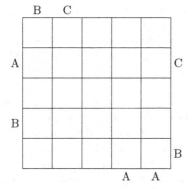

Fig. 3. An example of a 5×5 ABC End View puzzle with letters from the range A, B, and C (left) and its solution (right)

The construction of a ZKP protocol for ABC End View is very straightforward. We use a $\boxed{1}$ to encode letter A, a $\boxed{2}$ for letter B, a $\boxed{3}$ for letter C, and so on, and use a $\boxed{0}$ to encode an empty cell.

We can directly apply the FirstNonZero protocol to verify the second rule for each letter outside the edge of the grid. To verify the first rule, we apply the following *uniqueness verification protocol* for each row and column.

4.1 Uniqueness Verification Protocol

The uniqueness verification protocol [27] enables P to convince V that a sequence σ of n face-down cards is a permutation of different cards $a_1, a_2, ..., a_n$ in some order, without revealing their orders. It also preserves the orders of the cards in σ.

P performs the following steps.

$$\sigma: \boxed{?}\ \boxed{?}\ \cdots\ \boxed{?}$$
$$\boxed{1}\ \boxed{2}\ \cdots\ \boxed{n}$$

Fig. 4. A $2 \times n$ matrix constructed in Step 1

1. Publicly place cards $\boxed{1}$, $\boxed{2}$, ..., $\boxed{n}$ below the face-down sequence σ in this order from left to right to form a $2 \times n$ matrix of cards (see Fig. 4).
2. Turn over all face-up cards. Rearrange all columns of the matrix by a uniformly random permutation. (This can be implemented by putting both cards in each column into an envelope and scrambling all envelopes together.)
3. Turn over all cards in Row 1. V verifies that the sequence is a permutation of $a_1, a_2, ..., a_n$ (otherwise, V rejects).
4. Turn over all face-up cards. Rearrange all columns of the matrix by a uniformly random permutation.
5. Turn over all cards in Row 2. Rearrange the columns such that the cards in the bottom rows are $\boxed{1}$, $\boxed{2}$, ..., $\boxed{n}$ in this order from left to right. The sequence in the Row 1 now returns to its original state.

Our ZKP protocol for ABC End View uses $\Theta(n^2)$ cards and $\Theta(n)$ shuffles, in contrast to the similar protocol of Fukusawa and Manabe [6], which uses $\Theta(n^2)$ cards and $\Theta(n^2)$ shuffles. Moreover, as our protocol uses one card to encode each cell, the number of required cards is actually $n^2 + \Theta(n)$, while their protocol uses three cards to encode each cell, resulting in the total of $3n^2 + \Theta(n)$ cards.

We omit the proof of correctness and security of this protocol as it is a straightforward application of the FirstNonZero protocol, and a similar proof was already given in [6].

5 ZKP Protocol for Goishi Hiroi

Goishi Hiroi or *Hiroimono* is a variant of peg solitaire developed by Nikoli, a Japanese publisher famous for developing many popular logic puzzles including

Sudoku, Numberlink, and Slitherlink. In a Goishi Hiroi puzzle, m stones are placed in an $n \times n$ grid with each cell containing at most one stone. The player has to pick all m stones one by one according to the following rules [14].

1. The player can pick any stone as the first stone.
2. After picking a stone, the player has to travel horizontally or vertically to pick the next stone.
3. During the travel, if there is a stone on the path, the player must pick it. After that, that player may continue traveling in the same direction or turn left or right, but cannot go in the backward direction (see Fig. 5).

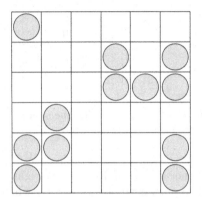 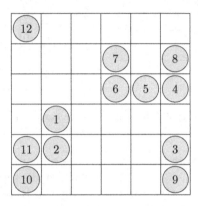

Fig. 5. An example of a 6×6 Goishi Hiroi puzzle with 12 stones (left) and its solution with each number i indicating the i-th stone that was picked (right)

Determining whether a given Goishi Hiroi puzzle has a solution has been proved to be NP-complete [1].

We will develop a ZKP protocol for Goishi Hiroi based on the FirstNonZero protocol.

5.1 Idea of the Protocol

The idea is that we can apply the chosen cut protocol to select a stone we want. After that, we take the $n - 1$ stones on each path from the selected stone in the direction to the north, east, south, and west (we extend the grid by $n - 1$ cells in all directions to support this), and apply the chosen cut protocol again to select the direction we want to travel. Then, we can apply the FirstNonZero protocol to select the first stone on that path.

Note that we also have to keep track of the direction we are traveling; the direction opposite to it will be the "forbidden direction" that we cannot travel in the next move. So, in each move, after selecting the direction, we have to also verify that it is not the forbidden direction.

5.2   Setup

First, publicly place a $\boxed{1}$ on each cell with a stone and a $\boxed{0}$ on each empty cell in the Goishi Hiroi grid. Also, extend the grid by $n-1$ cells in all directions by publicly placing "dummy cards" $\boxed{3}$s around the grid. Then, turn all cards face-down. We now have an $(3n-2) \times (3n-2)$ matrix of cards (see Fig. 6).

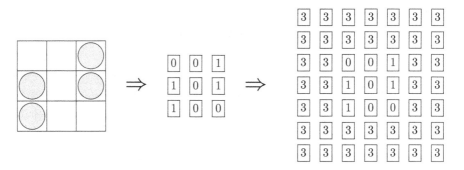

Fig. 6. The way we place cards on a 3×3 Goishi Hiroi grid during the setup

Note that if we arrange all cards in the matrix into a single sequence $C = (c_1, c_2, ..., c_{(3n-2)^2})$, starting at the top-left corner and going from left to right in Row 1, then from left to right in Row 2, and so on, we can locate exactly where the four neighbors of any given card are. Namely, the cards on the neighbor to the north, east, south, and west of a cell containing c_i are c_{i-3n+2}, c_{i+1}, c_{i+3n-2}, and c_{i-1}, respectively.

5.3   Main Protocol

To pick the first stone, P performs the following steps.

1. Apply the chosen cut protocol to select a card corresponding to the first stone.
2. Turn over the selected card to show that it is a $\boxed{1}$ (otherwise V rejects). Replace it with a $\boxed{2}$ and place it back to the grid.

To pick the second stone, P performs the following steps.

1. Apply the chosen cut protocol to select a card corresponding to the first stone.
2. Turn over the selected card to show that it is a $\boxed{2}$ (otherwise V rejects). Replace it with a $\boxed{0}$.
3. Take the $n-1$ cards on a path from the selected card in the direction to the north. Let $A_1 = (a_{(1,1)}, a_{(1,2)}, ..., a_{(1,n-1)})$ be the sequence of these cards in this order from the nearest to the farthest. Analogously, let A_2, A_3, and A_4 be the sequences of the $n-1$ cards on a path from the selected card in the direction to the east, south, and west. Stack each sequence into a single stack.

4. Place a $\boxed{0}$, called $a_{(i,0)}$, on top of A_i for each $i = 1,2,3,4$. Now we have $A_i = (a_{(i,0)}, a_{(i,1)}, ..., a_{(i,n-1)})$ for $i = 1,2,3,4$.
5. Apply the chosen cut protocol to select a stack A_k corresponding to the direction towards the second stone.
6. Apply the FirstNonZero protocol to the sequence $(a_{(k,1)}, a_{(k,2)}, ..., a_{(k,n-1)})$ to select a card corresponding to the second stone. V verifies that it is a $\boxed{1}$ (otherwise V rejects). Replace the selected card with a $\boxed{2}$.
7. Replace $a_{(k,0)}$ with a $\boxed{1}$. Also, replace each $a_{(i,0)}$ for $i \neq k$ with a $\boxed{0}$.
8. Place all cards back to the grid.

To pick each p-th stone for $p \geq 3$, the steps are very similar to picking the second stone. The only additional step is that, after P selects a direction, P has to show to V that it is not a forbidden direction. The formal steps are as follows.

1. Retain the cards $a_{(1,0)}, a_{(2,0)}, a_{(3,0)}, a_{(4,0)}$ from the previous iteration without revealing them. Swap $a_{(1,0)}$ and $a_{(3,0)}$. Swap $a_{(2,0)}$ and $a_{(4,0)}$. (The forbidden direction in this iteration is the direction opposite to the direction we were traveling in the previous iteration.)
2. Apply the chosen cut protocol to select a card corresponding to the $(p-1)$-th stone.
3. Turn over the selected card to show that it is a $\boxed{2}$ (otherwise V rejects). Replace it with a $\boxed{0}$.
4. Take the $n-1$ cards on a path from the selected card in the direction to the north. Let $A_1 = (a_{(1,1)}, a_{(1,2)}, ..., a_{(1,n-1)})$ be the sequence of these cards in this order from the nearest to the farthest. Analogously, let A_2, A_3, and A_4 be the sequences of the $n-1$ cards on a path from the selected card in the direction to the east, south, and west. Stack each sequence into a single stack.
5. Place the card $a_{(i,0)}$ on top of A_i for each $i = 1,2,3,4$. Now we have $A_i = (a_{(i,0)}, a_{(i,1)}, ..., a_{(i,n-1)})$ for $i = 1,2,3,4$.
6. Apply the chosen cut protocol to select a stack A_k corresponding to the direction towards the k-th stone.
7. Turn over the card $a_{(k,0)}$ to show that it is a $\boxed{0}$ (otherwise V rejects).
8. Apply the FirstNonZero protocol to the sequence $(a_{(k,1)}, a_{(k,2)}, ..., a_{(k,n-1)})$ to select a card corresponding to the k-th stone. V verifies that it is a $\boxed{1}$ (otherwise V rejects). Replace the selected card with a $\boxed{2}$. Replace the selected card with a $\boxed{2}$.
9. Replace $a_{(k,0)}$ with a $\boxed{1}$. Also, replace each $a_{(i,0)}$ for $i \neq k$ with a $\boxed{0}$.
10. Place all cards back to the grid.

If the verification passes for all $p = 1,2,...,m$, then V accepts.
Our ZKP protocol for Goishi Hiroi uses $\Theta(n^2)$ cards and $\Theta(m)$ shuffles.

6   Proof of Correctness and Security

We will prove the perfect completeness, perfect soundness, and zero-knowledge properties of our protocol for Goishi Hiroi.

Lemma 1 (Perfect Completeness). *If P knows a solution of the Goishi Hiroi puzzle, then V always accepts.*

Proof. Suppose P knows a solution of the puzzle. Consider when P picks the p-th stone $(p \geq 3)$ from the grid.

At the beginning of Step 1, the only card $a_{(i,0)}$ that is a $\boxed{1}$ is the one corresponding to the direction of travel in the previous iteration. So, after P swaps the cards, the only $\boxed{1}$ will be the one corresponding to the opposite direction of the direction of travel in the previous iteration.

In Step 3, the selected card was changed to $\boxed{2}$ in the previous iteration, so the verification will pass.

In Step 7, as the direction of travel cannot be the opposite direction of the previous iteration, the card must be a $\boxed{0}$, so the verification will pass.

In Step 8, as the stone has not been picked before, the card must be a $\boxed{1}$, so the verification will pass. Also, when invoking the FirstNonZero protocol, as the stone is the first one on the path, the FirstNonZero protocol will also pass.

As this is true for every $p \geq 3$, and the case $p = 2$ also works similarly, while the case $p = 1$ is trivial, we can conclude that V accepts. □

Lemma 2 (Perfect Soundness). *If P does not know a solution of the Goishi Hiroi puzzle, then V always rejects.*

Proof. We will prove the contrapositive of this statement. Suppose that V accepts, meaning that the verification passes for every iteration. Consider the p-th iteration $(p \geq 3)$.

In Step 3, the verification passes, meaning that the card is a $\boxed{2}$. As there is only one $\boxed{2}$ in the grid, which is the card selected in the previous iteration, the move in this iteration must start from that cell.

In Step 7, the verification passes, meaning that the card is a $\boxed{0}$, which means the current direction of travel is not the opposite direction of the direction of travel in the previous iteration, satisfying the rule of Goishi Hiroi.

In Step 8, the verification passes, meaning that the card is a $\boxed{1}$, which means there is a stone on the corresponding cell. Also, when invoking the FirstNonZero protocol, it also passes, meaning that the stone must be the first one on the path, satisfying the rule of Goishi Hiroi.

This means the p-th iteration corresponds to a valid move of picking a stone from the grid. As this is true for every $p \geq 3$, and the case $p = 2$ also works similarly, while the case $p = 1$ is trivial, we can conclude that P must know a valid solution of the puzzle. □

Lemma 3 (Zero-Knowledge). *During the verification, V obtains no information about P's solution.*

Proof. It is sufficient to show that all distributions of cards that are turned face-up can be simulated by a simulator S that does not know P's solution.

- In Steps 3 and 6 of the chosen cut protocol in Sect. 2.3, due to the pile-shifting shuffle, a $\boxed{1}$ has an equal probability to be at any of the n positions. Hence, these steps can be simulated by S.
- In Step 3 of the FirstNonZero protocol in Sect. 3, the cards $c_k, c_{k+1}, ..., c_{k+n-2}$ are all $\boxed{0}$s, and the card c_{k+n-1} is public information known to V. Hence, this step can be simulated by S.
- In the main protocol, there is only one deterministic pattern of the cards that are turned face-up, so the whole protocol can be simulated by S.

Therefore, we can conclude that V obtains no information about P's solution.   □

7 Future Work

We developed a card-based protocol to verify the first nonzero term of a sequence. We also construct card-based ZKP protocols for ABC End View and Goishi Hiroi puzzles. A possible future work is to explore methods to physically verify other interesting sequence-related functions, as well as developing ZKP protocols for other popular logic puzzles.

Acknowledgement. The author would like to thank Daiki Miyahara and Kyosuke Hatsugai for a valuable discussion on this research.

References

1. Andersson, D.: HIROIMONO is NP-complete. In: Crescenzi, P., Prencipe, G., Pucci, G. (eds.) FUN 2007. LNCS, vol. 4475, pp. 30–39. Springer, Heidelberg (2007). https://doi.org/10.1007/978-3-540-72914-3_5
2. Bultel, X., Dreier, J., Dumas, J.-G., Lafourcade, P.: Physical zero-knowledge proofs for Akari, Takuzu, Kakuro and KenKen. In: Proceedings of the 8th International Conference on Fun with Algorithms (FUN), pp. 8:1–8:20 (2016)
3. Bultel, X., Dreier, J., Dumas, J.-G., Lafourcade, P., Miyahara, D., Mizuki, T., Nagao, A., Sasaki, T., Shinagawa, K., Sone, H.: Physical zero-knowledge proof for Makaro. In: Izumi, T., Kuznetsov, P. (eds.) SSS 2018. LNCS, vol. 11201, pp. 111–125. Springer, Cham (2018). https://doi.org/10.1007/978-3-030-03232-6_8
4. Chien, Y.-F., Hon, W.-K.: Cryptographic and physical zero-knowledge proof: from Sudoku to Nonogram. In: Boldi, P., Gargano, L. (eds.) FUN 2010. LNCS, vol. 6099, pp. 102–112. Springer, Heidelberg (2010). https://doi.org/10.1007/978-3-642-13122-6_12
5. Dumas, J.-G., Lafourcade, P., Miyahara, D., Mizuki, T., Sasaki, T., Sone, H.: Interactive physical zero-knowledge proof for Norinori. In: Du, D.-Z., Duan, Z., Tian, C. (eds.) COCOON 2019. LNCS, vol. 11653, pp. 166–177. Springer, Cham (2019). https://doi.org/10.1007/978-3-030-26176-4_14
6. Fukusawa, T., Manabe, Y.: Card-based zero-knowledge proof for the nearest neighbor property: zero-knowledge proof of ABC end view. In: Batina, L., Picek, S., Mondal, M. (eds.) SPACE 2022. LNCS, vol. 13783, pp. 147–161. Springer, Cham (2022). https://doi.org/10.1007/978-3-031-22829-2_9

7. Goldreich, O., Micali, S., Wigderson, A.: Proofs that yield nothing but their validity and a methodology of cryptographic protocol design. J. ACM **38**(3), 691–729 (1991)
8. Goldwasser, S., Micali, S., Rackoff, C.: The knowledge complexity of interactive proof systems. SIAM J. Comput. **18**(1), 186–208 (1989)
9. Gradwohl, R., Naor, M., Pinkas, B., Rothblum, G.N.: Cryptographic and physical zero-knowledge proof systems for solutions of Sudoku puzzles. Theory Comput. Syst. **44**(2), 245–268 (2009). https://doi.org/10.1007/s00224-008-9119-9
10. Koch, A., Walzer, S.: Foundations for actively secure card-based cryptography. In: Proceedings of the 10th International Conference on Fun with Algorithms (FUN), pp. 17:1–17:23 (2020)
11. Lafourcade, P., Miyahara, D., Mizuki, T., Robert, L., Sasaki, T., Sone, H.: How to construct physical zero-knowledge proofs for puzzles with a "single loop" condition. Theor. Comput. Sci. **888**, 41–55 (2021)
12. Miyahara, D., et al.: Card-based ZKP protocols for Takuzu and Juosan. In: Proceedings of the 10th International Conference on Fun with Algorithms (FUN), pp. 20:1–20:21 (2020)
13. Miyahara, D., Sasaki, T., Mizuki, T., Sone, H.: Card-based physical zero-knowledge proof for Kakuro. IEICE Trans. Fundam. Electron. Commun. Comput. Sci. **E102.A**(9), 1072–1078 (2019)
14. Nikoli: Goishi Hiroi. https://www.nikoli.co.jp/ja/puzzles/goishi_hiroi/
15. Robert, L., Miyahara, D., Lafourcade, P., Libralesso, L., Mizuki, T.: Physical zero-knowledge proof and NP-completeness proof of Suguru puzzle. Inf. Comput. **285**(B), 104858 (2022)
16. Robert, L., Miyahara, D., Lafourcade, P., Mizuki, T.: Card-based ZKP for connectivity: applications to Nurikabe, Hitori, and Heyawake. New Gener. Comput. **40**(1), 149–171 (2022). https://doi.org/10.1007/s00354-022-00155-5
17. Robert, L., Miyahara, D., Lafourcade, P., Mizuki, T.: Card-based ZKP protocol for Nurimisaki. In: Devismes, S., Petit, F., Altisen, K., Di Luna, G.A., Fernandez Anta, A. (eds.) SSS 2022. LNCS, vol. 13751, pp. 285–298. Springer, Cham (2022). https://doi.org/10.1007/978-3-031-21017-4_19
18. Robert, L., Miyahara, D., Lafourcade, P., Mizuki, T.: Hide a liar: card-based ZKP protocol for Usowan. In: Du, D.Z., Du, D., Wu, C., Xu, D. (eds.) TAMC 2022. LNCS, vol. 13571, pp. 201–217. Springer, Cham (2022). https://doi.org/10.1007/978-3-031-20350-3_17
19. Ruangwises, S.: An improved physical ZKP for Nonogram. In: Du, D.-Z., Du, D., Wu, C., Xu, D. (eds.) COCOA 2021. LNCS, vol. 13135, pp. 262–272. Springer, Cham (2021). https://doi.org/10.1007/978-3-030-92681-6_22
20. Ruangwises, S.: Physical zero-knowledge proof for ball sort puzzle. In: Della Vedova, G., Dundua, B., Lempp, S., Manea, F. (eds.) CiE 2023. LNCS, vol 13967, pp. 246–257. Springer, Cham (2023). https://doi.org/10.1007/978-3-031-36978-0_20
21. Ruangwises, S.: Two standard decks of playing cards are sufficient for a ZKP for Sudoku. New Gener. Comput. **40**(1), 49–65 (2022). https://doi.org/10.1007/s00354-021-00146-y
22. Ruangwises, S., Itoh, T.: How to physically verify a rectangle in a grid: a physical ZKP for Shikaku. In: Proceedings of the 11th International Conference on Fun with Algorithms (FUN), pp. 24:1–24:12 (2022)
23. Ruangwises, S., Itoh, T.: Physical zero-knowledge proof for numberlink puzzle and k vertex-disjoint paths problem. New Gener. Comput. **39**(1), 3–17 (2021). https://doi.org/10.1007/s00354-020-00114-y

24. Ruangwises, S., Itoh, T.: Physical zero-knowledge proof for ripple effect. Theor. Comput. Sci. **895**, 115–123 (2021)

25. Ruangwises, S., Itoh, T.: Physical ZKP for connected spanning subgraph: applications to bridges puzzle and other problems. In: Kostitsyna, I., Orponen, P. (eds.) UCNC 2021. LNCS, vol. 12984, pp. 149–163. Springer, Cham (2021). https://doi.org/10.1007/978-3-030-87993-8_10

26. Ruangwises, S., Itoh, T.: Physical ZKP for Makaro using a standard deck of cards. In: Du, D.Z., Du, D., Wu, C., Xu, D. (eds.) TAMC 2022. LNCS, vol. 13571, pp. 43–54. Springer, Cham (2022). https://doi.org/10.1007/978-3-031-20350-3_5

27. Sasaki, T., Miyahara, D., Mizuki, T., Sone, H.: Efficient card-based zero-knowledge proof for Sudoku. Theor. Comput. Sci. **839**, 135–142 (2020)

28. Shinagawa, K., et al.: Card-based protocols using regular polygon cards. IEICE Trans. Fundam. Electron. Commun. Comput. Sci. **E100.A**(9), 1900–1909 (2017)

29. Ueda, I., Miyahara, D., Nishimura, A., Hayashi, Y., Mizuki, T., Sone, H.: Secure implementations of a random bisection cut. Int. J. Inf. Secur. **19**(4), 445–452 (2020). https://doi.org/10.1007/s10207-019-00463-w

Mechanism Design in Fair Sequencing

Zhou Chen[1], Yiming Ding[2], Qi Qi[2(✉)], and Lingfei Yu[3(✉)]

[1] School of Business, Hangzhou City University, Hangzhou, China
chenzhou@zucc.edu.cn
[2] Gaoling School of Artificial Intelligence, Renmin University of China,
Beijing, China
{dingym97,qi.qi}@ruc.edu.cn
[3] Zhejiang Gongshang University Hangzhou College of Commerce, Hangzhou, China
ylf@mail.zjgsu.edu.cn

Abstract. Sequencing agents to ensure fairness is a common issue in various domains, such as project presentations, job interview scheduling, and sports or musical game sequence arrangement. Since agents have their own positional preferences, we investigate whether extra credits can be assigned to some agents, so that all agents can choose their preferred positions. We propose an auction system to determine the fair number of credits that an agent may sacrifice for a position they like or request for a position they dislike. The system is modeled as a problem of pricing sequence positions, which demands budget-balanced and egalitarian conditions. We prove that deterministic protocols that guarantee DSIC (dominant-strategy incentive compatibility), budget-balance and egalitarianism do not exist. Furthermore, we design a randomized protocol that ensures being truthful is always the optimal response for every player. A particularly significant technical contribution we make is the establishment of the uniqueness of a randomized protocol with respect to the incentive compatible condition, which serves as the most suitable proxy when incentives are compatible.

Keywords: budget-balance · egalitarian · sequence positions · DSIC

1 Introduction

The sequence is crucial in various contexts, such as the athletic track in sprinting, the swimming lane, the diving and gymnastics competition order, the music or singing sequence, the project presentation order, and the job interview schedule. Previous studies indicate that people have varying preferences for music with different sequences [4,17]. Judges or audiences tend to be more impressed

This paper is supported by Beijing Outstanding Young Scientist Program NO. BJJWZYJH012019100020098, the Fundamental Research Funds for the Central Universities, and the Research Funds of Renmin University of China No. 22XNKJ07, and Major Innovation & Planning Interdisciplinary Platform for the "Double-First Class" Initiative, Renmin University of China.

M. Li et al. (Eds.): IJTCS-FAW 2023, LNCS 13933, pp. 184–194, 2023.
https://doi.org/10.1007/978-3-031-39344-0_14

with recent events than past ones due to the memory decay effect [1]. This phenomenon is extensively researched in various real-world scenarios, such as advertising campaigns [22], news coverage [22], and new product [13]. The impact of the presentation order on the final outcome of a selection process, particularly when information systems are evaluated, is widely acknowledged. The order effects have been demonstrated in various studies. Eisenberg and Barry [9] provide evidence that users' determination of document relevance in information systems is biased by the order and quantify the extent of this bias. Parker and Johnson [19], on the other hand, examine the impact of both the order and the number of presentations on reviewers' judgments. Hogarth and Einhorn [10] and Bruine de Bruin [5] identify the key task variables that account for the order effects in updating beliefs and build a belief-adjustment model through experimental examination and validation.

Studies on the order effect have a long history, dating back to Asch's [3] work on the change-in-meaning hypothesis, Anderson and Hubert's [2] research on attention decrement, and Clancy and Wachsler's [7] study on the fatigue factor's account of the order effect. More recently, Bruine de Bruin and Keren [6] examine the novelty effect of the direction of comparison on a judge's opinion, while Bruine de Bruin [5] further studies the effect in song contests and figure skating competitions. Xu and Wang [23] conceptualize the dynamics of judgment psychology by exploring the order effect forming mechanism. Page and Page [18] conduct a large-scale data analysis of bias in the case of idol series and show a systemic bias toward later contestants in sequential evaluation of candidate performance. Kondo et al. [12] extend the presence of the sequential effect to the domain of subjective decision-making through a face-attractiveness judgment task.

All these studies suggest that sequence arrangement plays a crucial role in subjective judgment across various applications. However, there have been few successful attempts in the literature to systematically resolve its bias. Random ordering is often used, such as in the Van Cliburn International Piano Competition, while diving follows a reversed order of presentation according to the players' early performance, echoing the observation that "last should be the first". In this study, we adopt a mechanism design approach to study the problem in terms of guiding principles that result in a fair solution.

Our goal is to develop a compensation scheme that charges players who gain an advantage to compensate players who would otherwise be disadvantaged by the presentation order, ensuring fairness such that every player obtains the same utility by their private evaluation. This is typically referred to as the *egalitarian* condition, as per Pazner and Schmeidler's [20] definition. Such a condition, if satisfied in the designed rule, guarantees that all parties have equal utility. Many studies have considered this egalitarian requirement for resource distribution problems. For example, Demange [8] proposed a "divide and choose" method for a set of agents to allocate a bundle of divisible goods, which could reach an efficient and egalitarian equilibrium. Olivier et al. [33] examine an egalitarian mechanism for a "one-sided" model, where the demanders do not act strategically and require their demands to be fulfilled precisely.

Another critical condition for the compensation scheme is that it must be budget-balanced, meaning any compensation for advantages must be paid by the players who gain them. Additionally, the scheme must not incentive any players to cheat to obtain a more favorable outcome. Ohseto [16] proves that even under some very strict conditions, there is still no deterministic allocation that satisfies DSIC, budget-balance, and egalitarianism. Thus, they provide the conditions for a mechanism that is DSIC and egalitarian without considering budget-balance. While our model is similar to Ohseto's for deterministic protocols, our work focuses on randomized protocols to find an mechanism that satisfies both egalitarianism and budget-balance while Ohseto only considers deterministic ones. Moulin [35], on the other hand, improves the VCG mechanism to make it both truthful and almost budget-balanced and applies it to solve the problem of assigning p identical objects to a group of n agents where $p < n$. Mishra and Sharma [26] design a mechanism for a private values single object auction model and prove that the mechanism is a dominant strategy incentive compatible and budget-balanced while satisfying equal treatment of equals.

Besides, there are also some other articles about designing mechanism for fair auction or allocation. Bei et al. [27] consider a variant problem of cake-cutting aiming to design a truthful and fair cake sharing mechanisms mechanism. Their research demonstrates that the leximin solution not only ensures truthfulness but also leads to the highest level of egalitarian welfare across all truthful and position oblivious mechanisms. Albert et al. [28] consider the multi-attribute resource allocation problem. They propose the multi-dimensional fairness taking into account all the attributes involved in the allocation problem. The results indicate that multi-dimensional fairness motivates agents to stay in the market, while improving the equity of wealth distribution without compromising the quality of allocation attributes. Ajay Gopinathan et al. [29] study the auctions for balancing social welfare and fairness in secondary spectrum markets. They incorporate randomization into the auction design to ensure the local fairness in that this way can guarantee each user a minimum probability of being assigned spectrum. And for global fairness, they adopt the max-min fairness criterion. They customize a new auction using linear programming techniques to achieve a balance between social welfare and max-min fairness, as well as to identify feasible channel allocations. What's more, mechanism design has been used to enhance fairness in ML algorithms. Finocchiaro et al. [30] introduce a mechanism towards algorithmic fairness. They developed a comprehensive framework that effectively links together the distinct frameworks of mechanism design and machine learning with the fairness guarantee. Menon and Williamson [34] propose choosing distinct thresholds for each group in a way that maximizes accuracy and minimizes demographic parity. Zafar et al. [31] suggest applying the preference-based fairness metrics that draw from concepts in economics and game theory to machine learning. They propose the preferred treatment and which can guarantee the envy-free fairness. More mechanism applied to enhance the fairness of the ML algorithms can be found in the survey of D. Pessach and E. Shmueli [32].

Contribution

We study the problem of finding a truthful auction protocol for a presentation ordering problem that is both budget-balanced and egalitarian.

- Firstly, we analyze the situation of deterministic mechanism. We present an analysis that shows that such mechanisms have a specific payment function. However, we also prove that no deterministic incentive compatible protocol exists under these conditions.
- To overcome this limitation, we move on to design a randomized mechanism that satisfies both egalitarianism and budget-balance. In this mechanism, truth-telling is the dominant strategy for all players, ensuring that no player can benefit from lying about their valuation. The randomized mechanism offers a viable solution for designing truthful mechanisms that satisfy the desired properties of being budget-balanced and egalitarian.
- However, the significance of our research does not end here. While previous research has established the uniqueness of randomized mechanisms for two-player case, the case of n-player (where $n \geq 3$) has never been proven. Therefore, we extend the existing results by Deng and Qi [36], who proved the uniqueness of randomized mechanisms for two-player case, to a more general scenario where $n \geq 3$, ensuring the completeness of the theorem.

In summary, we offer a viable solution in the form of a randomized mechanism that satisfies the desired properties of being budget-balanced, egalitarian, and incentive compatible. Moreover, we extend the existing results to the case of n-player case, which has never been proven before. Our findings can give implications for the design and implementation of truthful mechanisms in fair sequencing problem.

The paper is structured as follows. In Sect. 2, we define our model and introduce relevant concepts and definitions. In Sect. 3, we demonstrate that a deterministic protocol is not feasible and present our randomized protocol, which is proven to satisfy the properties of dominant strategy incentive compatibility (DSIC), budget-balance, and egalitarianism. Section 4 establishes the uniqueness condition and presents two cases where the theorem is proven. Section 5 concludes the paper.

2   Mathematical Modeling and Definitions

The sequencing problem is considered as an assignment problem, in which individuals are assigned to sequence positions. This assignment can be determined by an auction with predefined pricing rules and allocation rules.

Sequence positions can be considered heterogeneous commodities for sale in the auction: $n(n \geq 2)$ different goods are assigned to n players, each of whom has his/her own preferences. Suppose the valuation of these n goods by player $i, (1 \leq i \leq n)$ is $v_i = (v_{i1}, v_{i2}, \cdots, v_{in})$ and $v_{i1} + v_{i2} + \cdots + v_{in} = 0$. If $v_{ij} > 0$, this means good j is a position preferred by player i, with a larger number indicating a more preferable position. If $v_{ij} < 0$, this indicates a position disliked by the player. In this game, players bid a price for each position: $b_i = (b_{i1}, b_{i2}, \cdots, b_{in})$. In addition, the bids should satisfy $b_{i1} + b_{i2} + \cdots + b_{in} = 0$. We refer to this as the fair sequencing problem (called FSP for short).

Let $o_{ij}(\mathbf{b}) \in \{0, 1\}$ and $o_{ij}(b) = 1$ if player i receives the good j, $o_{ij}(b) = 0$ otherwise. One item j can only be assigned to one player: $\forall j, \sum_{i=1}^{n} o_{ij}(\mathbf{b}) = 1$. One player i also only receives one item: $\forall i, \sum_{j=1}^{n} o_{ij}(\mathbf{b}) = 1$. At the same time, player i is charged $t_i(\mathbf{b})$, which is positive for the winner and negative for the loser, and indicates the amount paid to the loser as compensation. With these notations, we consider the usual quasi-linear utility functions for players.

Definition 1. *Quasi-linear utility functions: the player i's , $i = 1, 2, \cdots, n$, utility function is*

$$u_i(\mathbf{b}) = \sum_{j=1}^{n} o_{ij}(\mathbf{b}) \times v_{ij} - t_i(\mathbf{b}).$$

In order to formally describe the transfer of payments from the winner to the loser, we introduce the following definition.

Definition 2. *Budget-balance: a protocol for sequence positions assignment is budget-balanced if the net charge to all players is zero, i.e.,*

$$\sum_{i=1}^{n} t_i(\mathbf{b}) = 0.$$

We introduce the egalitarian condition to even out differences in the players' private values. Pazner and Schmeidler [20] propose a general concept of egalitarian equivalence that equalizes two agents when their allocations are equivalent to a reference bundle comprised of both goods and money. However, for our purposes, we require a simple, purely monetary bundle. In other words, an egalitarian solution is one in which all players have equal utility.

Definition 3. *Egalitarian: a protocol for sequence positions assignment is egalitarian if the mechanism leads to equal utilities for all players: $u_1(\mathbf{b}) = u_2(\mathbf{b}) = \cdots = u_n(\mathbf{b})$.*

Vickrey et al. [24], Clarke [21] and Groves [25] lay down a fundamental principle in auction design to provide bidders with an incentive to submit their true value.

Definition 4. *Dominant-strategy incentive compatibility(DSIC): a protocol for sequence positions assignment is dominant-strategy incentive compatible if truthfulness is the dominant strategy among all players under the mechanism:*

$$u_i(v_i, b_{-i}) \geq u_i(v'_i, b_{-i}) \tag{1}$$

for every v'_i and b_{-i}.

Informally, in a dominant-strategy incentive compatible protocol, every player has an incentive to bid truthfully.

3 An Incentive Compatible Mechanism

In this section, we begin by providing a characterization of the pricing rule in our model. Specifically, we consider the conditions of egalitarianism and budget-balance, which allow us to derive an immediate corollary: the impossibility result for deterministic incentive compatible protocols in our model. This result has important implications for our further discussion on randomized protocols for the problem at hand.

3.1 Analysis for Egalitarian and Budget-Balanced Mechanisms

This subsection aims to provide a characterization of incentive compatible protocols that satisfy both the budget-balanced and the egalitarian conditions in our model. Specifically, we seek to identify protocols that not only ensure that the total payments made by all players are zero, but also that the resulting allocation is distributed as equally as possible among the players.

Lemma 1. *For a dominant-strategy incentive compatible mechanism with n players, it is both budget-balanced and egalitarian if and only if*

$$t_i = \sum_{j=1}^{n} o_{ij}(\mathbf{b})b_{ij} - \frac{1}{n}\sum_{j=1}^{n}\sum_{i=1}^{n} o_{ij}(\mathbf{b})b_{ij}$$

After analyzing the properties of dominant-strategy incentive compatibility, budget-balance, and egalitarianism in the n-player FSP problem, we arrive at a conclusion that may disappoint those seeking a deterministic solution. Our findings reveal that there is no deterministic mechanism that satisfies all three conditions simultaneously.

Theorem 1. *For n-player FSP, there does not exist a deterministic mechanism that satisfies dominant-strategy incentive compatibility, budget-balance, and egalitarianism.*

While this may seem discouraging at first, it underscores the importance of considering randomized protocols as a potential solution. By exploring such protocols, we can find a mechanism that is not only feasible but also fulfills the desired properties.

3.2 Design and Analysis of Randomized Incentive Compatible Mechanisms

As the impossibility of a deterministic mechanism to satisfy the necessary conditions in n-player FSP has been established, we turn our attention to exploring randomized mechanisms. We present the details of our randomized protocol for n-player FSP in the following section, including a description of the allocation mechanism, the payment scheme, and the satisfaction for incentive compatibility, budget-balance and egalitarianism. Through our analysis, we demonstrate that the randomized protocol is capable of achieving all three of these desirable properties.

Protocol 1.

1. *Each player submits its own bid:* $b_i = (b_{i1}, b_{i2}, \cdots, b_{in}), i = 1, \cdots, n$
2. *The system randomly select a winner sequence* $k_1, k_2, \cdots, k_n$, *i.e.* $o_{k_1 1} = o_{k_2 2} = \cdots = o_{k_n n} = 1$, *set price as* $b_{k_1 1}, b_{k_2 2}, \cdots, b_{k_n n}$.
3. *Every player earns a compensation of* $(b_{k_1 1} + b_{k_2 2} + \cdots + b_{k_n n})/n$.

In this protocol, each agents will submit its own bid for every positions. Then the protocol will allocate each position with equal probability and ask each agents to pay for their position. At last, each agent will earn a compensation which can guarantee the property of budget-balance. Also, this can be regarded as the compensation for other agents by the agent who has obtained better positions.

Theorem 2. *Protocol 1 is a dominant-strategy incentive compatible protocol and satisfies budget-balanced and ex-ante egalitarian conditions.*

4   Uniqueness for Randomized Incentive Compatible Protocols

In this section, we present a uniqueness theorem that establishes protocol 1 as the only auction protocol that satisfies the conditions of being egalitarian, budget-balanced, and dominant-strategy incentive compatible. The result of this theorem is split into two parts: the case of two players and the case of $n(n \geq 3)$ players. The two-player case has been previously established by Deng and Qi [36] in the GO game settings. However, their result only covers the two-player case and is not comprehensive enough for our purposes. Therefore, in this section, we provide a more comprehensive result that covers the case of $n \geq 3$. Additionally, for the sake of completeness, we also present Deng and Qi's results in Sect. 4.1.

Theorem 3. *The protocol 1 is the unique mechanism that satisfies DSIC, budget-balance and egalitarianism.*

4.1   Uniqueness of Randomized Mechanism in Two-Player FSP

First consider a two-player case. Suppose both players prefer the first position in a sequence. We can simplify the notation in this case: two players' valuation of the two items are $(v_1, -v_1)$ and $(v_2, -v_2)$ respectively. In this auction, they only need to bid for the first position with a bidding price b_i, $i = 1, 2$, and hence their bid for the other item is $-b_i$.

Then, the protocol 1 degenerates into the following form:

Protocol 2 (FSP for two-player).

1. *Each player submits its own bid:* b_i, $i = 1, 2$.
2. *The system sets its price at* $(b_1 + b_2)/2$ *and randomly selects a winner for the first position.*
3. *The winner compensates the opponent with* $(b_1 + b_2)/2$.

Proposition 1. *Any protocol with two players that satisfies DSIC, budget-balance and egalitarianism must choose the winner with equal probability and set the payment at $(b_1 + b_2)/2$.*

4.2   Uniqueness of Randomized Mechanism in n-Player FSP

We will now prove Theorem 3 for the case of n-player FSP, where n $\geq$ 3, by considering Protocol 1. We prove the following proposition, which forms the basis of the uniqueness theorem: First, we utilize the budget-balance and egalitarian conditions to derive the pricing rule, which is established in the conclusion of Lemma 1. Subsequently, we demonstrate the equal probability of each allocation by varying the true values in the incentive compatibility conditions.

It's important to note that our proof differs from the two-player case presented by Deng and Qi [36]. Specifically, we extend their findings to a more general scenario where $n \geq 3$, thus ensuring the completeness of the theorem. Our proof offers a comprehensive approach to the uniqueness theorem for randomized incentive compatible protocols in n-player FSPs.

Proposition 2. *The protocol 1 is the unique mechanism that satisfies DSIC, budget-balanced and egalitarian conditions for n-player FSP($n \geq 3$).*

We omit the details of the proof and only give the insight of this proof. Firstly, we can assume $w_{ij}(b_1, b_2, \cdots, b_n)$ is the probability player i wins position j when the bids are $b_1, b_2, \cdots, b_n$. Then, similar to lemma 1, we can get the payment as a function of w_{ij} under condition budget-balance and egalitarian. At last, with the condition of DSIC and some mathematical derivation, we can prove that $w_{ij} = \frac{1}{n}$ for all $i, j \in \{1, \cdots, n\}$ is the only possible solution.

5   Conclusion and Discussion

Our research is motivated by the challenge of allocating sequence positions in a competitive environment, such as determining the presentation order in a musical competition. We address this challenge by designing an incentive compatible mechanism to determine the compensation losers receive from the winner, while ensuring both egalitarianism and budget-balance.

We provide a non-existence theorem for designing a deterministic incentive compatible protocol that simultaneously satisfies budget-balanced and egalitarian conditions. Furthermore, we develop a randomized protocol that encourages honesty as an optimal response to opponents' strategies. While the protocol may be weak for risk-neutral players, for whom any bid profile is optimal, the uniqueness of the protocol supports it as a plausible solution.

In our model, each player assigns a value to the positions in a sequence. Protocols that are incentive compatible can induce players to speak honestly, such as those developed by Vickrey [24] , Clarke [21] and Groves [25] for auctions. Unfortunately, we prove that no deterministic incentive compatible protocol can satisfy budget-balanced and egalitarian conditions. So we proposed a randomized protocol. As we have mentioned in the introduction, our model is similar to

192 Z. Chen et al.

Ohseto's deterministic protocols. But, our model requires that losers have negative utility, which is not the case in Ohseto's model as it lacks a strategy-proof protocol. Positive results for any protocol designed using Ohseto's model would imply the same in our model while his non-existence result for a general model does not apply to our more restricted models.

Our study of randomized protocols is unique in this field, particularly due to its distinctiveness from existing research. We show that truth-telling is the optimal response strategy for risk-averse players. However, it is worth noting that Nisan and Ronen [15] proposed a definition of strong dominant-strategy incentive compatibility, stating that strongly DSIC requires a guarantee that there is at least one agent in Eq. 1 such that the inequality sign holds as following:

Definition 5. *(strong) Dominant-strategy incentive compatibility: a protocol for sequence positions assignment is strongly dominant-strategy incentive compatible if truthfulness is the weak dominant strategy among all players under the mechanism:*

$$u_i(v_i, b_{-i}) \geq u_i(v_i', b_{-i}), \quad \forall \ v_i' \ , \ b_{-i}$$

And $\exists j$, s.t.

$$u_j(v_j, b_{-j}) > u_j(v_j', b_{-j}), \quad \forall \ v_j' \ , \ b_{-j}$$

Our mechanism can only satisfy weak dominant-strategy incentive compatibility. It cannot guarantee the strong dominant-strategy incentive compatibility.

In a risk-averse scenario where there is a difference in utility, protocol 1 serves as a Nash implementation of budget-balanced and egalitarian conditions. The Nash implementation theory was specifically developed for situations in which impossibility hampers the achievement of social goals with certain properties ([11,14]). Our results can be considered as a positive contribution in the above cases.

References

1. Anderson, J.R.: Learning and Memory: An Integrated Approach. John Wiley & Sons Inc, Hoboken (2000)
2. Anderson, N.H., Hubert, S.: Effects of concomitant verbal recall on order effects in personality impression formation. J. Verbal Learn. Verbal Behav. **2**, 379–391 (1963)
3. Asch, S.E.: Forming impressions of personality. J. Abnorm. Soc. Psychol. **41**, 258–290 (1946)
4. Baucells, M., Smith, D., Weber, M.: Preferences over constructed sequences: empirical evidence from music (2016)
5. de Bruin, W.B.: Save the last dance for me: unwanted serial position effects in jury evaluations. Acta Psychol. **118**, 245–260 (2005)
6. Bruine de Bruin, W., Keren, G.: Order effects in sequentially judged options due to the direction of comparison. Organ. Behav. Hum. Decis. Process. **92**, 91–101 (2003)
7. Clancy, K.J., Wachsler, R.A.: Positional effects in shared-cost surveys. Public Opin. Q. **35**, 258–265 (1971)

8. Demange, G.: Implementing efficient egalitarian equivalent allocations. Economet-rica: J. Econometric Soc. **52**, 1167–1177 (1984)
9. Eisenberg, M., Barry, C.: Order effects: a study of the possible influence of pre-sentation order on user judgements of document relevance. J. Am. Soc. Inf. Sci. **39**(5), 293–300 (1988)
10. Hogarth, R., Einhorn, H.J.: Order effects in belief updating: the belief-adjustment model. Cogn. Psychol. **24**(1), 1–55 (1992)
11. Jackson, M.O.: A crash course in implementation theory. Soc. Choice Welfare **18**, 655–708 (2001)
12. Kondo, A., Takahashi, K., Watanabe, K.: Sequential effects in face-attractiveness judgment. Perception **41**(1), 43–49 (2012)
13. Mahajan, V., Muller, E., Sharma, S.: An empirical comparison of awareness fore-casting models of new product introduction. Mark. Sci. **3**(3), 179–197 (1984)
14. Maskin, E., Sjostrom, T.: Implementation theory. In: Arrow, K.J., Sen, A.K., Suzu-mura, K. (eds.) Handbook of Social Choice and Welfare 1st edn, vol. 1, pp. 237-.288. chapter 5. Elsevier (2002)
15. Nisan, N., Ronen, A.: Algorithmic mechanism design. In: Proceedings of the Thirty-First Annual ACM Symposium on Theory of Computing, Atlanta, Georgia, USA. ACM, pp. 129–140 (1999)
16. Ohseto, S.: Implementing egalitarian-equivalent allocation of indivisible goods on restricted domains. Econ. Theor. **23**, 659–670 (2004)
17. Rozin, A., Rozin, P., Goldberg, E.: The feeling of music past: how listeners remem-ber musical affect. Music Percept. Interdiscip. J. **22**(1), 15–39 (2004)
18. Page, L., Page, K.: Last shall be first: a field study of biases in sequential perfor-mance evaluation on the Idol series. J. Econ. Behav. Organ. **73**, 186–198 (2010)
19. Parker, L.M.P., Johnson, R.E.: Does order of presentation affect users' judgement of documents? J. Am. Soc. Inf. Sci. **41**, 493–494 (1990)
20. Pazner, E., Schmeidler, D.: Egalitarian-equivalent allocations: a new concept of economic equity. Q. J. Econ. **92**, 671–687 (1978)
21. Clarke, E.H.: Multipart pricing of public goods. Public Choice, **11**, 17–33 (1971)
22. Watt, J.H., Mazza, M., Snyder, L.: Agenda-setting effects of television news cov-erage and the effects decay curve. Commun. Res. **20**(3), 408–435 (1993)
23. Xu, Y., Wang, D.: Order effect in relevance judgment. J. Am. Soc. Inform. Sci. Technol. **59**, 1264–1275 (2008)
24. Vickrey, W.: Counter speculation, auctions, and competitive sealed tenders. J. Financ. **16**, 8–37 (1961)
25. Groves, T.: Incentives in teams. Econometrica: J. Econometric Soc. **41**, 617–631 (1973)
26. Mishra, D., Sharma, T.: A simple budget-balanced mechanism. Soc. Choice Welfare **50**, 147–170 (2018)
27. Bei, X., Lu, X., Suksompong, W.: Truthful cake sharing. In: Proceedings of the AAAI Conference on Artificial Intelligence, vol. 36(5), pp. 4809–4817 (2022)
28. Pla, A., Lopez, B., Murillo, J.: Multi-dimensional fairness for auction-based resource allocation. Knowl.-Based Syst. **73**, 134–148 (2015)
29. Gopinathan, A., Li, Z., Wu, C.: Strategyproof auctions for balancing social welfare and fairness in secondary spectrum markets. In: 2011 Proceedings IEEE INFO-COM, pp. 3020–3028 (2011)
30. Finocchiaro, J., et al.: Bridging machine learning and mechanism design towards algorithmic fairness. In: Proceedings of the 2021 ACM Conference on Fairness, Accountability, and Transparency, pp. 489–503 (2021)

31. Zafar, M.B., Valera, I., Rodriguez, M., Gummadi, K., Weller, A.: From parity to preference-based notions of fairness in classification. In: Advances in Neural Information Processing Systems, p. 30 (2017)
32. Pessach, D., Shmueli, E.: A review on fairness in machine learning. ACM Comput. Surv. (CSUR) **55**(3), 1–44 (2022)
33. Bochet, O., Ilkılıç, R., Moulin, H.: Egalitarianism under earmark constraints. J. Econ. Theory **148**(2), 535–562 (2013)
34. Menon, A.K., Williamson, R.C.: The cost of fairness in binary classification. In: Proceedings of the Conference on Fairness, Accountability, and Transparency, pp. 107–118 (2018)
35. Moulin, H.: Almost budget-balanced VCG mechanisms to assign multiple objects. J. Econ. Theory **144**(1), 96–119 (2009)
36. Deng, X., Qi, Q.: Priority right auction for Komi setting. In: Internet and Network Economics: 5th International Workshop. WINE 2009, pp. 521–528 (2009)

Red-Blue Rectangular Annulus Cover Problem

Sukanya Maji[1], Supantha Pandit[2], and Sanjib Sadhu[1(✉)]

[1] Department of CSE, National Institute of Technology, Durgapur, India
sm.20cs1102@phd.nitdgp.ac.in , sanjib.sadhu@cse.nitdgp.ac.in
[2] Dhirubhai Ambani Institute of Information and Communication Technology,
Gandhinagar, Gujarat, India

Abstract. We study the Generalized Red-Blue Annulus Cover problem for two sets of points, red (R) and blue (B). Each point $p \in R \cup B$ is associated with a positive penalty $\mathcal{P}(p)$. The red points have non-covering penalties and the blue points have covering penalties. The objective is to compute a rectangular annulus $\mathcal{A}$ such that the value of the function $\mathcal{P}(R^{out}) + \mathcal{P}(B^{in})$ is minimum, where $R^{out} \subseteq R$ is the set of red points not covered by $\mathcal{A}$ and $B^{in} \subseteq B$ is the set of blue points covered by $\mathcal{A}$. We study the problem for various types of rectangular annulus in one as well as two dimensions. We design a polynomial-time algorithm for each type of annulus.

Keywords: Annulus cover · Bichromatic point set · Rectangular annulus · Polynomial time algorithms

1 Introduction

An annulus is a closed region bounded by two geometric objects of the same type such as axis-parallel rectangles [9], circles [13], convex polygons [6], and many more. The minimum width circular annulus problem is related to the problem of finding a shape that fits in a given set of points in the plane [1]. Further, the largest empty annulus of different shapes has many potential applications [7,12].

One of the generalizations of "the problem of computing a minimum width annulus" is computing an annulus for a bichromatic point set where each point has a positive penalty. In this problem, a set of red and a set of blue points are given. Each red (resp. blue) point has a positive penalty called the non-covering (resp. covering) penalty, i.e. if a red point is not covered then its penalty is counted, and if a blue point is covered then its penalty is counted. The objective is to find an annulus $\mathcal{A}$ with minimum weight, where the weight of $\mathcal{A}$ is the total penalty of the red points not covered plus the total penalty of the blue points covered by $\mathcal{A}$. We define this problem as follows.

© Springer Nature Switzerland AG 2023
M. Li et al. (Eds.): IJTCS-FAW 2023, LNCS 13933, pp. 195–211, 2023.
https://doi.org/10.1007/978-3-031-39344-0_15

Generalized Red-Blue Annulus Cover (GRBAC) problem. Given a set of m blue points $B = \{p_1, p_2, \ldots, p_m\}$ and a set of n red points $R = \{q_1, q_2, \ldots, q_n\}$, and a penalty function $\mathcal{P} : R \cup B \to \mathbb{R}^+$. The objective is to find an annulus (of certain type and orientation) $\mathcal{A}$ such that the quantity $\delta = \sum_{q \in R^{out}} \mathcal{P}(q) + \sum_{p \in B^{in}} \mathcal{P}(p)$ is minimum, where $R^{out} \subseteq R$ is the set of points not covered by $\mathcal{A}$ and $B^{in} \subseteq B$ is the set of points covered by $\mathcal{A}$.

If the penalty of each red point is ∞ and a blue point is a **unit**, then the GRBAC problem becomes computing an annulus that covers all the red points while covering a minimum number of blue points. We refer to this as the *Red-Blue Annulus Cover (RBAC)* problem. We focus our study on the RBAC and GRBAC problems that compute an axis-parallel rectangular annulus. There exists multiple annuli covering the same set of blue and red points that minimizes δ. Throughout this paper, we report the one among these annuli, whose boundaries pass through some red points as an optimal solution.

Motivation: The oncologists remove the cancer cells while avoiding the healthy cells as much as possible, during the surgery or radiation therapy of the patients. We denote these two types of cells with two different colors. The excision of the tumor leads us to various geometric optimization problems, e.g. determining the smallest circle that encloses the red points or that separates the red points from the blue points. Instead of computing the smallest enclosing circle, the computation of a minimum width annulus which covers largest number of tumor cells with a minimum number of healthy cells, leads us to less wastage of healthy cells while performing surgery. This motivates us to formulate a problem of covering bichromatic point set by a minimum width annulus that covers a red colored point set completely while minimizing the blue colored point set covered by it. We can also assign different weights called penalties to each cancer cell depending on the stage of the cancer cells and the importance of the healthy cells, where removing a healthy cell and keeping cancer cells within the body during surgery leads to a penalty that is to be minimized. This motivates us to study the annulus problem by assigning penalties to each point.

2   Related Works

Finding the largest empty annuli of different shapes in a given point set is well-studied. This problem has many potential applications [7,12]. Díaz-Báñez et al. [8] first studied the problem for circular empty annulus, and proposed an $O(n^3 \log n)$-time and $O(n)$-space algorithm to solve it. Bae et al. [5] studied the problem of computing a maximum-width empty axis-parallel square and rectangular annulus in $O(n^3)$ and $O(n^2 \log n)$ time, respectively. Both of these algorithms require $O(n)$ space. Paul et al. [11] computed the maximum empty axis-parallel rectangular annulus that runs in $O(n \log n)$ time. Abellanas et al. [1] presented an $O(n)$-time algorithm for the rectangular annulus problem such that the annulus covers a given set of points. They considered several variations of this problem. Gluchshenko et al. [9] presented an $O(n \log n)$-time algorithm for the square annulus, and proved that this is optimal. Also, it is known that a minimum width annulus over all orientations can be found in $O(n^3 \log n)$ time for the square annulus [3] and in $O(n^2 \log n)$ time for the rectangular annulus [10]. Bae [4] addressed the problem of computing a minimum-width square or rectangular annulus that contains at least $(n - k)$ points out of n given points in the plane. The k excluded points are considered outliers of the n input points. Bae presented several fast algorithms for the problem.

Recently, Abidha and Ashok [2] studied the geometric separability problem. Given a bichromatic point set $P = R \cup B$ of red and blue points, a separator is an object of a certain type that separates R and B. They studied this problem when the separator is (i) a rectangular annulus of fixed orientation, (ii) a rectangular annulus of arbitrary orientation, (iii) a square annulus of fixed orientation, (iv) an orthogonal convex polygon. They constructed these separators in polynomial time and at the same time optimize given parameters.

3   Preliminaries and Notations

The x and y-coordinates of a point p are denoted by $x(p)$ and $y(p)$, respectively. The left, right, top, and bottom sides of a rectangle $\mathcal{R}$ are denoted by left($\mathcal{R}$), right($\mathcal{R}$), top($\mathcal{R}$) and bottom($\mathcal{R}$), respectively. A rectangular annulus $\mathcal{A}$ is defined by two parallel rectangles, one lying completely inside the other, and we denote these inner and outer rectangles of $\mathcal{A}$ by $\mathcal{R}_{in}$ and $\mathcal{R}_{out}$, respectively. Throughout this paper, we study only the axis-parallel rectangular annulus and hence, we use annulus to imply axis-parallel rectangular annulus unless otherwise stated. The rectangles $\mathcal{R}_{in}$ and $\mathcal{R}_{out}$ define four widths of $\mathcal{A}$: w_ℓ, w_r, w_t, and w_b, where the left width w_ℓ (resp. right width w_r) is the horizontal distance between the left (resp. right) boundaries of $\mathcal{R}_{in}$ and $\mathcal{R}_{out}$. Similarly, top width w_t (resp. bottom width w_b) is the vertical distance between the top (resp. bottom) sides of $\mathcal{R}_{in}$ and $\mathcal{R}_{out}$. A point p is said to be covered by an annulus $\mathcal{A}$ if p lies on or inside $\mathcal{R}_{out}$ but does not lie inside $\mathcal{R}_{in}$.

A rectangular annulus $\mathcal{A}$ is said to be *uniform* if $w_\ell = w_r = w_t = w_b$ (Fig. 1(a)). Otherwise, we say that $\mathcal{A}$ is *non-uniform* (Fig. 1(b),1(c)). It is observed that in a uniform annulus $\mathcal{A}$, the two rectangles $\mathcal{R}_{in}$ and $\mathcal{R}_{out}$ must be concentric. Further, in a *non-uniform concentric annulus*, $w_\ell = w_r$ and $w_t = w_b$ (Fig. 1(c)).

In this paper, we first study one-dimensional versions of the RBAC and GRBAC problems before proceeding to the two-dimensional versions. In a one-dimensional problem, the red and blue points lie on the real line L.

Our Contributions: Table 1 shows our contributions. We need to mention that, each solution needs $O((m+n) \log(m+n))$ preprocessing time and $O(m+n)$ space.

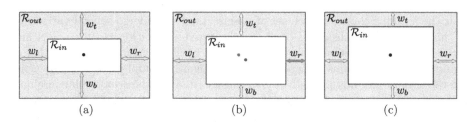

Fig. 1. (a) uniform annulus, (b) non-uniform annulus, and (c) non-uniform concentric annulus. Points depict the centers of the rectangles.

Table 1. Results obtained for different annulus covering problems

Problems	Dims	Type	Rectangular annulus	Time
RBACN-1D		RBAC	Non-uniform	$O(m+n)$
RBACU-1D	1D		Uniform	$O(m+n)$
GRBACN-1D		GRBAC	Non-uniform	$O(m+n)$
GRBACU-1D			Uniform	$O(n^2(m+n))$
RBACN-2D-non-concentric		RBAC	Non-uniform non-concentric	$O(n(m+n))$
RBACN-2D-concentric			Non-uniform concentric	$O(n(m+n))$
RBACU-2D-uniform	2D		Uniform	$O(n(m+n))$
GRBACN-2D-non-concentric		GRBAC	Non-uniform non-concentric	$O(n^6(m+n))$
GRBACN-2D-concentric			Non-uniform concentric	$O(n^5(m+n))$
GRBACU-2D-uniform			Uniform	$O(n^4(m+n))$

4   One-Dimensional Red-Blue Annulus Cover Problem

We consider the rectangular annulus covering problem in one dimension where the red-blue point set lies on a given straight line L (assuming L is horizontal). It is to be noted that, in one dimension, the rectangular annulus $\mathcal{A}$ defines two non-overlapping intervals, say the left interval $\mathcal{I}_L = [L_o, L_i]$ and the right interval $\mathcal{I}_R = [R_i, R_o]$, where L_o (resp. R_o) and L_i (resp. R_i) are the left (resp. right) outer and the left (resp. right) inner endpoints (see the Fig. 2). We use $p \in \mathcal{I}$ to denote that the point p is covered by the interval $\mathcal{I}$. Length of an interval $\mathcal{I}$ is denoted by $|\mathcal{I}|$. The distance between any two points a and b is $|ab|$.

In the rest of this section, a rectangular annulus $\mathcal{A}$ indicates a pair of non-overlapping intervals $\mathcal{I}_L$ and $\mathcal{I}_R$. We say a pair of intervals to be **uniform** if they are equal in length; otherwise, they are **non-uniform**. The $\mathcal{I}_L$ and $\mathcal{I}_R$ are said to be optimal, if they cover together all the red points and a minimum number of blue points in the $RBAC$ problem or have a minimum penalty in $GRBAC$ problem, among all possible pair of intervals. We discuss four different variations of this problem based on the choice of penalty for the red and blue points. As a pre-processing task, we sort and store all the blue points $(p_1, p_2, \ldots, p_m)$ as well as all the red points $(q_1, q_2, \ldots, q_n)$ in two separate arrays.

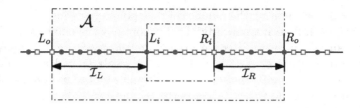

Fig. 2. Rectangular annulus $\mathcal{A}$ defines two non-overlapping intervals $\mathcal{I}_L$ and $\mathcal{I}_R$

Observation 1. *If the intervals are* **non-uniform**, *then each endpoint of the optimal pair* $(\mathcal{I}_L, \mathcal{I}_R)$ *must be on some red point in B, provided each interval* $(\mathcal{I}_L$ *and* $\mathcal{I}_R)$ *covers at least two red points. As a case of degeneracy, only one red point, say* q_i, *may exist in an interval, say* $\mathcal{I}_R$, *of* ϵ (> 0) *width whose one endpoint is at the* q_i.

Observation 2. *In case of* **uniform** *intervals* $\mathcal{I}_L$ *and* $\mathcal{I}_R$, *the endpoints of one of them must be any two red points.*

Justification of Observation 2. Suppose for the sake of contradiction, only one endpoint, say L_o of the interval $\mathcal{I}_L$ of the optimal pair $(\mathcal{I}_L$ and $\mathcal{I}_R)$ is a red point. If the endpoints of $\mathcal{I}_R$ do not lie on any point, then we shift $\mathcal{I}_R$ so that it's one endpoint, say $\mathcal{R}_i$ lies on a blue point without changing the penalty of $\mathcal{I}_R$. Now anchoring the $\mathcal{I}_L$ and $\mathcal{I}_R$ at their endpoint L_o and R_i, respectively, we decrease both the $\mathcal{I}_L$ and $\mathcal{I}_R$ by equal length until a red point is touched either by the endpoints L_i or R_o. Shortening of $\mathcal{I}_L$ and $\mathcal{I}_R$ may cause a reduction in their penalty if a blue point is eliminated from $\mathcal{I}_L$ or $\mathcal{I}_R$. If L_i touches a red point, then we are done; otherwise, anchoring $\mathcal{I}_R$ at R_o, we further decrease $\mathcal{I}_L$ and $\mathcal{I}_R$ until another red point is touched by R_i or L_i and thus contradicting our assumption, and hence the statement is proved.   □

4.1   The RBACN-1D Problem

This problem computes two non-overlapping non-uniform intervals $\mathcal{I}_L$ and $\mathcal{I}_R$ that cover together all the red points in R ($|R|=n$) and a minimum number of blue points in B ($|B|=m$). As per the Observation 1, we place the left endpoint of $\mathcal{I}_L$, i.e., L_o at q_1 and the right endpoint of $\mathcal{I}_R$, i.e., R_o at q_n so that none of the red points lie to the left of $\mathcal{I}_L$ and right of $\mathcal{I}_R$. Now we determine the other two endpoints L_i and R_i of $\mathcal{I}_L$ and $\mathcal{I}_R$, respectively so that all the red points in R and minimum number of blue points in B are covered by $\mathcal{I}_L \cup \mathcal{I}_R$. The two endpoints L_i and R_i must be any two consecutive red points q_i and q_{i+1}, $2 \leq i \leq n-2$. In a linear scan, we can compute the number of blue points lying inside each pair of consecutive red points. We select such a pair (q_i, q_{i+1}) that contains a maximum number of blue points, and choose these red points q_i and q_{i+1} as the right endpoint L_i of $\mathcal{I}_L$ and left endpoint R_i of $\mathcal{I}_R$, respectively. We report these two intervals $\mathcal{I}_L$ and $\mathcal{I}_R$ as a solution. We can also handle the degeneracy case by counting the number of blue points lying inside the pairs (q_1, q_2) and (q_{n-1}, q_n). If such a pair, say (q_1, q_2), contains the maximum number of blue points among all possible consecutive pairs (q_i, q_{i+1}), for $1 \leq i \leq (n-1)$, then we report $(q_1, q_1 + \epsilon)$ and (q_2, q_n) as the intervals $\mathcal{I}_L$ and $\mathcal{I}_R$, respectively. If the other pair (q_{n-1}, q_n) contains the maximum number of blue points, then we report (q_1, q_{n-1}) and $(q_n, q_n + \epsilon')$ as the two intervals. Note that we choose ϵ (resp. ϵ') such that the pair $(q_1, q_1 + \epsilon)$ (resp. $(q_n, q_n + \epsilon')$) does not contain any blue points.

Proof of Correctness: Our solution covers all the red points $\in R$. It discards all the blue points that lie before (resp. after) and the leftmost (resp. rightmost) red point. It also discards the maximum number of blue points lying between the two consecutive red points. Hence our solution is optimal, and we obtain the following result.

Theorem 1. *The RBACN-1D problem can be solved optimally in $O(m+n)$ time and $O(m+n)$ space, after $O((m+n)\log(m+n))$ preprocessing time.*

4.2   The RBACU-1D Problem

We compute two uniform intervals $\mathcal{I}_L$ and $\mathcal{I}_R$ that cover all red points and a minimum number of blue points. The Observation 2 holds for this problem. Depending on the color of the endpoints of each interval, the following two cases are possible.

Case (i): The Endpoints of $\mathcal{I}_L$ are Red. We choose a red point q_i such that $|q_1 q_i| > |q_{i+1} q_n|$ and $|q_1 q_{i-1}| < |q_i q_n|$. Take the interval $[q_1, q_i]$ as $\mathcal{I}_L$ and compute its corresponding interval $\mathcal{I}_R$ so that $|\mathcal{I}_R| = |\mathcal{I}_L|$, $\mathcal{I}_L \cap \mathcal{I}_R = \phi$, $\mathcal{I}_R$ covers the remaining red points those are not covered by $\mathcal{I}_L$ and $\mathcal{I}_R$ covers minimum number of blue points. We determine this $\mathcal{I}_R$ in the following way.

We sequentially search for a blue point, say p_j, that lie between q_i and q_{i+1}, so that $|p_j q_n|$ is as large as possible and $|p_j q_n| \leq |\mathcal{I}_L|$. We take $\mathcal{I}_R$ with its right endpoint placed at q_n and compute its left endpoint (which lies on or to the left of p_j) so that $|\mathcal{I}_R| = |\mathcal{I}_L|$. We compute the number of blue points covered by $\mathcal{I}_L \cup \mathcal{I}_R$. Next, we shift this $\mathcal{I}_R$ (keeping its length the same) toward the right so that one of its endpoints coincides with the blue point which occurs immediately next to the right of the corresponding previous endpoint, depending on whichever occurs earlier. Note that one endpoint of such $\mathcal{I}_R$ is a blue point, and so we shift $\mathcal{I}_R$ by a very small distance $\epsilon > 0$ toward left or right to discard that blue point and reduce one of the blue points covered by $\mathcal{I}_R$. We update the number of blue points covered by the two intervals. The above process continues until the left endpoint of $\mathcal{I}_R$ crosses the red point q_{i+1}. For this $\mathcal{I}_L = [q_1, q_i]$, we choose that position of $\mathcal{I}_R$ where the number of blue points covered by the $\mathcal{I}_R$ is minimized.

Now, in the next iteration, we increase the length of $\mathcal{I}_L$ by extending its right endpoint to its next red point q_{i+1} and repeat the above steps to choose its corresponding $\mathcal{I}_R$ while minimizing the number of blue points covered. This process is repeated until the right endpoint of $\mathcal{I}_L$ covers q_{n-1}.

Lemma 1. *The overall running time (amortized) to compute the $\mathcal{I}_R$ for all such $\mathcal{I}_L$ is $O(m+n)$.*

Proof. At each step, either the left endpoint or the right endpoint of $\mathcal{I}_R$ moves rightward through the blue or red points, and no point is reprocessed twice by the same endpoint of $\mathcal{I}_R$, and this proves the result. □

Case (ii): The Endpoints of $\mathcal{I}_R$ are Red. We can deal with this case similarly to Case (i). In this case, we take the right interval $\mathcal{I}_R = [q_i, q_n]$ with the red point q_i satisfying $|q_1q_{i-1}| < |q_iq_n|$ and $|q_1q_i| > |q_{i+1}q_n|$, and compute its corresponding left interval $\mathcal{I}_L$ while minimizing the number of blue points covered. Then we increase the length of $\mathcal{I}_R$ toward left only by placing its left endpoint on a red point lying before q_i, at each iteration and compute its corresponding $\mathcal{I}_L$. Such increase of $\mathcal{I}_R$ and computation of the corresponding $\mathcal{I}_L$, is repeated until $\mathcal{I}_R$ covers $\{q_n, q_{n-1}, \ldots, q_2\}$.

Finally, among all the pairs of intervals generated in the above two cases, we report the one with the minimum number of blue points covered.

Proof of Correctness: Our algorithm generates all possible uniform annulus covering all the red points and the minimum number of blue points following the Observation 2. This algorithm reports the pair covering minimum blue points among all such feasible solutions, and thus the solution is optimal.

The Lemma 1 leads to the following theorem.

Theorem 2. *The RBACU-1D problem can be solved optimally in $O(m+n)$ time and $O(m+n)$ space, with $O((m+n)\log(m+n))$ preprocessing time.*

5   One-Dimensional Generalized Red-Blue Annulus Cover Problem

5.1   The GRBACN-1D Problem

The two intervals $\mathcal{I}_L$ and $\mathcal{I}_R$ are non-uniform, and hence the Observation 1 also holds for this problem. We sequentially process all the red points rightwards starting from $q_1 \in R$. We associate a positive penalty to each blue or red point. Consider a red point q_i. Let $\mathcal{U}_1$ (resp. $\mathcal{U}_2$) be the set of red points $\in R$ (resp. blue points $\in B$) that lie on or to the left of $x(q_i)$. We compute two non-uniform non-overlapping optimal pair of intervals ($\mathcal{I}_L$ and $\mathcal{I}_R$) up to point q_i such that the following function is minimized.

$$\delta = \mathcal{P}(\mathcal{I}_L \cup \mathcal{I}_R) = \sum_{p \in B^{in}} \mathcal{P}(p) + \sum_{q \in R^{out}} \mathcal{P}(q)$$

where $B^{in} \subseteq \mathcal{U}_2$ is the set of blue points covered by $\mathcal{I}_L \cup \mathcal{I}_R$ and $R^{out} \subseteq \mathcal{U}_1$ is the set of red points not covered by $\mathcal{I}_L \cup \mathcal{I}_R$.

Note that $\sum_{q \in R^{out}} \mathcal{P}(q) = \mathcal{P}(R) - \sum_{q' \in R'} \mathcal{P}(q')$, where $\mathcal{P}(R)$ is the sum of penalties of the red points $\in \mathcal{U}_1$ and R' is the set of red points covered by $\mathcal{I}_L \cup \mathcal{I}_R$. Now, we compute this function up to q_n. We use an incremental approach to process each red point in R sequentially in increasing order of their x-coordinate. Before processing the point q_{i+1}, we maintain the four intervals $\mathcal{I}_1$, $\mathcal{I}_L$, $\mathcal{I}_R$ and $\mathcal{I}_{q_i}$ up to the red point q_i where the endpoints of each such interval are on some red points (see Observation 1).

(i) A single interval, say $\mathcal{I}_1 = [u, v]$ of minimum penalty among all possible intervals up to the point q_i, where $u, v \in R$ are the two endpoints of $\mathcal{I}_1$.

(ii) A pair of intervals $\mathcal{I}_L = [a, b]$ and $\mathcal{I}_R = [c, d]$ with minimum penalty among all the pair of intervals up to the point q_i, i.e. an optimal pair $(\mathcal{I}_L, \mathcal{I}_R)$ up to q_i.

(iii) An interval $\mathcal{I}_{q_i}$ having minimum penalty with its right endpoint at q_i.

The algorithm executes the following while processing the next red point q_{i+1}.

We first determine $\mathcal{I}_{q_{i+1}}$. The optimal pair of intervals $\mathcal{I}_L$ and $\mathcal{I}_R$ up to q_i either remains optimal or needs to be updated. We compute the penalties of the following pairs of intervals and return the optimal pair with a minimum penalty up to q_{i+1}.

$$(\mathcal{I}_1, \mathcal{I}_{q_{i+1}}), (\mathcal{I}_L, \mathcal{I}_{q_{i+1}}), (\mathcal{I}_R, \mathcal{I}_{q_{i+1}}), \text{ and } (\mathcal{I}_L, \mathcal{I} = [c, q_{i+1}]).$$

We must update the pair of intervals $(\mathcal{I}_L, \mathcal{I}_R)$ with the reported pair. We also update the single interval $\mathcal{I}_1$ to be used in the next iteration. Note that the width of one of the intervals can be ϵ (if the interval contains a single red point with a very large penalty) which corresponds to the degeneracy case.

Proof of Correctness: If the optimal pair $(\mathcal{I}_L, \mathcal{I}_R)$ up to point q_i, is to be updated while processing the point q_{i+1}, then the right endpoint of $\mathcal{I}_R$ (after update) must be the red point q_{i+1}. So we compute $\mathcal{I}_{q_{i+1}}$. If the left endpoint of $\mathcal{I}_{q_{i+1}}$ does not overlap with $\mathcal{I}_1$, $\mathcal{I}_L$ and $\mathcal{I}_R$ in the previous iteration (i.e. up to q_i), then one of these intervals becomes the left interval $\mathcal{I}_L$ in the current iteration (i.e. up to q_{i+1}), and the $\mathcal{I}_{q_{i+1}}$ becomes the right interval $\mathcal{I}_R$ up to point q_{i+1}. If the left endpoint of $\mathcal{I}_{q_{i+1}}$ overlaps, with $\mathcal{I}_R = [c, d]$, then the left endpoint of $\mathcal{I}_{q_{i+1}}$ will be c. So we compare $\mathcal{I} = [c, q_{i+1}]$ with $\mathcal{I}_L$. Note that $\mathcal{I}_{q_{i+1}}$ cannot overlap with $\mathcal{I}_L$ otherwise $\mathcal{I}_L$ and $\mathcal{I}_R$ would not have been optimal pair up to point q_i. Thus our algorithm produces the correct result up to q_{i+1} and, hence optimal solution after processing q_n.

To update the four intervals at each point $q_i \in R$, it needs $O(k_i + 1)$ time, where k_i is the number of blue points lying between q_{i-1} and q_i. Hence, we obtain the following result.

Theorem 3. *We can compute the non-uniform annulus of minimum penalty in the GRBACN-1D problem in $O(m + n)$ time using $O(m + n)$ space.*

5.2 The GRBACU-1D Problem

Here, we compute an optimal pair (i.e. of minimum penalty) of uniform intervals $\mathcal{I}_L = [L_o, L_i]$ and $\mathcal{I}_R = [R_i, R_o]$. In this case, $|\mathcal{I}_L| = |\mathcal{I}_R|$. Without loss of generality, we assume that the colors of both the endpoints of $\mathcal{I}_L$ are red (see the Observation 2). For a given such $\mathcal{I}_L$, we find an interval $\mathcal{I}_R$ lying to the right of $\mathcal{I}_L$ so that penalty of $\mathcal{I}_R$ is minimized. For this, first, we consider an interval $\mathcal{I}_L$ and then take another interval $\mathcal{I}$ of length $|\mathcal{I}_L|$ whose left endpoint coincides with a blue or red point that is not covered by $\mathcal{I}_L$. Then we shift this interval $\mathcal{I}$ rightward sequentially either with its left or right endpoint coinciding with a blue or red point lying immediately next to its previous left or right endpoint, depending on whichever occurs first. If one endpoint of such $\mathcal{I}$ is a blue point, then we shift it by a very small distance $\epsilon > 0$ toward the left or right to discard that blue point and thereby the penalty gets reduced. In this way, we compute the penalties of this interval $\mathcal{I}$ with one of its endpoints being at each different red or blue point (which are not covered by $\mathcal{I}_L$). We choose the one with minimum penalty as $\mathcal{I}_R$ for the given $\mathcal{I}_L$. Similarly, we can repeat the above tasks to search for a $\mathcal{I}_L$ by taking both the endpoints of $\mathcal{I}_R$ as red points. Finally, we choose the pair with minimum penalty.

Proof of Correctness: Our algorithm generates all possible pairs of intervals of equal length satisfying the Observation 2, and reports the pair with the minimum penalty, thus providing the optimal solution.

The above procedure needs $O(m + n)$ time. Since there are $O(n^2)$ distinct positions of $\mathcal{I}_L$, we obtain the following result

Theorem 4. *We can compute the uniform annulus with a minimum penalty in the GRBACU-1D problem in $O(n^2(m + n))$ time with $O(m + n)$ space.*

6   Two Dimensional Red-Blue Annulus Cover Problem

We compute an annulus $\mathcal{A}$ for a given set of bichromatic points lying on $\mathbb{R}^2$, which covers all the red points and the minimum number of blue points. We denote the left-most (resp. right-most) red point in R by q_ℓ (resp. q_r), and the bottom-most (resp. top-most) red point in R by q_b (resp. q_t). An annulus $\mathcal{A}$ is said to be *feasible* if it covers all the red points in R. Among all *feasible* annuli the one that covers the minimum number of blue points is called *minimum-$\mathcal{A}$*. Four points are sufficient to uniquely identify a rectangle. It can also be defined by its two opposite corner points. The two rectangles $\mathcal{R}_{out}$ and $\mathcal{R}_{in}$ of the annulus $\mathcal{A}$ may or may not be concentric. Depending on this, we study the two different variations (non-concentric and concentric) of the non-uniform annulus cover problem.

The four widths of a non-uniform non-concentric annulus are different (See Sect. 3), and we obtain the following result

Lemma 2. *Eight red points are sufficient to uniquely define the boundaries of a non-uniform non-concentric rectangular annulus.*

Proof. In non-concentric non-uniform annulus $\mathcal{A}$, the four widths w_ℓ, w_r, w_t and w_b are different. To define w_ℓ (resp. w_r), we need two red points on the left (resp. right) side of both $\mathcal{R}_{out}$ and $\mathcal{R}_{in}$. Similarly two red points on the top (resp. bottom) side of both $\mathcal{R}_{out}$ and $\mathcal{R}_{in}$ will be required to define w_t (resp. w_b). Therefore, a total of eight red points are sufficient to uniquely define a non-concentric non-uniform annulus $\mathcal{A}$ covering the minimum number of blue points.   □

The two horizontal (resp. vertical) widths of a non-uniform concentric annulus are the same, i.e. $w_\ell = w_r$ (resp. $w_t = w_b$) and we obtain the following result.

Lemma 3. *Six red points are sufficient to uniquely define the boundaries of a non-uniform concentric rectangular annulus.*

Proof. To define the horizontal (resp. vertical) width w_h (resp. w_v) of a concentric non-uniform annulus $\mathcal{A}$, two points must lie on the same vertical (resp. horizontal) side (either on the left side or on the right side) of $\mathcal{R}_{out}$ and $\mathcal{R}_{in}$. If these two points are on the left($\mathcal{R}_{out}$) and left($\mathcal{R}_{in}$) (resp. right($\mathcal{R}_{out}$) and right($\mathcal{R}_{in}$)) then a single point on either the right($\mathcal{R}_{out}$) (resp. left($\mathcal{R}_{out}$)) or right($\mathcal{R}_{in}$) (resp. left($\mathcal{R}_{in}$)) is needed to define the right (resp. left) boundaries of these rectangles. So, three points must lie on any three vertical sides of $\mathcal{R}_{out}$ and $\mathcal{R}_{in}$. Similarly, three points will be needed on any three of the horizontal sides of $\mathcal{R}_{out}$ and $\mathcal{R}_{in}$. Hence, a total of six red points are sufficient to uniquely define a concentric non-uniform annulus $\mathcal{A}$ covering the minimum number of blue points.   □

All the four widths of a uniform annulus are the same and, hence we obtain the following result.

Lemma 4. *Five red points are sufficient to uniquely define the boundaries of a uniform rectangular annulus.*

Proof. The four widths of a uniform rectangular annulus are equal. Two points lying on the same side of both $\mathcal{R}_{in}$ and $\mathcal{R}_{out}$ define the width of the annulus. So, another three points that lie on any of the six remaining boundaries of the annulus are sufficient to define the uniform rectangular annulus uniquely. So, in total five points are sufficient to define a uniform annulus uniquely. This proves the result.   □

6.1   The RBACN-2D-Non-Concentric Problem

In this problem, the annulus $\mathcal{A}$ is non-concentric and its width is non-uniform. Since $\mathcal{A}$ covers all the red points, we observe the following

Observation 3. *The left, right, top, and bottom sides of $\mathcal{R}_{out}$ are defined by q_ℓ, q_r, q_t, and q_b, respectively.*

We consider the set of red points $\{q_i\}$, ($q_i \in R \setminus \{q_\ell, q_r, q_t, q_b\}$) to generate $\mathcal{R}_{in}$ of all *feasible* annuli (as per Observation 3). For this, we take such a red point q_i to construct all *feasible* annuli with the left side of $\mathcal{R}_{in}$ being defined by that

q_i. For such a q_i, we consider all possible red points $q_j \in R \setminus \{q_\ell, q_r, q_t, q_b, q_i\}$ so that q_j defines the right side of all $\mathcal{R}_{in}$ which are generated as follows. We first construct a rectangle $\mathcal{R}_{in}$ being defined by two opposite corners q_i and q_j. Then we process the remaining red points sequentially in the increasing order of their x-coordinate to update the $\mathcal{R}_{in}$ and take each such red point as current q_j. We update the $\mathcal{R}_{in}$ so that its left and right sides pass through q_i and current q_j, respectively and $\mathcal{R}_{in}$ does not contain any red point inside it.

Lemma 5. *It needs a constant amount of time to update the $\mathcal{R}_{in}$.*

Proof. While updating $\mathcal{R}_{in}$ with current q_j, we need to check only the four boundary points of $\mathcal{R}_{in}$ constructed in the previous iteration by the red point q_{j-1}. If q_j lies above the top side of $\mathcal{R}_{in}$ (see Fig. 3(c)) or below the bottom side of $\mathcal{R}_{in}$ (see Fig. 3(d)), then no update of $\mathcal{R}_{in}$ is possible since a red point becomes inside $\mathcal{R}_{in}$. In other cases, we can update $\mathcal{R}_{in}$ as shown in Fig. 3(a) and Fig. 3(b). Thus it needs a constant amount of time to update $\mathcal{R}_{in}$. □

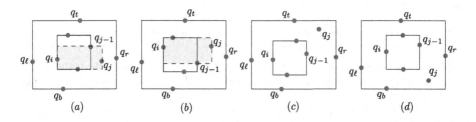

Fig. 3. $\mathcal{R}_{in}$ can be updated only in (a) and (b) as shown by the shaded rectangle.

We repeat the above procedure for all such q_i and report the *minimum-$\mathcal{A}$*.

Proof of Correctness: Our algorithm generates all possible annuli which are *feasible* and satisfy the Lemma 2. Finally, it chooses the annulus with minimum blue points covered, and thus the optimal result is obtained.

The above algorithm computes all possible rectangles $\mathcal{R}_{in}$ with q_i on its left side (containing no red points inside it) and counts the respective number of blue points covered by the annulus $\mathcal{A}$ in $O(m + n)$ amortized time (using the Lemma 5). Since there are $O(n)$ red points, we obtain the following result.

Theorem 5. *For a set of bichromatic points lying on $\mathbb{R}^2$, the RBACN-2D-non-concentric problem can be solved in $O(n(m + n))$ time using $O(m + n)$ space.*

6.2   The RBACN-2D-Concentric Problem

In the non-uniform concentric annulus $\mathcal{A}$, it's left (resp. top) width w_ℓ (resp. w_t) and right (resp. bottom) width w_r (resp. w_b) are the same and we say this width

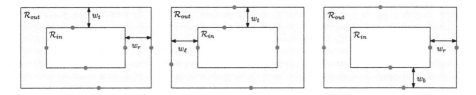

Fig. 4. Rectangular annulus defined by six red points

as the horizontal (resp. vertical) width, which is denoted by w_h (resp. w_v). The annulus $\mathcal{A}$ has either four points on the $\mathcal{R}_{in}$ and the remaining two points on $\mathcal{R}_{out}$ or vice versa, or three points on both $\mathcal{R}_{in}$ and $\mathcal{R}_{out}$ (See Fig. 4).

Fact 1. *Since the annulus $\mathcal{A}$ covers all the red points, the left (resp. right) side of the $\mathcal{R}_{out}$ cannot lie to the right (resp. left) of q_ℓ (resp. q_r), and the top (resp. bottom) side of the $\mathcal{R}_{out}$ cannot lie below (resp. above) the q_t (resp. q_b).*

Fact 1 states that one of the following four cases must be true for the annulus $\mathcal{A}$.

- **Case (i):** The left and bottom sides of $\mathcal{R}_{out}$ pass through q_ℓ and q_b, respectively.
- **Case (ii):** The left and top sides of $\mathcal{R}_{out}$ pass through q_ℓ and q_t, respectively.
- **Case (iii):** The right and bottom sides of $\mathcal{R}_{out}$ pass through q_r and q_b, respectively.
- **Case (iv):** The right and top sides of $\mathcal{R}_{out}$ pass through q_r and q_t, respectively.

We generate all possible annuli for Case (i) as follows (the annuli for the other three cases can be generated similarly):

We use two vertical sweep lines L_1 and L_2, where L_2 lies to the right of L_1 and these two lines generate the left and right sides of $\mathcal{R}_{in}$, respectively. These lines sweep in the rightward direction. The red points lying to the right of q_ℓ and above q_b are the event points for the sweep lines L_1 and L_2.

Fact 2. *Among six red points (see Lemma 3), two red points must lie on any two adjacent sides of the $\mathcal{R}_{in}$ and $\mathcal{R}_{out}$ to define both the widths w_h and w_v.*

Using Fact 2, we first generate all possible annuli with horizontal width w_h defined by q_ℓ and q_i, i.e. $w_h = |x(q_i) - x(q_\ell)|$, where the red point q_i lies to the right of q_ℓ. We take L_1 passing through q_i. Next, we take the vertical line L_2 at a distance w_h to the left of the rightmost point q_r. Next, we choose two red points q_u and q_v among the region bounded by the two lines L_1 and L_2 in such a way that q_u and q_v lies immediately above and below the horizontal line passing through q_i, respectively. Let $w_v = |y(q_v) - y(q_b)|$ (see Fig. 5(a)). We construct rectangular annulus of vertical width w_v. There are two cases:

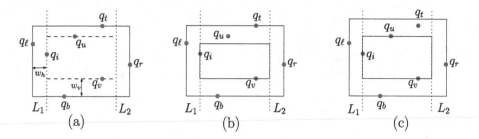

Fig. 5. Instance of Case(i) in RBACN-2D-concentric problem

(i) $|y(q_t) - y(q_u)| < |y(q_v) - y(q_b)|$: We create two concentric annuli as
  follows
    (a) top($\mathcal{R}_{out}$) passes through q_t and top($\mathcal{R}_{in}$) lies below it by a distance w_v
      (see Fig. 5(b)).
    (b) top($\mathcal{R}_{in}$) passes through q_u and top($\mathcal{R}_{out}$) lies above it by a distance w_v
      (see Fig. 5(c)).
  If any such annulus $\mathcal{A}$ is *feasible*, then we count the number of blue points
  covered by it.
(ii) $|y(q_t) - y(q_u)| > |y(q_v) - y(q_b)|$: In this case, no *feasible* annulus $\mathcal{A}$
  having vertical width w_v with its left($\mathcal{R}_{out}$), bottom($\mathcal{R}_{out}$), left($\mathcal{R}_{in}$) and
  right($\mathcal{R}_{in}$) being defined by q_ℓ, q_b, q_i and L_2, respectively, is possible.

Similarly, we can construct annulus of vertical width $w_v = |y(q_t) - y(q_u)|$. Note
that, if there are no such aforesaid red points q_u and q_v inside the region bounded
by L_1 and L_2, then we take the top (resp. bottom) side of the $\mathcal{R}_{in}$ at a small
distance $\epsilon > 0$ below (resp. above) the top (resp. bottom) side of $\mathcal{R}_{out}$.

In the next iteration, the line L_2 sweeps rightward to its next event point, say
a red point q_j, and also the right($\mathcal{R}_{out}$) shifts toward right keeping the horizontal
width of $\mathcal{A}$ same. If q_j lies below (resp. above) the bottom (resp. top) side of $\mathcal{R}_{in}$
in the previous iteration, then no new feasible rectangle $\mathcal{R}_{in}$ is possible using
q_i and q_j as the defining point of the left($\mathcal{R}_{in}$) and right($\mathcal{R}_{in}$), respectively;
otherwise, we construct a new $\mathcal{R}_{in}$ using the point q_j and $\mathcal{R}_{in}$ of the previous
iteration and make this $\mathcal{R}_{in}$ as large as possible without making it infeasible.
Then we create the corresponding annulus and count the number of blue points
inside it. In this way, the line L_2 iterates over all the red points q_j (lying to the
right of q_j and above q_b) to create all possible *feasible* annulus of the Case (i)
in $O(m + n)$ amortized time.

In a similar way, all possible annuli for the remaining three cases can be
generated. Among all possible annuli, we report the annulus that covers the
minimum number of blue points.

Note that the above algorithm for generating the annuli of Case (i) with
horizontal width w_h being defined by two red points on the left boundaries
of $\mathcal{R}_{out}$ and $\mathcal{R}_{in}$. So, it does not generate the annulus with the right width
defined by two red points. However such types of annuli will be generated while
determining the annuli for Case (iii) or Case (iv) (mentioned earlier). The above

four cases generate exhaustively all possible annuli that cover all the red points. We report the *minimum-𝒜*.

Proof of Correctness: Since the algorithm generates all possible feasible $\mathcal{R}_{in}$ (with no red points inside it) with its left side and the right side is defined by all possible pairs (q_i, q_j) (where $q_i, q_j \in R \setminus \{q_\ell, q_t, q_r, q_b\}$), it generates all possible *feasible* annuli following the Fact 1 and Fact 2. Thus, we have the following result.

Theorem 6. *For a set of bichromatic points in $\mathbb{R}^2$, the RBACN-2D-concentric problem can be solved in $O(n(m + n))$ time along with $O(m + n)$ space.*

6.3   The RBACU-2D Problem

We have six configurations (see Fig. 6) of uniform rectangular annulus defined by five red points (see the Lemma 4) which are as follows.

- The $\mathcal{R}_{out}$ (resp. $\mathcal{R}_{in}$) is defined by four points, and $\mathcal{R}_{in}$ (resp. $\mathcal{R}_{out}$) is defined by one point.
- The boundary of $\mathcal{R}_{in}$ (resp. $\mathcal{R}_{out}$) contains three points and that of $\mathcal{R}_{out}$ (resp. $\mathcal{R}_{in}$) contains two points on its two adjacent sides or vice versa.

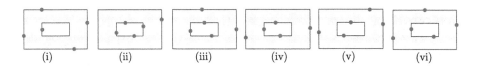

(i)   (ii)   (iii)   (iv)   (v)   (vi)

Fig. 6. Six possible configurations of uniform annulus defined by five red points

First, we consider the configuration where $\mathcal{R}_{out}$ is defined by four points, and these points must be the red points q_ℓ, q_t, q_r and q_b. We take a vertical sweep line L, that sweeps from left to right (resp. right to left) over the remaining $(n - 4)$ red points to search a red point q_i, for which a *feasible* annulus of uniform width, say w, exists. If such *feasible* annulus exists, then q_i defines the left (resp. right) side of the $\mathcal{R}_{in}$. Similarly, we can sweep a horizontal line downward (resp. upward) to find a feasible annulus where q_i defines the top (resp. bottom) side of $\mathcal{R}_{in}$. Among all such annuli, we choose the one that covers the minimum number of blue points. This procedure needs $O(m + n)$ time.

We can generate the annuli of other configurations also. Suppose the width, say w, is defined by the q_ℓ and q_i. To generate the annuli of the width w, we need to move the boundaries of $\mathcal{R}_{in}$ and $\mathcal{R}_{out}$ as follows:

(i) Maintaining the horizontal distance between the right sides of R_{in} and R_{out} as w, we shift both these sides toward the right until the right side of R_{in} hits the first red point that lies to the right of q_i and inside an open-sided rectangular zone whose top, bottom and left side is defined by the top, bottom and left side of R_{in}, respectively. Thus we generate the right boundaries of the annulus with width w by sequentially scanning all the points once. Similarly, we can construct the upper and lower boundaries of the annulus with the same width w.

(ii) On the other hand, if q_i defines the right width w $(=|x(q_r) - x(q_i)|)$ of the annulus, then we shift the left, top, and bottom sides of both R_{in} and R_{out} in the leftward, upward, and downward direction, respectively, to generate the left, top and bottom boundaries of the annulus with same width w.

Depending on the configurations to be generated, we need to move any three sides or two adjacent sides or two non-adjacent sides, or one side of the aforesaid annulus $\mathcal{A}$. Among all the *feasible* annuli, we report the *minimum-$\mathcal{A}$*.

Proof of Correctness: We choose all the red points as the aforesaid point $q_i \in R$ and then generate *feasible* annuli of all possible configurations (mentioned above) following the Lemma 4. Finally, we choose the one that covers the minimum number of blue points. Thus, our algorithm gives the correct result. Since $|R| = n$, we obtain the following result.

Theorem 7. *For a set of bichromatic points lying on $\mathbb{R}^2$, the RBACU-2D problem can be solved optimally in $O(n(m + n))$ time and $O(m + n)$ space.*

7 Two Dimensional Generalized Red-Blue Annulus Cover Problem

The bichromatic points in $\mathbb{R}^2$ are associated with different penalties. We compute the annulus $\mathcal{A}$ of non-uniform width with the minimum penalty and there are two versions (non-concentric and concentric) of this problem depending on the positions of the $\mathcal{R}_{out}$ and $\mathcal{R}_{in}$.

7.1 The GRBACN-2D-Non-Concentric Problem

Lemma 2 holds for this problem. We take any four red points, say q_1, q_2, q_3 and q_4 to construct $\mathcal{R}_{out}$ (see Fig. 7). We choose any two red points, say q_i and q_j inside $\mathcal{R}_{out}$. We construct $\mathcal{R}_{in}$ with its left side and right side passing through q_i and q_j, respectively. Consider two vertical lines V_i and V_j passing through q_i and q_j, respectively. We denote the horizontal line passing through any point, say q_i by H_i. Suppose that $y(q_i) > y(q_j)$. The top (resp. bottom) side of $\mathcal{R}_{in}$ cannot lie below (resp. above) the H_i (resp. H_j). We sequentially scan all the points lying within the yellow (resp. green) colored region (see Fig. 7) bounded by the intersection of V_i, V_j, H_i (resp. H_j) and top (resp. bottom) side of $\mathcal{R}_{out}$. We choose two red points, say q_a and q_b within the yellow and green shaded region, to define the annulus having minimum penalty in $O(m + n)$ time. We can choose the six points ($q_1, q_2, q_3, q_4, q_i, q_j$) in $\binom{n}{6}$ ways, and hence we obtain the following result.

Theorem 8. *The GRBACN-2D-non-concentric problem can be solved optimally in $O(n^6(m + n))$ time using $O(m + n)$ space.*

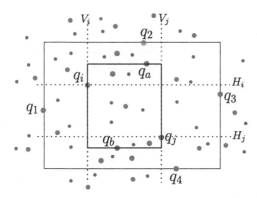

Fig. 7. The GRBACN-2D-Non-Concentric Problem.

7.2   The GRBACN-2D-Concentric Problem

Lemma 3 holds good for this problem, and hence, we obtain the following result

Theorem 9. *The GRBACN-2D-concentric problem can be solved in $O(n^5(m + n))$ time and $O(m + n)$ space.*

7.3   The GRBACU-2D-Uniform Problem

Lemma 4 holds good for this problem, and hence, we obtain the following result.

Theorem 10. *The GRBACU-2D-uniform problem can be solved optimally in $O(n^4(m + n))$ time and $O(m + n)$ space.*

References

1. Abellanas, M., Hurtado, F., Icking, C., Ma, L., Palop, B., Ramos, P.A.: Best fitting rectangles. In: EuroCG (2003)
2. Abidha, V.P., Ashok, P.: Geometric separability using orthogonal objects. Inf. Process. Lett. **176**, 106245 (2022)
3. Bae, S.W.: Computing a minimum-width square annulus in arbitrary orientation. Theor. Comput. Sci. **718**, 2–13 (2018)
4. Bae, S.W.: Computing a minimum-width square or rectangular annulus with outliers. Comput. Geom. **76**, 33–45 (2019)
5. Bae, S.W., Baral, A., Mahapatra, P.R.S.: Maximum-width empty square and rectangular annulus. Comput. Geom. **96**, 101747 (2021)
6. Barequet, G., Goryachev, A.: Offset polygon and annulus placement problems. Comput. Geom. **47**(3), 407–434 (2014)
7. de Berg, M., Cheong, O., van Kreveld, M.J., Overmars, M.H.: Computational Geometry: Algorithms and Applications. Springer, Berlin, Heidelberg (2008). https://doi.org/10.1007/978-3-540-77974-2
8. Díaz-Báñez, J.M., Hurtado, F., Meijer, H., Rappaport, D., Sellarès, J.A.: The largest empty annulus problem. Int. J. Comput. Geom. Appl. **13**(4), 317–325 (2003)

9. Gluchshenko, O., Hamacher, H.W., Tamir, A.: An optimal $O(n \log n)$ algorithm for finding an enclosing planar rectilinear annulus of minimum width. Oper. Res. Lett. **37**(3), 168–170 (2009)

10. Mukherjee, J., Mahapatra, P.R.S., Karmakar, A., Das, S.: Minimum-width rectangular annulus. Theor. Comput. Sci. **508**, 74–80 (2013)

11. Paul, R., Sarkar, A., Biswas, A.: Finding the maximum empty axis-parallel rectangular annulus. In: IWCIA, pp. 139–146 (2020)

12. Preparata, F.P., Shamos, M.I.: Computational Geometry - An Introduction. Texts and Monographs in Computer Science. Springer, New York (1985). https://doi.org/10.1007/978-1-4612-1098-6

13. Roy, U., Zhang, X.: Establishment of a pair of concentric circles with the minimum radial separation for assessing roundness error. Comput. Aided Des. **24**(3), 161–168 (1992)

Applying Johnson's Rule in Scheduling Multiple Parallel Two-Stage Flowshops

Guangwei Wu[1,2(✉)], Fu Zuo[1], Feng Shi[2,3], and Jianxin Wang[2]

[1] College of Computer and Information Engineering, Central South University of Forestry and Technology, Changsha, People's Republic of China
will199031827@hotmail.com
[2] School of Computer Science and Engineering, Central South University, Changsha, People's Republic of China
[3] Xiangjiang Laboratory, Changsha, People's Republic of China

Abstract. It is well-known that the classical Johnson's Rule leads to optimal schedules on a two-stage flowshop. However, it is still unclear how Johnson's Rule would help in scheduling multiple parallel two-stage flowshops with the objective of minimizing the makespan. Thus within the paper, we study the problem and propose a new efficient algorithm that incorporates Johnson's Rule applied on each individual flowshop with a carefully designed job assignment process to flowshops. The algorithm is successfully shown to have a runtime $O(n \log n)$ and an approximation ratio $7/3$, where n is the number of jobs. Compared with the recent PTAS result for the problem, our algorithm has a larger approximation ratio, but it is more efficient in practice from the perspective of runtime.

Keywords: scheduling · two-stage flowshops · approximation algorithm · cloud computing

1 Introduction

Recently, there have been increasing interests in the study of scheduling multiple parallel two-stage flowshops [3–6, 11, 14, 17, 19–22], due to its wide applications in the area of cloud computing, industrial manufacturing, transportation, etc. [1, 2, 11]. Our study was partially motivated by the research in cloud computing and data centers [19]. In certain applications of cloud computing, a client request can be regarded as a two-stage job, consisting of a disk-reading stage and a network-transformation stage, where the network-transformation stage cannot start until the disk-reading stage brings the required data into main memory, and each

This work is supported by the National Natural Science Foundation of China under Grants 62072476; Natural Science Foundation of Hunan Province under Grant 2020JJ4949 and 2021JJ40791; Excellent Youth Project of Scientific Research of Hunan Provincial Education Department under Grant 19B604; the Open Project of Xiangjiang Laboratory (No. 22XJ03005).

M. Li et al. (Eds.): IJTCS-FAW 2023, LNCS 13933, pp. 212–224, 2023.
https://doi.org/10.1007/978-3-031-39344-0_16

server can be regarded as a two-stage flowshop that can handle both the disk-reading and network-transformation for each client request. It has been observed that the costs of the two stages of a request are comparable, not necessarily closely correlated, and depend on different cloud services, the involved servers in the cloud, and bandwidth of local I/O (disk or flash) and of network [23].

The current paper studies algorithms that schedule two-stage jobs on multiple parallel two-stage flowshops with the objective of minimizing the scheduling makespan (i.e., the completion time of all jobs). It is clear that the problem is at least as hard as the classical MAKESPAN problem that has been studied extensively (for more details, please refer to the survey [10]), which can be regarded as scheduling one-stage jobs on multiple one-stage flowshops. As a result, the problem in our study is NP-hard even for two flowshops and becomes strongly NP-hard when the number of flowshops in the system is part of the input [7]. Our aim is to present an approximation algorithm for the problem, thus we introduce the related approximation algorithm work in the following.

Scheduling on a fixed number of two-stage flowshops has been studied. Kovalyov seems the first one to study the problem [14]. Vairaktarakis and Elhafsi [18] proposed a formulation that leads to a pseudo-polynomial time algorithm for the problem on two flowshops. Zhang and van de Velde [22] proposed approximation algorithms with approximation ratios 3/2 and 12/7, respectively, when the number of flowshops are 2 and 3. Using a formulation similar to that in [18], Dong et al. [4] gave a pseudo-polynomial time algorithm and a fully polynomial-time approximation scheme for the problem on a fixed number of flowshops. Wu et al. [19] proposed a new formulation different from that given in [4,18], which leads to a fully polynomial-time approximation scheme with an improved runtime for the problem. The problem where the number of flowshops is part of the input has also been studied. Wu et al. [20] studied two restricted cases of the problem and presented approximation algorithms with ratio 11/6 for both cases. An approximation algorithm with ratio 2.6 for general case was developed [21]. Dong et al. presented a PTAS for the problem [6], using the classic scaling technique for the classical flowshop scheduling problem [16]. Recently, the multiple parallel two-stage flowshops scheduling problem with a deadline got lots of attentions, where the objective is to achieve the maximum total profit of selected jobs that can be finished by the deadline. For the case that the number of flowshops is part of the input, Chen et al. [3] first gave an efficient algorithm with approximation ratio 4, then improved the ratio to 3 but at the cost of a more expensive time complexity. For the case that the number of flowshops is fixed, Chen et al. [3] gave a 2-approximation algorithm, then Tong et al. [17] presented a PTAS.

Scheduling on a single two-stage flowshop is the classical TWO-STAGE FLOWSHOP problem, which has been studied extensively in the literature. It is well-known that the classical Johnson's Rule can give a schedule with the *optimal* completion time on the flowshop [13]. However, it has been unclear how Johnson's Rule would help to schedule on multiple parallel two-stage flowshops with the objective of minimizing the makespan. None of the previous research [4,6,14,18,21] was able to take advantage of Johnson's Rule. In fact, Johnson's

Rule seems to somehow conflict with the optimization goal in the process of assigning jobs to flowshops: for jobs whose first stage is less costful than their second stage, Johnson's Rule sorts the assigned jobs in *increasing order* in terms of the first stage costs in flowshop, while in the study of the classical MAKESPAN problem, it has been well-known that assigning the jobs to machines in increasing order in terms of costs can result in schedules with poor makespan [8,9].

Thus it is meaningful and necessary to study the relationship between Johnson's Rule and scheduling in multiple two-stage flowshops, which can further deepen our understanding on the problem of two-stage flowshop scheduling and lead to better algorithms for the problem. In this paper, we make a step toward this direction. We first pre-arrange the given set of two-stage jobs into a sequence, then propose rules of assigning the jobs to flowshops based on the pre-arranged order, finally apply Johnson's Rule on the jobs assigned to each flowshop. The pre-arranged order is very different from the one specified by Johnson's Rule but it, plus the rules of assigning the jobs to the flowshops, eliminates the anomalies of assigning the jobs to the flowshops in a very unbalanced manner. By a thorough analysis, we show that Johnson's Rule applied on each flowshop can nicely be incorporated with the pre-arranged order and the job assignment rules, which provides effective methods to derive upper bounds on the completion time of a flowshop in terms of the optimal solution of the problem instance. Consequently, an approximation algorithm is successfully given, based on the method given above for the problem in which the number of flowshops is part of the input. The algorithm has an approximation ratio $7/3$ and runs in time $O(n \log n)$, indicating that it can be effectively implemented in practical applications.

2   Preliminaries

For a set G of two-stage jobs to be scheduled on m identical two-stage flowshops, we make the following assumptions:

1. each job consists of an R-operation and a T-operation;
2. each flowshop has an R-processor and a T-processor that can run in parallel and can process the R-operations and the T-operations, respectively, of the assigned jobs;
3. the R-operation and the T-operation of a job must be executed in the R-processor and the T-processor, respectively, of the same flowshop, in such a way that the T-operation cannot start unless the R-operation is completed;
4. there is no precedence constraint among the jobs; and
5. preemption is not allowed.

Under this model, each job J can be given as a pair $J = (r, t)$ of non-negative integers, where r, the *R-time*, is the time for processing the R-operation of J by an R-processor, and t, the *T-time*, is the time for processing the T-operation of J by a T-processor. A *schedule* $\mathcal{S}$ of a set of jobs on m flowshops consists of an *assignment* that assigns each job to a flowshop, and for each flowshop, the *execution orders* of the R- and T-operations of the jobs assigned to that flowshop.

The *completion time* of a flowshop M under a schedule $\mathcal{S}$ is the time when M finishes all R- and T-operations of the jobs assigned to M. The *makespan* $C_{\max}$ of $\mathcal{S}$ is the largest flowshop completion time under the schedule $\mathcal{S}$ over all flowshops. Following the 3-field notation $\alpha|\beta|\gamma$ suggested in [10], we refer to the scheduling model studied in this paper as $P|2\mathrm{FL}|C_{\max}$, formally defined as follows:

$P|2\mathrm{FL}|C_{\max}$
    Given a set of n two-stage jobs and an integer m, construct a schedule of the jobs on m identical two-stage flowshops whose makespan is minimized.

Note that the number m of two-stage flowshops is given as part of the input. Like most previous work on flowshop scheduling, our algorithm is based on *permutation scheduling* [15]. Thus, we will use an ordered sequence $\langle J_1, J_2, \ldots, J_t \rangle$ to represent a schedule $\mathcal{S}$ of the two-stage jobs on a single two-stage flowshop if both executions of the R- and the T-operations of the jobs, by the R- and the T-processors of the flowshops, respectively, strictly follow the given order. If our objective is to minimize the makespan of schedules, then we can make the following assumptions (where $\bar{\tau}_0 = t_0 = 0$) (see [19] for detailed discussions).

Lemma 1 ([19]). *Let* $\mathcal{S} = \langle J_1, J_2, \ldots, J_t \rangle$ *be a two-stage job schedule on a single two-stage flowshop, where* $J_i = (r_i, t_i)$, *for* $1 \le i \le t$. *Let* $\bar{\rho}_i$ *and* $\bar{\tau}_i$, *respectively, be the times at which the R-operation and the T-operation of job* J_i *are started. Then for all* i, $1 \le i \le t$, *we can assume:*

(1) $\bar{\rho}_i = \sum_{k=1}^{i-1} r_k;$ *and*   (2) $\bar{\tau}_i = \max\{\bar{\rho}_i + r_i, \bar{\tau}_{i-1} + t_{i-1}\}.$

The problem of scheduling two-stage jobs on a single two-stage flowshop is the classical TWO-STAGE FLOWSHOP problem, which can be solved *optimally* in time $O(n \log n)$ by sorting the given jobs using Johnson's Rule [13], which gives a schedule of the given jobs with the *minimum completion time* on the flowshop. Johnson's Rule can be formally stated as follows:

Johnson's Rule [13].
    On a set of two-stage jobs (r_i, t_i), $1 \le i \le n$, divide the jobs into two disjoint groups G_1 and G_2, where G_1 contains all jobs (r_h, t_h) with $r_h \le t_h$, and G_2 contains all jobs (r_g, t_g) with $r_g > t_g$. Order the jobs in a sequence such that the first part consists of the jobs in G_1, sorted in non-decreasing order of R-times, and the second part consists of the jobs in G_2, sorted in non-increasing order of T-times.

Johnson's order of a set of two-stage jobs is to order the jobs into a sequence that satisfies the conditions given in Johnson's Rule above.

3   The Algorithm

We start with a description of the algorithm. Given a set G of n two-stage jobs and an integer m representing the number of two-stage flowshops, our algorithm involves three main steps. The first *sorting step* sorts the jobs in G into a job sequence $S^* = \langle J_1^*, \ldots, J_d^*, J_{d+1}^*, \ldots, J_n^* \rangle$, where the subsequence $S_1^* = \langle J_1^*, \ldots, J_d^* \rangle$ contains the jobs $J_h^* = (r_h^*, t_h^*)$ with $r_h^* \leq 3t_h^*/2$, in non-increasing order by T-time, and the following job subsequence $S_2^* = \langle J_{d+1}^*, \ldots, J_n^* \rangle$ contains the rest jobs in G, in non-increasing order by R-time. The second *assignment step* assigns the jobs, in the order of the job sequence S^*, to the m flowshops, in such a way that each job in the subsequence S_1^* is assigned to the flowshop with the minimum ψ-value, while each job in the subsequence S_2^* is assigned to the flowshop with the minimum ρ-value (the ψ-value and ρ-value of a flowshop will be defined precisely in the algorithm). Finally, in the third *permutation step*, for each flowshop M_q, the algorithm sorts the jobs assigned to the flowshop M_q into Johnson's order. The algorithm is given in Fig. 1.

Algorithm Approx
INPUT: a set G of n two-stage jobs and an integer m.
OUTPUT: a schedule $\mathcal{S}$ of the jobs in G on m two-stage flowshops $M_1, \ldots, M_m$.

1.   sort the jobs in G into a sequence $S^* = \langle J_1^*, \ldots, J_d^*, J_{d+1}^*, \ldots, J_n^* \rangle$, where $S_1^* = \langle J_1^*, \ldots, J_d^* \rangle$ are the jobs $J_i^* = (r_i^*, t_i^*)$ satisfying $r_i^* \leq 3t_i^*/2$, sorted in non-increasing order by T-time, and $S_2^* = \langle J_{d+1}^*, \ldots, J_n^* \rangle$ are the jobs satisfying $r_i^* > 3t_i^*/2$, sorted in non-increasing order by R-time;
2.   **for** $q = 1$ **to** m **do** $\{\rho_q = 0;\ \psi_q = 0;\ T_q = \emptyset;\}$
3.   **for** $i = 1$ **to** d **do**
3.1   find the flowshop M_q with the minimum ψ-value ψ_q;
3.2 $T_q = T_q \cup \{J_i^*\};\ \psi_q = \psi_q + t_i^*;\ \rho_q = \rho_q + r_i^*;$ \\ assign J_i^* to M_q
4.   **for** $i = d + 1$ **to** n **do**
4.1   find the flowshop M_q with the minimum ρ-value ρ_q;
4.2 $T_q = T_q \cup \{J_i^*\};\ \rho_q = \rho_q + r_i^*;$ \\ assign J_i^* to M_q
5.   **for** $q = 1$ **to** m **do**
         sort the jobs of T_q in the flowshop M_q into Johnson's order S_q;
6.   return $\mathcal{S} = \{S_1, S_2 \ldots, S_m\}$.

Fig. 1. An approximation algorithm for $P|2FL|C_{\max}$

The sorting and assignment steps of the algorithm **Approx** are the same as that in the algorithm proposed in [21]. The difference is the additional permutation step in the algorithm, which sorts the jobs on each flowshop into Johnson's order. The permutation step is obviously natural in practice, because Johnson's order is an optimal permutation for scheduling on a single two-stage flowshop [13]. However, it is unclear how such a step benefits scheduling algorithms for multiple two-stage flowshops, which is the focus of the current paper.

Theorem 1. *The algorithm* **Approx** *runs in time $O(n \log n)$.*

Proof. As showed in [21], the sorting and assignment steps (steps 1–4) take time $O(n \log n)$. The permutation step of sorting jobs on each flowshop into Johnson's order runs in time $O(n_1 \log n_1) + O(n_2 \log n_2) + \ldots + O(n_m \log n_m) = O(n \log n)$, where n_q $(1 \leq q \leq m)$ is the number of jobs assigned to the flowshop M_q, and $n_1 + \ldots + n_m = n$. Thus, the algorithm runs in time $O(n \log n)$. □

We use τ_{opt} as the optimal makespan for the job set G scheduled on m two-stage flowshops. The following lemma gives lower bounds on the value τ_{opt}.

Lemma 2. *For any job* $J_i = (r_i, t_i)$ *in the set* G, $r_i + t_i \leq \tau_{opt}$. *Thus,* $\min\{r_i, t_i\} \leq \tau_{opt}/2$. *Moreover, if* J_i *belongs to the sequence* S_2^*, *then* $t_i < 2\tau_{opt}/5$.

Proof. It is obvious that $r_i + t_i \leq \tau_{opt}$ because both R- and T-operations of the job J_i are executed by the same flowshop, and the T-operation cannot start until the R-operation is completed. In case the job J_i is in S_2^*, the inequality $t_i < 2\tau_{opt}/5$ follows from $r_i + t_i \leq \tau_{opt}$ and $r_i > 3t_i/2$. □

For the convenience of discussion, we introduce the following notations:

- M_h: the flowshop where the schedule $\mathcal{S}$ by **Approx** achieves its makespan;
- τ^*: the makespan of the schedule $\mathcal{S}$, i.e., the completion time of M_h under $\mathcal{S}$;
- G_h: the set of jobs assigned to M_h by the schedule $\mathcal{S}$;
- S_h: the sequence $\langle J_1, \ldots, J_z \rangle$ for G_h on M_h constructed by step 5 of **Approx**;
- H_1: the set of jobs $J_i = (r_i, t_i)$ in G_h satisfying $r_i \leq t_i$;
- H_2: the set of jobs $J_i = (r_i, t_i)$ in G_h satisfying $t_i < r_i \leq 3t_i/2$;
- H_3: the set of jobs $J_i = (r_i, t_i)$ in G_h satisfying $r_i > 3t_i/2$.

According to the algorithm **Approx**, the sequence S_h is in Johnson's order. Thus, S_h is an optimal schedule of the job set G_h on the flowshop M_h, and can be written as a concatenation of two subsequences: $S_h = \langle S_{h,1}, S_{h,2} \rangle$, where $S_{h,1}$ consists of the jobs in H_1, sorted in non-decreasing order by R-time, and $S_{h,2}$ consists of the jobs in $H_2 \cup H_3$, sorted in non-increasing order by T-time. Note that the sequence S_h is very different from the one given by the assignment step of the algorithm (steps 3–4) to the flowshop M_h, which first assigns the jobs in the set $H_1 \cup H_2 \subseteq S_1^*$ to M_h, in non-increasing order by T-time, then assigns the jobs in the set $H_3 \subseteq S_2^*$ to M_h, in non-increasing order by R-time.

Lemma 3 ([21]). (1) *The sum of the T-times of the jobs in the set $H_1 \cup H_2$ is bounded by $4\tau_{opt}/3$; (2) for any job J_i in $H_1 \cup H_2$, the sum of the T-times of the jobs in the set $(H_1 \cup H_2) \setminus \{J_i\}$ is bounded by τ_{opt}; and (3) there are at most two jobs in $H_1 \cup H_2$ whose T-time is larger than $\tau_{opt}/3$.*

4   The Analysis of the Algorithm

In this section, we study the approximation ratio of the algorithm **Approx**. We divide the analysis into two cases, based on whether the set H_3 is empty. The main result of this section, which is also the main result of the paper, is

Theorem 2. *The algorithm* **Approx** *is a (7/3)-approximation algorithm.*

The theorem is a direct consequence of Theorem 3 and Theorem 4, which will be shown in the following subsections.

4.1   Case 1. $H_3 = \emptyset$

We first consider the case where $H_3 = \emptyset$, i.e., no job $J = (r, t)$ assigned to the flowshop M_h satisfies the condition $r > 3t/2$.

In this case, consider the schedule (i.e., the sequence) $S_h = \langle J_1, J_2, \ldots, J_z \rangle$, which consists of the jobs in G_h sorted in Johnson's order, and has completion time τ^*. Let x $(1 \le x \le z)$ be the minimum job index in the sequence S_h such that the T-operations of the jobs $J_x, J_{x+1}, \ldots, J_z$ are executed continuously with no idle time by the T-processor of M_h. Let

$$a_0 = \sum_{i=1}^{x-1} r_i, \quad b_0 = \sum_{i=1}^{x-1} t_i, \quad a_1 = \sum_{i=x+1}^{z-1} r_i, \quad b_1 = \sum_{i=x+1}^{z-1} t_i.$$

The analysis for **Case 1** is further divided into two subcases.

Subcase 1.1. $H_3 = \emptyset$, $J_x \in H_1$, and $J_z \in H_1 \cup H_2$; and
Subcase 1.2. $H_3 = \emptyset$, and both J_x and J_z are in H_2.

Note that the case where J_x is in H_2 while J_z is in H_1 is impossible because in Johnson's order S_h, all jobs $J_i = (r_i, t_i)$ in H_1 (with $r_i \le t_i$) appear before any jobs $J_j = (r_j, t_j)$ in H_2 (with $r_j > t_j$).

The configuration of the schedule S_h in **Subcase 1.1** is given in Fig. 2. The execution of the jobs in the block "for b_0" may not be continuous, but this has no impact on our analysis.

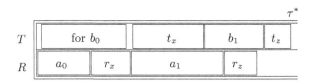

Fig. 2. The configuration of the flowshop M_h in **Subcase 1.1**

Lemma 4. *In* **Subcase 1.1**, *we have* $\tau^* \le 11\tau_{opt}/6$.

Proof. By Lemma 1, the R-processor of the flowshop M_h runs with no idle time. By the definition of the index x, the T-operations of the jobs $J_x, \ldots, J_z$ are executed with no idle time. The operation t_x must start right after the operation r_x is completed: otherwise by Lemma 1, the execution of t_x would be waiting for the completion of t_{x-1}, and would start right after t_{x-1} is completed. Thus the T-operations of the jobs $J_{x-1}, J_x, \ldots, J_z$ would be executed with no idle time, contradicting the minimality of the index x. Combining all these facts gives:

$$\tau^* = a_0 + r_x + t_x + b_1 + t_z.$$

Because S_h is in Johnson's order, the assumption that J_x belongs to the set H_1 implies that all jobs $J_1, \ldots, J_{x-1}$ are also in the set H_1, where each job $J_i = (r_i, t_i)$ satisfies $r_i \le t_i$. Therefore, $a_0 \le b_0$, which gives:

$$\tau^* \le b_0 + r_x + t_x + b_1 + t_z = r_x + (b_0 + t_x + b_1 + t_z) \le \frac{\tau_{opt}}{2} + \frac{4\tau_{opt}}{3} = \frac{11\tau_{opt}}{6},$$

where in the last inequality, we have used $r_x \le \tau_{opt}/2$ because of Lemma 2 and $J_x \in H_1$, and $b_0 + t_x + b_1 + t_z \le 4\tau_{opt}/3$ by Lemma 3(1).   □

We now consider **Subcase 1.2**, where both jobs J_x and J_z are in the set H_2. Thus, the job sequence $S_h = \langle S_{h,1}, S_{h,2} \rangle$, where $S_{h,1}$ consists of the jobs in H_1 sorted in non-decreasing order by R-time, and $S_{h,2}$ consists of the jobs in H_2 sorted in non-increasing order by T-time, has both J_x and J_z in the subsequence $S_{h,2}$. Thus, the sum $a_0 = \sum_{i=1}^{x-1} r_i$ of R-times of the jobs in $\{J_1, \ldots, J_{x-1}\}$ can be written as $a_0 = a_{01} + a_{02}$, where a_{01} is the sum of R-times for the jobs in $H_1 \cap \{J_1, \ldots, J_{x-1}\}$ while a_{02} is the sum of R-times for the jobs in $H_2 \cap \{J_1, \ldots, J_{x-1}\}$. Similarly, the sum $b_0 = \sum_{i=1}^{x-1} t_i$ is partitioned into $b_0 = b_{01} + b_{02}$ based on the sets H_1 and H_2. By the definitions of the sets H_1 and H_2, we have $a_{01} \le b_{01}$ and $b_{02} \le a_{02} \le 3b_{02}/2$ (where we use $b_{02} \le a_{02}$ instead of $b_{02} < a_{02}$ because a_{02} could be 0). Moreover, by Lemma 2, and because both J_x and J_z are in H_2, we have $t_x \le \tau_{opt}/2$ and $t_z \le \tau_{opt}/2$.

Lemma 5. *In* **Subcase 1.2***, we have* $\tau^* \le 7\tau_{opt}/3$.

The proof of Lemma 5 will be given in a complete version. Combining Lemmas 4 and 5, we derive

Theorem 3. *In* **Case 1** *where the set* H_3 *is empty, the makespan* τ^* *of the schedule* S *constructed by the algorithm* **Approx** *is bounded by* $7\tau_{opt}/3$.

4.2   Case 2. $H_3 \neq \emptyset$

Now we consider the case where the set H_3 is not empty. The schedule $S_h = \langle S_{h,1}, S_{h,2} \rangle$ on the flowshop M_h follows Johnson's Rule, where $S_{h,1}$ consists of the jobs in H_1, sorted in non-decreasing order by R-time, and $S_{h,2}$ consists of the jobs in $H_2 \cup H_3$, sorted in non-increasing order by T-time.

Suppose that the sequence S_h is $\langle J_1, \ldots, J_x, \ldots, J_z \rangle$, where again we let x be the smallest index such that in the sequence S_h, the T-operations of the jobs $J_x, J_{x+1}, \ldots, J_z$ are executed continuously. Let $J_i = (r_i, t_i)$ for all i. Slightly different from that in **Case 1**, let $a_0 = \sum_{i=1}^{x-1} r_i$, $b_0 = \sum_{i=1}^{x-1} t_i$, $a_1 = \sum_{i=x+1}^{z} r_i$, and $b_1 = \sum_{i=x+1}^{z} t_i$. Let J_c be the last job assigned to M_h by the algorithm **Approx**. Then J_c belongs to H_3 thus is in the subsequence $S_{h,2}$ because H_3 is not empty. Note the schedule S_h is in Johnson's order, so J_c may not be J_z.

Lemma 6. *For any job* $J_y = (r_y, t_y)$ *in the set* H_3, $t_y < 2\tau_{opt}/5$. *In* **Case 2**, *for the job* J_c *last assigned to* M_h *by the algorithm* **Approx**, $\sum_{i=1}^{z} r_i - r_c \le \tau_{opt}$.

Proof. The inequality $t_y < 2\tau_{opt}/5$ follows directly from Lemma 2 since all jobs in H_3 are in the sequence S_2^*.

Since $J_c = (r_c, t_c)$ is the last job assigned to M_h by the algorithm and H_3 is not empty, J_c is in the sequence S_2^*. When J_c was being assigned, the flowshop

M_h had the minimum ρ-value, which is $\sum_{i=1}^{z} r_i - r_c$, among all flowshops. Thus the total sum $\sum_{i=1}^{n} r_i^*$ of R-times of the input job set G is at least $m(\sum_{i=1}^{z} r_i - r_c)$, which implies that $\sum_{i=1}^{z} r_i - r_c \leq \tau_{opt}$. □

Now we are ready to study **Case 2**. Our analysis is based on the schedule

$$S_h = \langle S_{h,1}, S_{h,2} \rangle = \langle J_1, \ldots, J_x, \ldots, J_z \rangle.$$

Case 2 is divided into four subcases.

Subcase 2.1. $J_x \in H_2 \cup H_3$, thus in $S_{h,2}$;
Subcase 2.2. $J_x \in H_1$ is the last job in $S_{h,1}$;
Subcase 2.3. $J_x \in H_1$ is the first job in $S_{h,1}$; and
Subcase 2.4. $J_x \in H_1$ is in $S_{h,1}$, but neither the first nor the last in $S_{h,1}$.

We start with **Subcase 2.1**. The related proofs will be given in a complete version.

Lemma 7. *In* **Subcase 2.1**, *we have* $\tau^* \leq 2\tau_{opt}$.

Now we consider **Subcase 2.2**, where $J_x \in H_1$ is the last job in $S_{h,1}$.

Lemma 8. *In* **Subcase 2.2**, *we have* $\tau^* \leq 98\tau_{opt}/45$.

The proof of Lemma 8 will be given in a complete version.
Now consider **Subcase 2.3**, where J_x is the first job in $S_{h,1}$ so $a_0 = b_0 = 0$.

Lemma 9. *In* **Subcase 2.3**, *we have* $\tau^* \leq 7\tau_{opt}/3$.

Proof. **Subcase 2.3** is further divided into two cases, based on whether $H_2 = \emptyset$.

First consider the case where $H_2 \neq \emptyset$. The jobs for a_1 and b_1 are given in two consecutive sequences: a sequence for the jobs in H_1 (let a_{11} and b_{11} be the sum of the R-times and T-times of these jobs, respectively), followed by a sequence for the jobs in $H_2 \cup H_3$ (let $a_{1,23}$ and $b_{1,23}$ be the sum of the R-times and T-times of these jobs, respectively). The set of jobs for a_{11} and b_{11} can be assumed to be not empty – otherwise, the case would become **Subcase 2.2**. See Fig. 3. Let a_{12} and a_{13} be, respectively, the sum of R-times for the jobs in $H_2 \cap \{J_{x+1}, \ldots, J_{c-1}, J_{c+1}, \ldots, J_z\}$ and in $H_3 \cap \{J_{x+1}, \ldots, J_{c-1}, J_{c+1}, \ldots, J_z\}$, and let b_{12} and b_{13} be for the sums of T-times for these jobs. We have $a_{1,23} = a_{12} + a_{13} + r_c$ and $b_{1,23} = b_{12} + b_{13} + t_c$. Since $H_2 \neq \emptyset$, the set of jobs for a_{12} and b_{12} is not empty. We have

$$
\begin{aligned}
\tau^* &= r_x + t_x + b_{11} + b_{1,23} = r_x + t_x + b_{11} + b_{12} + b_{13} + t_c \\
&= r_x + t_x + b_{11} + \frac{1}{3}b_{12} + \frac{2}{3}b_{12} + b_{13} + t_c \\
&\leq r_x + t_x + b_{11} + \frac{1}{3}b_{12} + \frac{2}{3}a_{12} + \frac{2}{3}a_{13} + t_c \qquad (1) \\
&= \frac{2}{3}(r_x + a_{12} + a_{13}) + \frac{1}{3}r_x + t_x + b_{11} + \frac{1}{3}b_{12} + t_c \\
&\leq \frac{2}{3}(\tau_{opt} - a_{11}) + \frac{1}{3}r_x + t_x + b_{11} + \frac{1}{3}b_{12} + t_c, \qquad (2)
\end{aligned}
$$

where in inequality (1), we have used the relations $a_{12} \geq b_{12}$ and $a_{13} \geq 3b_{13}/2$ because a_{12} and b_{12} are for jobs in H_2 while a_{13} and b_{13} are for jobs in H_3. In inequality (2), we have used the relation $r_x + a_{11} + a_{12} + a_{13} \leq \tau_{opt}$ by Lemma 6.

As explained above, the set of jobs for the value a_{11} is not empty. Thus, $a_{11} \geq r_x$ because the job J_x and the jobs for the value a_{11} are in the sequence $S_{h,1}$ that is sorted in non-decreasing order by R-time. Thus,

$$\tau^* \leq \frac{2}{3}(\tau_{opt} - r_x) + \frac{1}{3}r_x + t_x + b_{11} + \frac{1}{3}b_{12} + t_c \leq \frac{2}{3}\tau_{opt} + t_x + b_{11} + \frac{1}{3}b_{12} + t_c$$
$$= \frac{2}{3}\tau_{opt} + \frac{1}{3}(t_x + b_{11} + b_{12}) + \frac{2}{3}(t_x + b_{11}) + t_c$$
$$\leq \frac{2}{3}\tau_{opt} + \frac{1}{3} \cdot \frac{4}{3}\tau_{opt} + \frac{2}{3}\tau_{opt} + \frac{2}{5}\tau_{opt} = \frac{98}{45}\tau_{opt}, \tag{3}$$

where in inequality (3), we have used the following relations: (i) by Lemma 3(1), $t_x + b_{11} + b_{12} \leq 4\tau_{opt}/3$, because the job J_x and the jobs for b_{11} and b_{12} are in the set $H_1 \cup H_2$; (ii) by Lemma 3(2), $t_x + b_{11} \leq \tau_{opt}$, because $H_2 \neq \emptyset$ so the set of jobs for b_{12} is not empty; and (iii) by Lemma 6, $t_c \leq 2\tau_{opt}/5$.

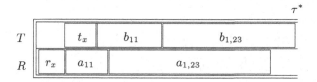

Fig. 3. The configuration of the flowshop M_h for **Subcase 2.3**.

Now consider the case where $H_2 = \emptyset$. Again we can assume that the set of jobs for a_{11} and b_{11} is not empty (to distinguish from **Subcase 2.2**). We have

$$\tau^* = r_x + t_x + b_{11} + b_{1,23} = r_x + t_x + b_{11} + b_{13} + t_c$$
$$= r_x + t_x + b_{11} + \frac{1}{4}b_{13} + \frac{3}{4}b_{13} + t_c \leq r_x + t_x + b_{11} + \frac{1}{4}b_{13} + \frac{1}{2}a_{13} + t_c \tag{4}$$
$$= \frac{1}{2}(r_x + a_{13}) + \frac{1}{2}r_x + t_x + b_{11} + \frac{1}{4}b_{13} + t_c$$
$$\leq \frac{1}{2}(\tau_{opt} - r_x) + \frac{1}{2}r_x + t_x + b_{11} + \frac{1}{4}b_{13} + t_c \tag{5}$$
$$= \frac{1}{2}\tau_{opt} + t_x + b_{11} + \frac{1}{4}b_{13} + t_c \leq \frac{11}{6}\tau_{opt} + \frac{1}{4}b_{13} + t_c, \tag{6}$$

where in inequality (4), we have used the relation $a_{13} \geq 3b_{13}/2$ because the jobs for a_{13} and b_{13} are in the set H_3. For inequality (5), observe that by Lemma 6, $r_x + a_{11} + a_{13} \leq \tau_{opt}$ so $r_x + a_{13} \leq \tau_{opt} - a_{11}$. Moreover, since the set of jobs for a_{11} is not empty, we have $a_{11} \geq r_x$. Combining all these gives $r_x + a_{13} \leq \tau_{opt} - r_x$. Inequality (6) has used the relation $t_x + b_{11} \leq 4\tau_{opt}/3$ by Lemma 3(1).

If the set of jobs for a_{13} and b_{13} has at most one job, which is in the set H_3, then by Lemma 6, $b_{13} \leq 2\tau_{opt}/5$. By the same lemma, since the job J_c is in the set H_3, $t_c \leq 2\tau_{opt}/5$. Thus, from (6) we have

$$\tau^* \leq \frac{11}{6}\tau_{opt} + \frac{1}{4} \cdot \frac{2}{5}\tau_{opt} + \frac{2}{5}\tau_{opt} = \frac{7}{3}\tau_{opt}.$$

If the set of jobs for a_{13} and b_{13} has at least two jobs, then by Lemma 6, $a_{13} \leq \tau_{opt}$. Since the job J_c is the last job assigned on M_h thus has the smallest R-time over all jobs in H_3, and there are at least two jobs for a_{13}, we have $r_c \leq a_{13}/2$. Thus, $a_{13}+r_c \leq 3a_{13}/2 \leq 3\tau_{opt}/2$. Since the job J_c and all jobs for a_{13} and b_{13} are in H_3, we have $b_{13}+t_c \leq 2(a_{13}+r_c)/3 \leq \tau_{opt}$, and $t_c \leq 2r_c/3 \leq \tau_{opt}/3$. Bringing all these in (6) gives

$$\tau^* \leq \frac{11}{6}\tau_{opt} + \frac{1}{4}(b_{13} + t_c) + \frac{3}{4}t_c \leq \frac{11}{6}\tau_{opt} + \frac{1}{4}\tau_{opt} + \frac{3}{4} \cdot \frac{\tau_{opt}}{3} = \frac{7}{3}\tau_{opt}.$$

This completes the proof that in **Subcase 2.3**, $\tau^* \leq 7\tau_{opt}/3$.   □

Finally, we consider **Subcase 2.4**, in which $J_x \in H_1$ is neither the first nor the last job in the sequence $S_{h,1}$. The related proofs will be given in a complete version.

Lemma 10. *In* **Subcase 2.4,** *we have* $\tau^* \leq 98\tau_{opt}/45$.

Based on Lemmas 7–10, we derive directly

Theorem 4. *In* **Case 2** *where the set H_3 is not empty, the makespan τ^* of the schedule S constructed by the algorithm* **Approx** *is bounded by $7\tau_{opt}/3$.*

5   Conclusion and Final Remarks

We proposed an approximation algorithm for scheduling multiple parallel two-stage flowshops where the number of flowshops is part of the input. We proved that the algorithm runs in time $O(n \log n)$ and has an approximation ratio bounded by 7/3. Although the known PTAS for the problem [6] has a better approximation ratio, it and the algorithms adopting the similar framework with it always have impractical runtime. Just take the first step of the PTAS for example. The PTAS needs to enumerate all distinct configurations of flow-shop before applying mixed integer linear program. The number of all possible distinct configurations reaches $((43/\epsilon)^2 + 2)^{2((43/\epsilon)^2+2)+((43/\epsilon)^2+1)^2 \cdot ((43/\epsilon)^4+1)^2}$, where $\epsilon > 0$ is an error parameter, which prevents the PTAS from practical applications even when $\epsilon = 4/3$ (the reason considering the setting $\epsilon = 4/3$ is that $1 + 4/3$ is our approximation ratio). Therefore, one of our contributions is using the traditional algorithmic techniques such as sorting and searching to ensure the feasibility of the algorithm in practice. Another contribution of the paper is showing how the classical Johnson's Rule can be used in the design and analysis of approximation algorithms for scheduling multiple parallel two-stage flowshops.

A direct future study is to design algorithms with better approximation ratio while keeping the feasibility in practice. Since optimal schedules for a single two-stage flowshop can be constructed using Johnson's Rule, the most difficult part of the problem of scheduling multiple parallel two-stage flowshops is to decide how to assign jobs among flowshops. Techniques studied in the classical MAKESPAN problem for one-stage machines do not seem directly applicable because there both jobs and machines are measured by a single parameter, i.e., the cost of a job and the load of a machine that is the sum of the costs of the jobs assigned to the machine, while in scheduling multiple parallel two-stage flowshops, both jobs and flowshop load are measured by at least two parameters, i.e., the R-time and T-time of a job, and the ψ-value and the ρ-value of a flowshop (see the algorithm **Approx** in Fig. 1). We point out that the difference of the R-time and T-time of a job, and the configuration of a flowshop seem to also have significant impact in the scheduling. The sum of R-time and T-time of a job does not seem to be a good measure for the cost of the job because their difference can also impact the completion time of a flowshop. Moreover, well-known techniques in the study of one-stage machine scheduling, such as list scheduling and LPT (largest processing time) schedule [10] seem to somehow conflict with Johnson's Rule. For instance, an LPT schedule for one-stage machine scheduling assigns jobs to machines in non-increasing order in job cost, while Johnson's Rule requires assigning two-stage jobs whose T-time is larger than its R-time in non-decreasing order in R-time. In summary, it seems to require development of new techniques for scheduling multiple parallel two-stage flowshops.

References

1. Artiba, A., Tahon, C.: Production planning knowledge-based system for pharmaceutical manufacturing lines. Eur. J. Oper. Res. **61**(1–2), 18–29 (1992)
2. Blazewicz, J., Ecker, K.H., Schmidt, G., Weglarz, J.: Scheduling in Computer and Manufacturing Systems. Springer, Berlin (2012)
3. Chen, J., Huang, M., Guo, Y.: Scheduling multiple two-stage flowshops with a deadline. Theor. Comput. Sci. **921**, 100–111 (2022)
4. Dong, J., et al.: An FPTAS for the parallel two-stage flowshop problem. Theor. Comput. Sci. **657**, 64–72 (2017)
5. Dong, J., et al.: Corrigendum to "An FPTAS for the parallel two-stage flowshop problem". Theor. Comput. Sci. **687**, 93–94 (2017)
6. Dong, J., Jin, R., Luo, T., Tong, W.: A polynomial-time approximation scheme for an arbitrary number of parallel two-stage flow-shops. Eur. J. Oper. Res. **218**(1), 16–24 (2020)
7. Garey, M.R., Johnson, D.S.: Computers and Intractability: A Guide to the Theory of NP-completeness. W.H. Freeman and Company, New York (1979)
8. Graham, R.L.: Bounds for certain multiprocessing anomalies. Bell Labs Tech. J. **45**(9), 1563–1581 (1966)
9. Graham, R.L.: Bounds on multiprocessing timing anomalies. SIAM J. Appl. Math. **17**(2), 416–429 (1969)

10. Graham, R.L., Lawler, E.L., Lenstra, J.K., Kan, A.R.: Optimization and approximation in deterministic sequencing and scheduling: a survey. Ann. Discret. Math. **5**, 287–326 (1979)

11. He, D.W., Kusiak, A., Artiba, A.: A scheduling problem in glass manufacturing. IIE Trans. **28**(2), 129–139 (1996)

12. Hochbaum, D.S., Shmoys, D.B.: Using dual approximation algorithms for scheduling problems: theoretical and practical results. J. ACM **34**(1), 144–162 (1987)

13. Johnson, S.M.: Optimal two- and three-stage production schedules with setup times included. Nav. Res. Logist. Q. **1**(1), 61–68 (1954)

14. Kovalyov, M.Y.: Efficient epsilon-approximation algorithm for minimizing the makespan in a parallel two-stage system. Vesti Academii navuk Belaruskai SSR, Ser. Phiz.-Mat. Navuk **3**, 119 (1985). (in Russian)

15. Ruiz, R., Maroto, C.: A comprehensive review and evaluation of permutation flowshop heuristics. Eur. J. Oper. Res. **165**(2), 479–494 (2005)

16. Schuurman, P., Woeginger, G.J.: A polynomial time approximation scheme for the two-stage multiprocessor flow shop problem. Theor. Comput. Sci. **237**, 105–122 (2000)

17. Tong, W., Xu, Y., Zhang, H.: A polynomial-time approximation scheme for parallel two-stage flowshops under makespan constraint. Theor. Comput. Sci. **922**, 438–446 (2022)

18. Vairaktarakis, G., Elhafsi, M.: The use of flowlines to simplify routing complexity in two-stage flowshops. IIE Trans. **32**(8), 687–699 (2000)

19. Wu, G., Chen, J., Wang, J.: Scheduling two-stage jobs on multiple flowshops. Theor. Comput. Sci. **776**, 117–124 (2019)

20. Wu, G., Chen, J., Wang, J.: On scheduling inclined jobs on multiple two-stage flowshops. Theor. Comput. Sci. **786**, 67–77 (2019)

21. Wu, G., Chen, J., Wang, J.: On scheduling multiple two-stage flowshops. Theor. Comput. Sci. **818**, 74–82 (2020)

22. Zhang, X., van de Velde, S.: Approximation algorithms for the parallel flow shop problem. Eur. J. Oper. Res. **216**(3), 544–552 (2012)

23. Zhang, Y., Zhou, Y.: TransOS: a transparent computing-based operating system for the cloud. Int. J. Cloud Comput. **4**(1), 287–301 (2012)

The Fair k-Center with Outliers Problem: FPT and Polynomial Approximations

Xiaoliang Wu[1,2], Qilong Feng[1,2(✉)], Jinhui Xu[3], and Jianxin Wang[2,4]

[1] School of Computer Science and Engineering, Central South University,
Changsha 410083, China
wuxiaoliang@csu.edu.cn, csufeng@mail.csu.edu.cn
[2] Xiangjiang Laboratory, Changsha 410205, China
[3] Department of Computer Science and Engineering, State University of New York
at Buffalo, New York 14260-1660, USA
jinhui@cse.buffalo.edu
[4] Hunan Provincial Key Lab on Bioinformatics, Central South University,
Changsha 410083, China
jxwang@mail.csu.edu.cn

Abstract. The fair k-center and k-center with outliers problems are two important variants of the k-center problem in computer science, which have attracted lots of attention. The previous best results for the above two problems are a 3-approximation (ICML 2020) and a 2-approximation (ICALP 2016), respectively. In this paper, we consider a common generalization of the two mentioned variants of the k-center problem, denoted as the fair k-center with outliers (FkCO) problem. For the FkCO problem, we are given a set X of points in a metric space and parameters k and z, where the points in X are divided into several groups, and each point is assigned a color to denote which group it belongs to. The goal is to find a subset $C \subseteq X$ of k centers and a set Z of at most z outliers such that C satisfies fairness constraints, and the objective $\max_{x \in X \setminus Z} \min_{c \in C} d(x, c)$ is minimized. In this paper, we study the Fixed-Parameter Tractability (FPT) approximation algorithm and polynomial approximation algorithm for the FkCO problem. Our main contributions are: (1) we propose a $(1 + \epsilon)$-approximation algorithm in FPT time for the FkCO problem in a low-dimensional doubling metric space; (2) we achieve a polynomial 3-approximation algorithm for the FkCO problem with the reasonable assumptions that all optimal clusters are well separated and have size greater than z.

Keywords: k-center · fair k-center · approximation algorithm

This work was supported by National Natural Science Foundation of China (62172446), Open Project of Xiangjiang Laboratory (22XJ02002), and Central South University Research Programme of Advanced Interdisciplinary Studies (2023QYJC023).

© Springer Nature Switzerland AG 2023
M. Li et al. (Eds.): IJTCS-FAW 2023, LNCS 13933, pp. 225–238, 2023.
https://doi.org/10.1007/978-3-031-39344-0_17

1   Introduction

Clustering is one of the most popular problems in machine learning, and has lots of applications in data mining, image classification, etc. Given a set of points, the goal of clustering is to partition the point set into several disjoint clusters such that the points in the same cluster are close to each other, and the points in different clusters are far away from each other. Several classic clustering models have been extensively studied, such as k-center, k-median, and k-means. In this paper, we focus on the k-center problem that is known to be NP-hard [22], and admits a 2-approximation algorithm [23,28]. Many variations of the k-center problem have been studied in the literature, including fair k-center [5,16,29,33], k-center with outliers [10,12,19,36], capacitated k-center [4,17], fault-tolerant k-center [13,20,31], lower-bounded k-center [1,3], etc.

The fair k-center problem has received lots of attention recently. For the fair k-center problem, we are given a set X of n points in a metric space, where X comprises m disjoint groups $X_1, \ldots, X_m$, and the points in X_h are colored with color $h \in \{1, \ldots, m\}$. We are also given a parameter k and a vector $\gamma = (k_1, \ldots, k_m)$ with $\sum_{h=1}^{m} k_h = k$. The goal is to find a subset $C \subseteq X$ of k centers such that $|C \cap X_h| = k_h$ for any $h \in \{1, \ldots, m\}$, and the objective $\max_{x \in X} \min_{c \in C} d(x, c)$ is minimized. Chen *et al.* [14] studied the matroid center problem, which generalizes the fair k-center problem, and proposed a 3-approximation with running time $\Omega(n^2 \log n)$. For the fair k-center problem, there was a $(3 \cdot 2^{m-1} - 1)$-approximation with running time $O(nkm^2 + km^4)$ based on a swap technique [33]. Jones, Lê Nguyên, and Nguyen [29] improved the time complexity to $O(nk)$, and maintained the approximation factor 3 using the maximum matching method. Chiplunkar, Kale and Ramamoorthy [16] considered the distributed algorithm for the fair k-center problem under the massively parallel computation model. Recently, Angelidakis *et al.* [5] considered the lower-bounded version of the fair k-center problem, and proposed a 15-approximation algorithm that runs in time $O(nk^2 + k^5)$. Additionally, there are lots of works on other definitions of fairness, including group fairness [2,7,8,26], proportional fairness [15,34,37], individual fairness [35,38], etc.

For the k-center with outliers problem, we are given a set X of n points in a metric space and parameters k and z. The goal is to find a subset $C \subseteq X$ of size k and a set Z of at most z outliers such that the objective $\max_{x \in X \setminus Z} \min_{c \in C} d(x, c)$ is minimized. Charikar *et al.* [12] first considered the k-center with outliers problem, and proposed a 3-approximation using a greedy algorithm. Recently, the approximation ratio was improved to 2 [10,27] using the linear programming method. In [19], a bi-criteria approximation algorithm was given, which achieves a 2-approximation by admitting $(1 + \epsilon)z$ outliers to be removed, where ϵ is a parameter used to control the number of outliers. Malkomes *et al.* [36] considered the distributed algorithm for the k-center with outliers, and proposed a 4-approximation algorithm. The approximation ratio was improved to $(2+\epsilon)$ [9].

In this paper, we consider a common generalization of the two mentioned variants of the k-center problem, denoted as the fair k-center with outliers (FkCO) problem. For the FkCO problem, we are given a set X of n points in a metric space, where X comprises m disjoint groups $X_1, \ldots, X_m$, and the points in X_h are colored with color $h \in \{1, \ldots, m\}$. We are also given parameters k and z, and a vector $\gamma = (k_1, \ldots, k_m)$ with $\sum_{h=1}^{m} k_h = k$. The goal is to find a subset $C \subseteq X$ of k centers and a set Z of at most z outliers such that $|C \cap X_h| = k_h$ for any $h \in \{1, \ldots, m\}$, and the objective $\max_{x \in X \setminus Z} \min_{c \in C} d(x, c)$ is minimized.

There exist some obstacles to obtain FPT approximation algorithm and polynomial approximation algorithm for the FkCO problem.

- The coreset construction and sampling are commonly used techniques in designing the FPT approximation algorithms for the clustering problems. However, for the FkCO problem, the coreset technique fails because there are no results of coreset working for the problem, and the sampling-based algorithms rely heavily on the properties of Euclidean space resulting in the hardness of extending the technique to metric space. Goyal and Jaiswal [24] considered the FPT approximation algorithm for the k-center with outliers problem under fairness constraints. However, the definition of fairness in [24] is different to ours. Therefore, the method in [24] is hard to work for the FkCO problem.

- For the k-center with outliers and fair k-center problems, the best known results have approximation ratios 2 [11] and 3 [29], respectively. One natural idea of solving the FkCO problem is firstly to apply the method in [11], and then adjust the obtained result to satisfy the fairness constraints. However, the major obstacle of the above method is that the result returned by the algorithm in [11] is hard to be converted a solution satisfying fairness constraints, since a feasible solution may not exist due to the process of removing outliers without considering the colors of points. Similarly, another natural idea is firstly to use the algorithm in [29] to obtain a set of k centers satisfying fairness constraints, and then remove outliers based on the above k centers. However, how to compute the loss in approximation guarantee remains troublesome. Therefore, none of the algorithms for the k-center with outliers and fair k-center problems works for the FkCO problem.

- In [39], a 3-approximation algorithm with polynomial time was presented for the FkCO problem. The approximation algorithm starts with a set of candidate points without outliers by using the algorithm in [12]. In fact, some points that are not outliers may be removed in the above process of obtaining candidate points, resulting in the hardness on the finding of feasible solution. Han et al. [25] gave a 4-approximation algorithm in polynomial time for the individual fair k-center with outliers problem. However, the method in [25] is not workable for the FkCO problem due to the different definitions of fairness.

1.1   Our Results and Techniques

In this paper, we obtain the following results for the FkCO problem.

Theorem 1. *Given an instance $\mathcal{I} = (X, d, k, z, \mathcal{G}, H, \gamma)$ of the Fk CO problem and a parameter ϵ, assume that D is the doubling dimension of X. Then, there exists an algorithm that obtains a $(1+\epsilon)$-approximate solution of $\mathcal{I}$ in FPT time. i.e., in $f(k, z, \epsilon) \cdot n^{O(1)}$ time.*

We now give the general idea of our algorithm. Given an instance $\mathcal{I} = (X, d, k, z, \mathcal{G}, H, \gamma)$ of the FkCO problem and a parameter ϵ, our algorithm starts with the reduction of the number of points in X. We make use of the method given in [9] to obtain a set $T \subseteq X$ with $O((k + z) \cdot (16/\epsilon)^D)$ points, where D is the doubling dimension of C. The set T has a property that for any point in X, the distance between the point and T is at most $\epsilon\tau_o$, where τ_o is the cost of optimal solution of $\mathcal{I}$. Note that many real world datesets often have lower intrinsic dimensions [6]. As in [19], we assume that the set X has a low doubling dimension D. Then, based on T, we prove theoretically that there must exist a feasible solution $(\hat{C}, Z)$ of $\mathcal{I}$, where $\hat{C} \subseteq X$ satisfies fairness constraints, and Z is a subset of X with at most z outliers, respectively. Moreover, we prove that for any point in $X \backslash Z$, the distance between the point and $\hat{C}$ is at most $(1+\epsilon)\tau_o$. To find such a solution $(\hat{C}, Z)$, we consider each subset C of T with size k, use the matching method to convert C into $\hat{C}$, and output the set satisfying fairness constraints with minimal clustering cost.

We also consider a polynomial approximation algorithm for the FkCO problem under some practical assumptions. In practice, compared with the outliers, the number of points in each optimal cluster is usually not too small, i.e., it is a rare to have an optimal cluster with size less than z. Moreover, we also consider another practical assumption that the optimal clusters are well separated. This property has been studied in practical applications for other clustering problems [18,30]. Based on the above assumptions, we have the following result.

Theorem 2. *Given an instance $\mathcal{I} = (X, d, k, z, \mathcal{G}, H, \gamma)$ of the Fk CO problem, assume that each optimal cluster of $\mathcal{I}$ has size greater than z, and the distance between any two optimal centers is greater than $4\tau_o$, where τ_o is the cost of optimal solution of $\mathcal{I}$. Then, there exists an algorithm that achieves a 3-approximate solution of $\mathcal{I}$ in polynomial time.*

We now give the general idea of our algorithm. Given an instance $\mathcal{I} = (X, d, k, z, \mathcal{G}, H, \gamma)$ of the FkCO problem, assume that τ_o is the cost of optimal solution of $\mathcal{I}$. Otherwise, we can guess the cost τ_o of optimal solution by considering a binary search over all the possible distances between points, resulting in at most a factor $O(n^2)$ of the running time, where n is the number of points in X. The above method has been widely used in literature [16,26,32] to solve the k-center problem and related problems. Our main idea is to select a point from each optimal cluster by a greedy strategy, which ensures the existence of a feasible solution. To find such a feasible solution, we use a matching method to get k centers satisfying fairness constraints based on the obtained centers.

2   Preliminaries

In this section, we give some formal definitions of related problems. Given a set X of points in a metric space $(\mathcal{X}, d)$, for a point $x \in X$ and a subset $C \subseteq X$, let $d(x, C) = \min_{c \in C} d(x, c)$ denote the minimum distance of x to any point in C. For any positive integer m, let $[m] = \{1, \ldots, m\}$. For any nonempty subset $C \subseteq X$ of centers and any $c \in C$, a ball $B(c, r)$ is the set of points that are within a distance r from c, i.e., $B(c, r) = \{x \in X \mid d(c, x) \leq r\}$.

Definition 1 (the k-center problem). *Given a set X of points in a metric space $(\mathcal{X}, d)$ and an integer k, the goal is to find a subset $C \subseteq X$ of k centers such that the cost $\max_{x \in X} d(x, C)$ is minimized.*

Definition 2 (the k-center with outliers problem). *Given a set X of points in a metric space $(\mathcal{X}, d)$ and two integers k, z, the goal is to find a subset $C \subseteq X$ of k centers and a set Z of at most z outliers such that the cost $\max_{x \in X \setminus Z} d(x, C)$ is minimized.*

Given an instance (X, d, k, z) of the k-center with outliers problem, a pair (C, Z) is called a feasible solution of the instance if C is a subset of X with k centers, and Z is a set with at most z outliers.

Definition 3 (the FkCO problem). *Given a set X of points in a metric space $(\mathcal{X}, d)$, two integers k, z, a set of colors $H = \{1, \ldots, m\}$, m disjoint groups $\mathcal{G} = \{X_1, \ldots, X_m\}$ with $\cup_{h=1}^{m} X_h = X$, and a vector $\gamma = \{k_1, \ldots, k_m\}$ with $\sum_{h=1}^{m} k_h = k$, where the points in X_h are colored with color $h \in H$, the goal is to find a subset $C \subseteq X$ of k centers and a set Z of at most z outliers such that the cost $\max_{x \in X \setminus Z} d(x, C)$ is minimized, and C satisfies the following fairness constraints.*

$$|C \cap X_h| = k_h, \forall h \in H \tag{1}$$

Given an instance $(X, d, k, z, \mathcal{G}, H, \gamma)$ of the FkCO problem, a pair (C, Z) is called a feasible solution of the instance if $C \subseteq X$ satisfies constraint (1), and Z is a set with at most z outliers. Moreover, we use $\text{cost}(C, Z) = \max_{x \in X \setminus Z} d(x, C)$ to denote the cost of (C, Z).

Definition 4 (doubling dimension). *Given a set X of points in a metric space $(\mathcal{X}, d)$, the doubling dimension of X is the smallest number D such that for any radius r and a point $x \in X$, all points in the ball $B(x, r)$ are always covered by the union of at most 2^D balls with radius $r/2$.*

Since many real world datesets often have lower intrinsic dimensions [6], in this paper, we assume that the set X has a low doubling dimension D as in [19]. Throughout this paper, let $|X| = n$. Given an instance $\mathcal{I}$ of the FkCO problem, let (C^*, Z^*) denote an optimal solution of $\mathcal{I}$, where $C^* = \{c_1^*, \ldots, c_k^*\}$ is the set of k optimal centers, and Z^* is the set of z optimal outliers, respectively. Thus, for any $h \in H$, we have $|C^* \cap X_h| = k_h$. Let $O^* = \{O_1^*, \ldots, O_k^*\}$ be the set of k optimal clusters. Thus, we have $O^* = \bigcup_{i=1}^{k} O_i^* = X \setminus Z^*$.

Algorithm 1. An FPT approximation algorithm for the FkCO problem

---

Input: An instance $\mathcal{I} = (X, d, k, z, \mathcal{G}, H, \gamma)$ of the FkCO problem and parameters ϵ, τ_o
Output: A feasible solution of $\mathcal{I}$
1: $T \leftarrow$ reduce the number of points in X using method in [9];
2: $C_{min} \leftarrow \emptyset, Z_{min} \leftarrow \emptyset$;
3: **for** each subset $C = \{c_1, \ldots, c_k\} \subseteq T$ with size k **do**
4:    $B \leftarrow \emptyset$;
5:    **for** $i = 1$ to k **do**
6:       $B_i \leftarrow \{x \in X \mid d(x, c_i) \leq (1 + \epsilon)\tau_o\}$;
7:       $B \leftarrow B \cup \{B_i\}$;
8:    **end for**
9:    $(\hat{C}, Z) \leftarrow$ GET-SOLUTION$(X, C, d, k, \mathcal{G}, H, \gamma, \tau_o, \epsilon, B)$;
10:    **if** $\hat{C} = \emptyset$ or $|Z| > z$ **then**
11:       **continue**;
12:    **end if**
13:    **if** cost$(\hat{C}, Z) <$ cost(C_{min}, Z_{min}) **then**
14:       $C_{min} \leftarrow \hat{C}, Z_{min} \leftarrow Z$;
15:    **end if**
16: **end for**
17: **return** (C_{min}, Z_{min}).

---

3 An FPT Approximation Algorithm for the FkCO Problem in Doubling Metric Space

In this section, we present a $(1 + \epsilon)$-approximation algorithm for the FkCO problem with FPT time in a low-dimensional doubling metric space. The running time of the algorithm is $f(k, z, \epsilon) \cdot n^{O(1)}$, where $f(k, z, \epsilon) = ((k + z) \cdot \epsilon^{-1})^{O(k)}$. We now give the general idea of solving the FkCO problem. For a given instance $\mathcal{I} = (X, d, k, z, \mathcal{G}, H, \gamma)$ of the FkCO problem and parameters ϵ, τ_o, our algorithm starts with the reduction of the number of points in X by using the method in [9] (see Subsect. 3.1). Then, based on the reduced points, we prove theoretically the existence of a $(1 + \epsilon)$-approximate solution of $\mathcal{I}$. To find such a solution, we consider all subsets with size k of the above reduced points. For each subset, we try to convert it to one satisfying fairness constraints (see Subsect. 3.2). Finally, we output the set satisfying fairness constraints with minimum clustering cost. The specific process is given in Algorithm 1.

3.1 Reduce the Number of Points

In this section, we show how to reduce the number of points. For a given instance $\mathcal{I} = (X, d, k, z, \mathcal{G}, H, \gamma)$ of the FkCO problem and a parameter ϵ, we apply the method proposed in [9] to obtain a set $T \subseteq X$ with size $O((k + z) \cdot (\frac{24}{\epsilon})^D)$, where ϵ is a parameter used to control the size of T, and D is the doubling dimension of X, respectively. As in [19], we assume that D is a low doubling dimension of X. Moreover, by Lemma 4 and Lemma 6 from [9], we have the following theorem.

Theorem 3 ([9]). *Given an instance $\mathcal{I} = (X, d, k, z)$ of the k-center with outliers problem and a parameter ϵ, assume that D is the doubling dimension of X. Then, we can obtain a subset $T \subseteq X$ with size $O((k + z) \cdot (\frac{24}{\epsilon})^D)$ in polynomial time such that for any $x \in X$, $d(x, T) \le \epsilon\tau$, where τ is the cost of optimal solution of $\mathcal{I}$.*

Note that for a given instance $(X, d, k, z, \mathcal{G}, H, \gamma)$ of the FkCO problem and a parameter ϵ, Theorem 3 still holds due to the fact $\tau \le \tau_o$, where τ_o is the optimal cost of the FkCO problem instance. Therefore, for any $x \in X$, we have $d(x, T) \le \epsilon\tau_o$.

Lemma 1. *Given an instance $\mathcal{I} = (X, d, k, z, \mathcal{G}, H, \gamma)$ of the FkCO problem and parameters ϵ, assume that Z^* is the set of optimal z outliers of $\mathcal{I}$, and τ_o is the cost of optimal solution of $\mathcal{I}$, respectively. Let T be the set of points returned by step 1 in Algorithm 1. Then, there must exist a subset $C \subseteq T$ with size k such that for any $x \in X \backslash Z^*$, $d(x, C) \le (1 + \epsilon)\tau_o$.*

Proof. Let (C^*, Z^*) be an optimal solution of $\mathcal{I}$, where $C^* = \{c_1^*, \ldots, c_k^*\}$ is the set of k optimal centers, and Z^* is the set of z optimal outliers, respectively. Let $O^* = \{O_1^*, \ldots, O_k^*\}$ be the set of k optimal clusters. Thus, we have $O^* = \cup_{i=1}^k O_i^* = X \backslash Z^*$. Assume that τ_o is the cost of optimal solution of $\mathcal{I}$. Let T be the set of points returned by step 1 in Algorithm 1. For any $i \in [k]$, let $\pi(c_i^*) = \arg\min_{x \in T} d(x, c_i^*)$ denote the closest point in T to c_i^*. By Theorem 3, we have that there exists a point in T with distance at most $\epsilon\tau_o$ to c_i^*. Thus, we have $d(c_i^*, \pi(c_i^*)) \le \epsilon\tau_o$, because $\pi(c_i^*)$ is the closest point in T to c_i^*. Then, for any $x \in O_i^*$ ($i \in [k]$), by the triangle inequality, we have

$$d(x, \pi(c_i^*)) \le d(x, c_i^*) + d(c_i^*, \pi(c_i^*)) \le \tau_o + \epsilon\tau_o \le (1 + \epsilon)\tau_o.$$

Thus, there exists a subset $C = \{\pi(c_1^*), \ldots, \pi(c_k^*)\} \subseteq T$ such that for any $x \in \cup_{i=1}^k O_i^* = X \backslash Z^*$, $d(x, C) \le (1 + \epsilon)\tau_o$. □

Lemma 1 implies that there must exist a subset C of T that induces a $(1 + \epsilon)$-approximation without satisfying fairness constraints by removing the points in X with distance greater than $(1 + \epsilon)\tau_o$ from the centers in C. More formally, we call C a set of ϵ-optimal clustering centers of $\mathcal{I}$. In fact, there are $|T|^k$ subsets with size k of T, and one of them induces a $(1 + \epsilon)$-approximation without satisfying fairness constraints of $\mathcal{I}$. To determine which one is the set of ϵ-optimal clustering centers, we need to consider each subset with size k of T, and output the subset with the minimum clustering cost.

3.2   Find a Feasible Solution

Recall that in the previous section, for the FkCO problem, we prove the existence of a $(1 + \epsilon)$-approximate solution without satisfying the fairness constraints. In the section, we first prove that for the FkCO problem, there must exist a $(1 + \epsilon)$-approximate solution (note that the solution satisfies fairness constraints).

Lemma 2. *Given an instance $\mathcal{I} = (X, d, k, z, \mathcal{G}, H, \gamma)$ of the FkCO problem and a parameter ϵ, let T be the set of points returned by step 1 in Algorithm 1, and let $C = \{c_1, \ldots, c_k\} \subseteq T$ be the set of ϵ-optimal clustering centers of $\mathcal{I}$, respectively. Then, there must exist a $(1 + \epsilon)$-approximate solution of $\mathcal{I}$.*

Proof. Let (C^*, Z^*) be an optimal solution of $\mathcal{I}$, where $C^* = \{c_1^*, \ldots, c_k^*\}$ is the set of k optimal centers, and Z^* is the set of z optimal outliers, respectively. Let $O^* = \{O_1^*, \ldots, O_k^*\}$ be the set of k optimal clusters. Thus, we have $O^* = \cup_{i=1}^{k} O_i^* = X \backslash Z^*$. Assume that τ_o is the cost of optimal solution of $\mathcal{I}$. Let T be the set of points returned by step 1 of Algorithm 1.

We now prove the existence of a $(1+\epsilon)$-approximate solution of $\mathcal{I}$. Recall that the set C of ϵ-optimal clustering centers has the property that for each $c_i^* \in C^*$ ($i \in [k]$), there exists a center in C from c_i^* with distance at most $\epsilon\tau_o$. Thus, for each $c_i \in C$ ($i \in [k]$), there must exist a point, denoted by $f(c_i)$, that has the same color as c_i^* (possibly c_i^* itself), with distance at most $\epsilon\tau_o$ from c_i. Let $\hat{C} = \{f(c_1), \ldots, f(c_k)\}$. Then, for any $x \in O_i^*$ ($i \in [k]$), by the triangle inequality and Lemma 1, we have

$$d(x, f(c_i)) \leq d(x, c_i) + d(c_i, f(c_i)) \leq (1 + \epsilon)\tau_o + \epsilon\tau_o \leq (1 + 2\epsilon)\tau_o.$$

Thus, for each $x \in \cup_{i=1}^{k} O_i^* = X \backslash Z^*$, we have $d(x, \hat{C}) \leq (1 + 2\epsilon)\tau_o$. Let $\epsilon' = 2\epsilon$. Therefore, $(\hat{C}, Z^*)$ is a $(1 + \epsilon')$-approximate solution of $\mathcal{I}$.   □

Lemma 2 implies that the existence of a $(1 + \epsilon)$-approximate solution. To find such a $(1 + \epsilon)$-approximate solution, we present an algorithm, called GET-SOLUTION, that gives a feasible solution in polynomial time. Assume that we have found the set of ϵ-optimal clustering centers $C = \{c_1, \ldots, c_k\}$, and τ_o is the cost of optimal solution of $\mathcal{I}$, respectively. For each $i \in [k]$, let $B_i = \{x \in X \mid d(x, c_i) \leq (1 + \epsilon)\tau_o\}$ be the set of points with distance at most $(1 + \epsilon)\tau_o$ to c_i. Let $B = \{B_1, \ldots, B_k\}$. The general idea of GET-SOLUTION is as follows. For a given instance $\mathcal{I} = (X, d, k, z, \mathcal{G}, H, \gamma)$ of the FkCO problem, parameters ϵ, τ_o, a set C of k centers, and a set B of k balls, GET-SOLUTION starts with a bipartite graph G based on C and B. Then, it finds a maximum matching on G to obtain a set $\hat{C}$ of k centers satisfying fairness constraints.

Lemma 3. *Given an instance $\mathcal{I} = (X, d, k, z, \mathcal{G}, H, \gamma)$ of the FkCO problem and parameters ϵ, τ_o, let $C = \{c_1, \ldots, c_k\}$ be the set of ϵ-optimal clustering centers of $\mathcal{I}$, and let $B = \{B_1, \ldots, B_k\}$ be the set of k balls corresponding to C, respectively. Then, GET-SOLUTION gives a $(1 + \epsilon)$-approximate solution of $\mathcal{I}$ in polynomial time.*

Proof. Recall that Lemma 2 shows the existence of a $(1 + \epsilon)$-approximate solution of $\mathcal{I}$. We now prove that algorithm GET-SOLUTION can give such a solution. GET-SOLUTION starts with the construction of a bipartite graph $G = (A_1 \cup A_2, E)$. The left vertex set A_1 contains k vertices in total, where for each center $c_i \in C$ ($i \in [k]$), it contains one vertex u_i. The right vertex set $A_2 = \cup_{h=1}^{m} V_h$, where for each color $h \in H$, V_h contains k_h identical vertices.

Algorithm 2. GET-SOLUTION

Input: An instance $\mathcal{I} = (X, d, k, z, \mathcal{G}, H, \gamma)$ of the FkCO problem, parameters ϵ, τ_o, a
set $C = \{c_1, \ldots, c_k\}$ of k centers, and a set $B = \{B_1, \ldots, B_k\}$ of k balls
Output: A feasible solution of $\mathcal{I}$
1: Let $A_1 = A_2 = \emptyset$, and $E = \emptyset$;
2: **for** $i = 1$ to k **do**
3:     Construct a vertex u_i, and add it to A_1;
4: **end for**
5: **for** $h = 1$ to m **do**
6:     Construct a set V_h of k_h identical vertices, and add the vertices in V_h to A_2;
7: **end for**
8: Let $G = (A_1 \cup A_2, E)$;
9: For any vertices $a \in A_1$ and $b \in A_2$, let (a, b) denote the edge between a and b;
10: **for** $i = 1$ to k **do**
11:     **for** $h = 1$ to m **do**
12:         **if** $\exists x \in X_h$ and $d(c_i, x) \le \epsilon\tau_o$ **then**
13:             **for** each vertex w in V_h **do**
14:                 Add edge (u_i, w) to E;
15:             **end for**
16:         **end if**
17:     **end for**
18: **end for**
19: Find the maximum matching M of G;
20: **if** $M = \emptyset$ **then**
21:     **return** $(\emptyset, X\backslash \cup_{i=1}^k B_i)$.
22: **end if**
23: $\hat{C} \leftarrow \emptyset$;
24: **for** each edge $(a, b) \in M$ **do**
25:     Let p be a point with color h such that $d(c_i, p) \le \epsilon\tau_o$, where a is the corresponding
        vertex of center $c_i \in C$, and b is in V_h, respectively;
26:     $\hat{C} \leftarrow \hat{C} \cup \{p\}$;
27: **end for**
28: **return** $(\hat{C}, X\backslash \cup_{i=1}^k B_i)$.

For each $i \in [k]$ and $h \in H$, if there exists a point $x \in X_h$ with $d(x, c_i) \le \epsilon\tau_o$,
then the corresponding vertex u_i is connected to all vertices in V_h. Let M be
the maximum matching returned by the Ford-Fulkerson algorithm [21] on G.
Then, M immediately induces a set $\hat{C}$ of k centers as follows. For each edge
(a, b) in M, assume that vertex a corresponds to center $c_i \in C$, and vertex b
is in V_h. We add a point $p \in X_h$ to $\hat{C}$. Since $|V_h| = k_h$, $\hat{C}$ contains k_h points
with color h. Therefore, $\hat{C}$ is a set of k centers satisfying fairness constraints. Let
$Z = X\backslash \cup_{i=1}^k B_i$. By Lemma 1, it is easy to get that $|Z| \le z$. Therefore, $(\hat{C}, Z)$
a $(1 + \epsilon)$-approximate solution of $\mathcal{I}$.

The remaining task is to bound the running time of GET-SOLUTION. It is
easy to get that steps 1–9 can be executed in time $O(k) + O(m)$. The running
time of steps 10–18 is $O(km)$. Since Ford-Fulkerson algorithm can be executed
in polynomial time, step 19 has a polynomial running time. Steps 24–27 can be

done in time $O(nk)$ due to the fact $|M| = k$. Therefore, the total running time of GET-SOLUTION is polynomial. □

To complete the proof of Theorem 1, we now analyze the running time of Algorithm 1. By Theorem 3, the running time of obtaining the set T is polynomial. Since $|T| = O((k+z) \cdot (24/\epsilon)^D)$, there are at most $|T|^k = O((k+z) \cdot (\frac{24}{\epsilon})^D)^k$ subsets of size k. Therefore, for constant D, the number of iterations in step 3 of Algorithm 1 can be bounded by

$$O((k+z) \cdot (\frac{24}{\epsilon})^D)^k \leq ((k+z) \cdot \epsilon^{-1})^{O(k)}.$$

For each subset, by Lemma 3, the running time of GET-SOLUTION can be bounded in time polynomial in n, i.e., $n^{O(1)}$ time. Therefore, the total running time of Algorithm 1 can be bounded by $n^{O(1)} \cdot ((k+z) \cdot \epsilon^{-1})^{O(k)}$, which is FPT. By the above discussion, Theorem 1 can be proved.

4   A Polynomial Approximation Algorithm for the FkCO Problem

In this section, we present a polynomial time 3-approximation algorithm (see Algorithm 3) for the FkCO problem under some reasonable assumptions. we assume that the optimal clusters are far away from each other, and have size greater than z. The first assumption implies that the optimal clusters are well separated, and it has been used in literature to solve the k-center with outliers problem [19] and other clustering problems as well [18,30]. The second assumption is practical, because in most case the optimal clusters are not too small compared with the number of outliers.

We now give the general idea of our algorithm, which is inspired by the method presented in [12] for solving the k-center with outliers problem. For a given $\mathcal{I} = (X, d, k, z, \mathcal{G}, H, \gamma)$ of the FkCO problem and a parameter τ_o, our algorithm starts by obtaining a solution (C, Z) (steps 1–7 of Algorithm 3), where C is a set of k centers without satisfying fairness constraints, and Z is a set of at most z outliers, respectively. Specifically, we execute the following process k times to obtain (C, Z). Initially, let $Z = X$. Then, we select a point c from Z such that the ball $B(c, \tau_o)$ contains the most number of points in Z, add the point c to C, and delete the points in $B(c, 2\tau_o)$ from Z. Based on the above assumptions, we can prove that the selected k centers in C fall into each optimal cluster separately. Finally, based on (C, Z), we call algorithm GET-SOLUTION to obtain a feasible solution $(\hat{C}, Z)$, where $\hat{C}$ satisfies fairness constraints. The specific process is given in Algorithm 3.

Let (C^*, Z^*) be an optimal solution, where $C^* = \{c_1^*, \ldots, c_k^*\}$ is the set of k optimal centers. Let $O^* = \{O_1^*, \ldots, O_k^*\}$ be the set of k optimal clusters. Formally, we have the following lemma.

Lemma 4. *Given an instance $\mathcal{I} = (X, d, k, z, \mathcal{G}, H, \gamma)$ of the FkCO problem and a parameter τ_o, assume that each optimal cluster O_i^* $(i \in [k])$ has size greater*

Algorithm 3. A polynomial 3-approximation algorithm for the FkCO problem

Input: An instance $\mathcal{I} = (X, d, k, z, \mathcal{G}, H, \gamma)$ of the FkCO problem and a parameter τ_o
Output: A feasible solution of $\mathcal{I}$

1: $C \leftarrow \emptyset$, $Z \leftarrow X$, $B \leftarrow \emptyset$;
2: **while** $|C| < k$ **do**
3:     $c \leftarrow \arg\max_{c \in Z} |B(c, \tau_o) \cap Z|$;
4:     $C \leftarrow C \cup \{c\}$;
5:     $B \leftarrow B \cup \{B(c, 2\tau_o)\}$;
6:     $Z \leftarrow Z \backslash B(c, 2\tau_o)$;
7: **end while**
8: $(\hat{C}, Z) \leftarrow$ GET-SOLUTION$(X, C, d, k, \mathcal{G}, H, \gamma, \tau_o, 1, B)$;
9: **return** $(\hat{C}, Z)$.

---

than z, and for any $j, j' \in [k]$ $(j \neq j')$, $d(c_j^*, c_{j'}^*) > 4\tau_o$. Then, Algorithm 3 gives a 3-approximate solution of $\mathcal{I}$ in polynomial time.

Proof. Let (C, Z) be the output of Algorithm 3 after steps 1–7 are executed. Let $C = \{c_1, \ldots, c_k\}$. Recall that in each iteration of the while-loop, we select a point $c \in Z$ such that the ball $B(c, \tau_o)$ contains the most number of points in Z, and then delete the points in $B(c, 2\tau_o)$ from Z. We now prove that in each iteration, c falls into one optimal cluster in O^*, and does not fall into the set Z^* of optimal outliers. If c falls into one optimal cluster in O^*, denoted by O_i^* $(i \in [k])$. For any $x \in O_i^*$, by triangle inequality, we have

$$d(x, c) \leq d(x, c_i^*) + d(c_i^*, c) \leq 2\tau_o.$$

Thus, the ball $B(c, 2\tau_o)$ can cover all points in O_i^*. Since for any $j, j' \in [k]$ $(j \neq j')$, $d(c_j^*, c_{j'}^*) > 4\tau_o$, $B(c, 2\tau_o)$ will not cover any point from other optimal cluster in O^* different from O_i^*. Moreover, since in step 3 we select a point c such that $B(c, \tau_o)$ contains the most number of points in Z, and each optimal cluster has size greater than z, c does not fall into the set Z^* of outliers. By the above discussion, we have that $c_1, \ldots, c_k$ fall into the k optimal clusters $O_1^*, \ldots, O_k^*$ separately. Moreover, we have that $\cup_{i=1}^k B(c_i, 2\tau_o)$ covers all points in O^*. Therefore, after steps 1–7 are executed, the number of outliers in Z is at most z, i.e., $|Z| \leq z$.

We now prove that there exists a subset $\hat{C} \subseteq X \backslash Z$ satisfying fairness constraints. Recall that the process of selecting centers to C, we have that $c_1, \ldots, c_k$ fall into each optimal clusters separately. Therefore, for each $c_i^* \in C^*$ $(i \in [k])$, we have $d(c_i, c_i^*) \leq \tau_o$. Moreover, for each $c_i \in C$ $(i \in [k])$, there must exist a point, denoted by $f(c_i)$, that has the same color as c_i^* (possibly c_i^* itself), with distance at most τ_o from c_i. Let $\hat{C} = \{f(c_1), \ldots, f(c_k)\}$. Then, for any $x \in O_i^*$ $(i \in [k])$, by the triangle inequality, we have

$$d(x, f(c_i)) \leq d(x, c_i) + d(c_i, f(c_i)) \leq 2\tau_o + \tau_o \leq 3\tau_o.$$

Thus, for each $x \in \cup_{i=1}^k O_i^* \subseteq X \backslash Z$, we have $d(x, \hat{C}) \leq 3\tau_o$. Therefore, $(\hat{C}, Z)$ is a 3-approximate solution of $\mathcal{I}$. By Lemma 3, we have that GET-

SOLUTION$(X, C, d, k, \mathcal{G}, H, \gamma, \tau_o, 1, B)$ can give such a solution $(\hat{C}, Z)$, where ϵ is set to 1. The proof is similar to Lemma 3.

The remaining task is to prove that the running time of Algorithm 3 is polynomial. It is easy to get that step 1 and steps 4–6 can be executed in time $O(1)$. Step 3 can be executed in time $O(n^2)$, because for each point in the current set Z, we compute all the points in Z within distance τ_o from the point. Therefore, the running time of steps 1–7 can be bounded by $O(n^2 k)$. By Lemma 3, GET-SOLUTION can be executed in polynomial time. Therefore, the total running time of Algorithm 3 is polynomial. □

Theorem 2 follows from Lemma 4.

5 Conclusions

In this paper, we consider the FkCO problem, which generalizes the well-studied fair k-center and k-center with outliers problems. We present several algorithms to solve the FkCO problem from the perspective of FPT approximation and polynomial approximation. An open question is that whether it is possible to design a constant-factor polynomial approximation algorithm for the FkCO problem without any assumptions. We believe that solving this question would be of great interest.

References

1. Aggarwal, G., et al.: Achieving anonymity via clustering. ACM Trans. Algorithms **6**(3), 49:1–49:19 (2010)
2. Ahmadian, S., Epasto, A., Kumar, R., Mahdian, M.: Clustering without over-representation. In: Proceedings of the 25th ACM SIGKDD International Conference on Knowledge Discovery and Data Mining, pp. 267–275 (2019)
3. Ahmadian, S., Swamy, C.: Approximation algorithms for clustering problems with lower bounds and outliers. In: Proceedings of the 43rd International Colloquium on Automata, Languages, and Programming, pp. 69:1–69:15 (2016)
4. An, H., Bhaskara, A., Chekuri, C., Gupta, S., Madan, V., Svensson, O.: Centrality of trees for capacitated k-center. Math. Program. **154**(1–2), 29–53 (2015)
5. Angelidakis, H., Kurpisz, A., Sering, L., Zenklusen, R.: Fair and fast k-center clustering for data summarization. In: Proceedings of the 39th International Conference on Machine Learning, pp. 669–702 (2022)
6. Belkin, M.: Problems of learning on manifolds. The University of Chicago (2004)
7. Bera, S.K., Chakrabarty, D., Flores, N., Negahbani, M.: Fair algorithms for clustering. In: Proceedings of the 33rd International Conference on Neural Information Processing Systems, pp. 4955–4966 (2019)
8. Bercea, I.O., Groß, M., Khuller, S., Kumar, A., Rösner, C., Schmidt, D.R., Schmidt, M.: On the cost of essentially fair clusterings. In: Proceedings of the 22nd International Conference on Approximation Algorithms for Combinatorial Optimization Problems and 23rd International Conference on Randomization and Computation, pp. 18:1–18:22 (2019)

9. Ceccarello, M., Pietracaprina, A., Pucci, G.: Solving k-center clustering (with outliers) in mapreduce and streaming, almost as accurately as sequentially. Proc. VLDB Endowment **12**(7), 766–778 (2019)

10. Chakrabarty, D., Goyal, P., Krishnaswamy, R.: The non-uniform k-center problem. In: Proceedings of the 43rd International Colloquium on Automata, Languages, and Programming, pp. 67:1–67:15 (2016)

11. Chakrabarty, D., Goyal, P., Krishnaswamy, R.: The non-uniform k-center problem. ACM Trans. Algorithms **16**(4), 1–19 (2020)

12. Charikar, M., Khuller, S., Mount, D.M., Narasimhan, G.: Algorithms for facility location problems with outliers. In: Proceedings of the 12th Annual ACM-SIAM Symposium on Discrete Algorithms, pp. 642–651 (2001)

13. Chechik, S., Peleg, D.: The fault-tolerant capacitated k-center problem. Theoret. Comput. Sci. **566**, 12–25 (2015)

14. Chen, D.Z., Li, J., Liang, H., Wang, H.: Matroid and knapsack center problems. Algorithmica **75**(1), 27–52 (2016)

15. Chen, X., Fain, B., Lyu, L., Munagala, K.: Proportionally fair clustering. In: Proceedings of the 36th International Conference on Machine Learning, pp. 1032–1041 (2019)

16. Chiplunkar, A., Kale, S., Ramamoorthy, S.N.: How to solve fair k-center in massive data models. In: Proceedings of the 37th International Conference on Machine Learning, pp. 1877–1886 (2020)

17. Cygan, M., Hajiaghayi, M., Khuller, S.: LP rounding for k-centers with non-uniform hard capacities. In: Proceedings of the 53rd Annual Symposium on Foundations of Computer Science, pp. 273–282 (2012)

18. Daniely, A., Linial, N., Saks, M.: Clustering is difficult only when it does not matter. arXiv preprint arXiv:1205.4891 (2012)

19. Ding, H., Yu, H., Wang, Z.: Greedy strategy works for k-center clustering with outliers and coreset construction. In: Proceedings of the 27th Annual European Symposium on Algorithms, pp. 40:1–40:16 (2019)

20. Fernandes, C.G., de Paula, S.P., Pedrosa, L.L.C.: Improved approximation algorithms for capacitated fault-tolerant k-center. Algorithmica **80**(3), 1041–1072 (2018)

21. Ford, L.R., Fulkerson, D.R.: Maximal flow through a network. Can. J. Math. **8**, 399–404 (1956)

22. Garey, M.R., Johnson, D.S.: Computers and Intractability: A Guide to the Theory of NP-Completeness. WH Freeman (1979)

23. Gonzalez, T.F.: Clustering to minimize the maximum intercluster distance. Theoret. Comput. Sci. **38**, 293–306 (1985)

24. Goyal, D., Jaiswal, R.: Tight FPT approximation for constrained k-center and k-supplier. Theoret. Comput. Sci. **940**, 190–208 (2023)

25. Han, L., Xu, D., Xu, Y., Yang, P.: Approximation algorithms for the individually fair k-center with outliers. J. Glob. Optim. 1–16 (2022)

26. Harb, E., Lam, H.S.: KFC: A scalable approximation algorithm for k-center fair clustering. In: Proceedings of the 34th International Conference on Neural Information Processing Systems, pp. 14509–14519 (2020)

27. Harris, D.G., Pensyl, T., Srinivasan, A., Trinh, K.: A lottery model for center-type problems with outliers. ACM Trans. Algorithms **15**(3), 1–25 (2019)

28. Hochbaum, D.S., Shmoys, D.B.: A best possible heuristic for the k-center problem. Math. Oper. Res. **10**(2), 180–184 (1985)

29. Jones, M., Lê Nguyên, H., Nguyen, T.: Fair k-centers via maximum matching. In: Proceedings of the 37th International Conference on Machine Learning, pp. 4940–4949 (2020)
30. Kanungo, T., Mount, D.M., Netanyahu, N.S., Piatko, C.D., Silverman, R., Wu, A.Y.: An efficient k-means clustering algorithm: analysis and implementation. IEEE Trans. Pattern Anal. Mach. Intell. **24**(7), 881–892 (2002)
31. Khuller, S., Pless, R., Sussmann, Y.J.: Fault tolerant k-center problems. Theoret. Comput. Sci. **242**(1–2), 237–245 (2000)
32. Khuller, S., Sussmann, Y.J.: The capacitated k-center problem. SIAM J. Discret. Math. **13**(3), 403–418 (2000)
33. Kleindessner, M., Awasthi, P., Morgenstern, J.: Fair k-center clustering for data summarization. In: Proceeding of the 36th International Conference on Machine Learning, pp. 3448–3457 (2019)
34. Li, B., Li, L., Sun, A., Wang, C., Wang, Y.: Approximate group fairness for clustering. In: Proceedings of the 38th International Conference on Machine Learning, pp. 6381–6391 (2021)
35. Mahabadi, S., Vakilian, A.: Individual fairness for k-clustering. In: Proceedings of the 37th International Conference on Machine Learning, pp. 6586–6596 (2020)
36. Malkomes, G., Kusner, M.J., Chen, W., Weinberger, K.Q., Moseley, B.: Fast distributed k-center clustering with outliers on massive data. In: Advances in Neural Information Processing Systems, vol. 28 (2015)
37. Micha, E., Shah, N.: Proportionally fair clustering revisited. In: Proceedings of the 47th International Colloquium on Automata, Languages, and Programming, pp. 85:1–85:16 (2020)
38. Negahbani, M., Chakrabarty, D.: Better algorithms for individually fair k-clustering. In: Proceedings of the 35th International Conference on Neural Information Processing Systems, pp. 13340–13351 (2021)
39. Yuan, F., Diao, L., Du, D., Liu, L.: Distributed fair k-center clustering problems with outliers. In: Proceedings of the 22nd International Conference on Parallel and Distributed Computing, Applications and Technologies, pp. 430–440 (2022)

Constrained Graph Searching on Trees

Lusheng Wang[1,3], Boting Yang[2(✉)], and Zhaohui Zhan[1]

[1] Department of Computer Science, City University of Hong Kong, Kowloon Tong,
Hong Kong, Special Administrative Region of China
{lusheng.wang,zhaohui.zhan}@cityu.edu.hk
[2] Department of Computer Science, University of Regina, Regina, SK, Canada
boting.yang@uregina.ca
[3] City University of Hong Kong Shenzhen Research Institution, Shenzhen, China

Abstract. Megiddo et al. (1988) introduced the edge searching problem, which is to find the minimum number of searchers to capture the robber in the edge searching model. Dyer et al. (2008) introduced the fast searching problem that is to find the minimum number of searchers to capture the robber in the fast searching model. In this paper, we consider these two graph searching problems under some constraints. One constraint is that a subset of vertices, called start vertices, are initially occupied by searchers before we place additional searchers on the graph. Another constraint is that some of the searchers must end their search at certain vertices called halt vertices. We focus on trees with n vertices. Let k be the number of times to move searchers from start vertices. For the edge searching problem, we give an $O(kn)$-time algorithm for computing the edge search number of a tree that contains only start vertices or only halt vertices. For a tree that contains both start vertices and halt vertices, we present an $O(n^2)$-time algorithm to compute the edge search number. We show that all these problems are monotonic. For the fast searching problem, we propose a linear-time algorithm for computing the fast search number of a tree that contains only start vertices or only halt vertices. For a tree with n vertices that contains s start vertices and h halt vertices, we give an $O((s+h)n)$-time algorithm to compute the fast search number.

Keywords: edge search number · fast search number · graph searching · tree

1 Introduction

Graph searching was first studied in 1976 by Parsons [14], who introduced a continuous version of the graph searching problem. Megiddo et al. [13] discretized this model. They showed that deciding the edge search number of a graph is NP-hard. They also provided a linear time algorithm for computing the edge search number of trees. The edge search number of a graph is closely related to some other graph parameters. Let G be a graph and let $es(G)$ denote the edge search

ⓒ Springer Nature Switzerland AG 2023
M. Li et al. (Eds.): IJTCS-FAW 2023, LNCS 13933, pp. 239–251, 2023.
https://doi.org/10.1007/978-3-031-39344-0_18

number of G. Ellis et al. [8] showed that $vs(G) \leq es(G) \leq vs(G) + 2$, where $vs(G)$ is the vertex separation number of graph G. Kinnersley [10] showed that $pw(G) = vs(G)$, where $pw(G)$ is the pathwidth of G. Hence $pw(G) \leq es(G) \leq pw(G) + 2$. LaPaugh [11] and Beinstock and Seymour [3] proved that the edge searching problem is monotonic, which implies that the problem of finding the edge search number is in NP, and thus finding the edge search number of a graph is NP-complete.

Dyer et al. [7] introduced the fast searching problem. They considered trees and complete bipartite graphs and presented results on the fast search number of these graphs. Yang [19] proved that the problem of finding the fast search number of a graph is NP-complete. The fast searching problem remains NP-complete for Eulerian graphs. Even the problem of deciding whether the fast search number of G is a half of the number of odd vertices in G is NP-complete; it remains NP-complete for planar graphs with maximum degree 4. Dereniowski et al. [6] showed that the fast searching problem is NP-hard for multigraphs and for graphs. Stanley and Yang [15] gave a linear time algorithm to compute the fast search number of Halin graphs. They also gave a quadratic time algorithm for computing the fast search number of cubic graphs. Note that the problem of finding the edge search number of cubic graphs is NP-complete [12]. Xue et al. [17] presented lower bounds and upper bounds on the fast search number of complete k-partite graphs. They also solved the open problem of determining the fast search number of complete bipartite graphs. Xue and Yang [16] proved an explicit formula for computing the fast search number of the Cartesian product of an Eulerian graph and a path. Very recently, Xue et al. [18] propose a new method for computing the fast search number of k-combinable graphs. Using this method, they gave a linear time algorithm for computing the fast search number of cactus graphs.

Besides the edge searching and the fast searching mentioned above, many other graph searching models have been studied; see [1,2,4,5,9] for surveys.

For the problems that we will investigate in this paper, we have two special subsets of vertices of a given graph: the set of *start vertices*, denoted V_s, and the set of the *halt vertices*, denoted V_h. The start vertices and halt vertices are also simply called *starts* and *halts* respectively. A constrained graph searching with respect to starts and halts satisfies the following three conditions:

1. Before the searching process, every start vertex is initially occupied by exactly one searcher.
2. During the entire searching process, each start vertex remains clear, that is, either it is occupied by a searcher or all its incident edges are clear.
3. Once a halt vertex is occupied by a searcher, it must be occupied by at least one searcher for the remainder of the searching process.

Such constrained graph searching problems are referred to as *Graph Searching with Starts and Halts* (briefly, GSSH). In a GSSH problem, the searchers that occupy start vertices initially are called *starting searchers*. All other searchers that are placed on the graph during the searching process are called *additional searchers*.

Note that our algorithms for GSSH problems can be used as subroutines for searching algorithms that has the divide-and-conquer style. This is one of our major motivations to study GSSH problems. In this kind of algorithms, we first decompose a given graph into two or more subgraphs; we then solve the problem on these subgraphs; and finally we compose the solutions of subgraphs to solve the original problem on the given graph. When we search a subgraph, some vertices may have been occupied by searchers because these searchers stopped on these vertices at the end of the searching process when we clear previous subgraphs. So these vertices can be considered as start vertices of this subgraph and can also be considered as halt vertices of the previous subgraphs. As an application, our algorithms presented in this paper can be used to search the cycle-disjoint graphs [20], where there are many induced subgraphs that are trees.

Another motivation for studying GSSH problems comes from real applications. For example, suppose there is a robber hiding in an area of the city which is modelled by a graph. In this scenario, the starting searchers (or cops) are deployed at street intersections connecting the area containing the robber with the rest of the city. The cops can divided the area into neighborhoods, and then they search for the robber neighborhood by neighborhood. In this case, the cops first search a particular neighborhood, during which if a cop arrives at an intersection that leads to another neighborhood, this cop stays at the intersection until the particular neighborhood is cleared. These intersections can be modelled by the halt vertices in an instance of the GSSH problem. So GSSH problems can be seen as a particular type of subgraph searching.

All the GSSH problems discussed in this paper take place on trees. We will investigate these problems on two graph searching models: the edge searching model and the fast searching model.

2   Preliminaries

Let $G = (V, E)$ be a graph with vertex set V and edge set E. We also use $V(G)$ and $E(G)$ to denote the vertex set and edge set of G respectively. A *leaf* is a vertex that has degree one. A vertex is *odd* (resp. *even*) when its degree is an odd (resp. even) number.

Given a graph G that contains a robber hiding on vertices or along edges, in the edge searching problem introduced by Megiddo et al. [13], a team of searchers want to capture the invisible robber. The robber can move at a great speed at any time from one vertex to another vertex along a searcher-free path between the two vertices. Initially, G contains no searchers. So all edges are *dirty* initially. In order to clear all edges, the searchers have the following three actions:

1. *placing* a searcher on a vertex in the graph,
2. *removing* a searcher from a vertex, and
3. *sliding* a searcher along an edge from one endpoint to the other.

There are two ways to *clear* an edge: (1) at least two searchers are located on one endpoint of the edge, and one of them slides along the edge from this endpoint to the other; (2) a searcher is located on one endpoint of the edge, and all edges incident on this endpoint, except the edge itself, are clear and the searcher slides along the edge from this endpoint to the other.

A *search strategy* for a graph G is a sequence of searchers' actions such that all edges of G are cleared after the last action is carried out. The *edge search number* of G, denoted by $es(G)$, is the minimum number of searchers required to clear G. An edge search strategy is *optimal* if at each step it uses at most $es(G)$ searchers. A search strategy $S = \{a_1, a_2, \ldots, a_m\}$ is called *monotonic* if $A_{i-1} \subseteq A_i$, for all $1 \le i \le m$, where A_i is the set of edges that are clear after the action a_i and $A_0 = \emptyset$. A searching problem is *monotonic* if there exists an optimal search strategy that is monotonic for each instance of the searching problem. When a searcher is removed or slides from a vertex v in a search strategy, if the graph has a path containing v that connects a dirty edge to a clear edge and this path is not occupied by any searchers, then the clear edge becomes *recontaminated* immediately because the robber moves exceedingly fast. Such a strategy is not monotonic.

The edge searching problem has a strong connection with the fast searching problem, which was first introduced by Dyer et al. [7]. The fast searching model has the same setting as the edge searching model except that (1) every edge is traversed exactly once by a searcher and (2) a searcher cannot be removed from a vertex and then placed on another vertex later. The minimum number of searchers required to clear G in the fast searching model is the *fast search number* of G, denoted by $fs(G)$. A *fast search strategy* is a sequence of actions such that the final action leaves all edges of G cleared. Since every edge is traversed exactly once by a searcher, we know that a fast search strategy for a connected graph with m edges contains exactly $fs(G)$ placing actions and m sliding actions. A fast search strategy is called *optimal* if at each step it uses at most $fs(G)$ searchers. Note that all fast search strategies are monotonic because searchers are allowed to slide along each edge exactly once. So the fast searching problem is monotonic.

3   Edge Searching with Starts

In this section we study the edge searching with start vertices, which is formally described as follows.

Edge Searching with Starts (ESS)
Instance: A tree T and a set of start vertices $V_s \subseteq V(T)$, such that each $v \in V_s$ is occupied by exactly one starting searcher.
Question: What is the minimum number of additional searchers for clearing T in the edge searching model such that each vertex from V_s remains clear during the entire search?

We define the *ESS number* of T with respect to V_s, denoted by $ess(T, V_s)$, as the minimum number of additional searchers required to clear T in the ESS model. An **ESS** search strategy is *optimal* if it uses $ess(T, V_s)$ searchers to clear T in the **ESS** model. We say that a vertex v is *clear* if all edges incident with v are clear or v is occupied by a searcher. A searcher on a vertex $v \in V_s$ is called *moveable* if there is only one contaminated edge incident with v and all other edges incident with v are clear.

Given a tree T and a subset of vertices $V' \subseteq V(T)$, define $T \ominus V'$ to be the set of maximal induced subtrees $\{T_1, T_2, \ldots, T_k\}$ of T such that $\bigcup_{1 \leq i \leq k}\{T_i\} = T$ and for each $v \in V' \cap V(T_i)$ where $1 \leq i \leq k$, v is a leaf in T_i.

During the progression of an edge search strategy for a tree T, a searcher can be removed from a vertex without causing recontamination if all its incident edges are clear. A searcher is called *free* when it is not currently occupying a vertex of T.

A formal description of our algorithm EDGESEARCHS for computing $ess(T, V_s)$ is given in Algorithm 1.

Algorithm 1. EDGESEARCHS(T, V_s) (Edge Search with Starts)

Input: A tree T with a nonempty set of start vertices V_s.
Output: $ess(T, V_s, E_c)$.

1: $e_s \leftarrow 0$; $f \leftarrow 0$; $E_c \leftarrow \emptyset$.
2: While there exists a moveable searcher on a vertex $u \in V_s$: let uv be the only dirty edge incident with u; slide this searcher from u to v to clear the edge uv; $V_s \leftarrow (V_s \setminus \{u\}) \cup \{v\}$; $E_c \leftarrow E_c \cup \{uv\}$.
3: While there exists a searcher on a clear vertex $v \in V_s$: remove the searcher from v, $V_s \leftarrow V_s \setminus \{v\}$ and $f \leftarrow f + 1$.
4: Compute $T \ominus V_s$; let $\{T_1, T_2, \ldots, T_k\}$ be the set of the completely dirty trees in $T \ominus V_s$. Find a subtree, say T_1, with the smallest edge search number, that is, $es(T_1) \leq es(T_i)$, for $2 \leq i \leq k$.
5: Clear all edges of T_1 using $es(T_1)$ searchers; $E_c \leftarrow E_c \cup E(T_1)$; $e_s \leftarrow \max\{e_s, es(T_1) - f\}$.
6: If $E_c = E(T)$ then *return* e_s; otherwise, go to Step 2

We now consider the correctness of Algorithm 1. We will first prove that the actions performed in Steps 2 and 3, are part of some optimal **ESS** search strategy. We then prove that $ess(T, V_s)$ is at least as large as the edge search number of the subtree T_1 which is cleared in Step 5. Once we prove the above in Lemmas 1 and 2, the correctness of Algorithm 1 follows. We begin with the following property of the algorithm.

Lemma 1. *For an instance (T, V_s) of the* **ESS** *problem, there is an optimal* **ESS** *strategy for (T, V_s) that contains the actions performed in Steps 2 and 3 of Algorithm 1.*

From Lemma 1, we can prove the following result.

Lemma 2. *For an instance* (T, V_s) *of the* ESS *problem, let* T_1 *be the subtree computed in Step 4 of Algorithm 1. Then*

$$ess(T, V_s) \geq es(T_1) - f,$$

where f *is the number of free searchers removed from* T *in Step 3.*

From Lemma 2, we can show one of our main results.

Theorem 1. *For an instance* (T, V_s) *of the* ESS *problem, the value of* e_s *at the termination of Algorithm 1 is equal to* $ess(T, V_s)$.

Theorem 2. *Let* (T, V_s) *be an instance of the* ESS *problem with* $|V(T)| = n$. *Then* $ess(T, V_s)$ *can be computed in* $O(kn)$ *time, where* k *is the number of times when Step 4 is run in Algorithm 1.*

Now we turn our attention to considering the monotonicity.

Theorem 3. *The* ESS *problem is monotonic.*

4   Fast Searching with Starts

In this section we consider the following fast searching problem with start vertices.

Fast Searching with Starts (FSS)
Instance: A tree T and a set of start vertices $V_s \subseteq V(T)$, such that each $v \in V_s$ is occupied by exactly one starting searcher.
Question: What is the minimum number of additional searchers for clearing T in the fast searching model such that each vertex from V_s remains clear during the entire search?

We define the *FSS number* of T with respect to V_s, denoted $fss(T, V_s)$, as the minimum number of additional searchers required to clear T in the FSS model. An FSS search strategy is *optimal* if it uses $fss(T, V_s)$ searchers to clear T in the FSS model.

A description of our algorithm FASTSEARCHS for computing $fss(T, V_s)$ is given in Algorithm 2.

Now we turn our attention to proving the correctness of our algorithm in the following theorem.

Theorem 4. *Let* (T, V_s) *be an instance of the* FSS *problem which is the input of Algorithm 2. Then the output* f_s *at the termination of the algorithm is equal to* $fss(T, V_s)$.

Next, we show the running time of Algorithm 2.

Theorem 5. *Let* (T, V_s) *be an instance of the* FSS *problem with* $|V(T)| = n$. *Algorithm 2 can be implemented in* $O(n)$ *time.*

Algorithm 2. FASTSEARCHS(T, V_s) (Fast Search with Starts)

Input: A tree T, and a set of start vertices V_s.

Output: $fss(T, V_s)$.

1: Initially $f_s \leftarrow 0$.
2: If V_s contains a vertex u that is a leaf in T such that the component in T that contains u has an odd vertex $v \in V(T) \setminus V_s$, then: update T by deleting all edges on the path between u and v from T and deleting isolated vertices from T; update V_s by deleting the start vertices that are not on the current T; repeat Step 2 until the condition is not satisfied.
3: If there is a component T' in T such that all odd vertices of T' belong to V_s, then: update T by deleting all edges and vertices of T' from T; update V_s by deleting the vertices that are not on the current T; repeat Step 3 until the condition is not satisfied.
4: If $E(T) \neq \emptyset$, then:
  (4.1) arbitrarily select two leaves u and v in the same component of T;
  (4.2) update T by deleting all edges on the path between u and v from T and deleting isolated vertices from T; update V_s by deleting the vertices that are not on the current T; $f_s \leftarrow f_s + 1$; go to Step 2.
5: *Return* f_s.

5   Searching Trees with Halts

In this section we consider the searching problems where we have only *halt vertices*. We formally state our problems as follows.

Edge/Fast Searching with Halts (ESH/FSH)

Instance: A tree T with a set of halt vertices $V_h \subseteq V(T)$.

Question: What is the minimum number of searchers for clearing T in the edge/fast searching model such that once a searcher occupies a vertex $v \in V_h$, v remains occupied by at least one searcher for the remainder of the search?

The *ESH number* (resp. *FSH number*) of T with respect to V_h, denoted by $esh(T, V_h)$ (resp. $fsh(T, V_h)$), is defined as the minimum number of searchers required to clear T under the ESH model (resp. FSH model).

Note that the ESH problem can be considered as an inversion of the ESS problem. In the ESS problem, recall that initially all edges of the graph are dirty, each $v \in V_s$ is occupied by exactly one starting searcher and none of the vertices in $V \setminus V_s$ contains a searcher. Notice that a search strategy is a sequence of searchers' actions that result in every edge in the graph being cleared. Let s be an action of searchers. The *reverse* of s, denote s^{-1}, is defined as follows:

- If s is "sliding a searcher from v to u", then s^{-1} is "sliding a searcher from u to v";
- if s is "removing a searcher from v", then s^{-1} is "placing a searcher on v";
- if s is "placing a searcher on v", then s^{-1} is "removing a searcher from v".

For a search strategy $S = (s_1, s_2, \ldots, s_k)$, the *inverse* of S, denoted by S^{-1}, is defined as $S^{-1} = (s_k^{-1}, \ldots, s_2^{-1}, s_1^{-1})$.

Let S be an ESS strategy satisfying the following condition: during the progression of S, if there is a vertex v such that all edges incident with v are cleared and v is occupied by searcher(s), then the searcher(s) are removed from v in the next action(s). Such an ESS strategy S is called *standard*. It is not hard to see that a standard ESS strategy has the following property.

Lemma 3. *For any instance* (T, V_s) *of the* ESS *problem, there is a monotonic standard* ESS *strategy that clears* T *using* $ess(T, V_s)$ *searchers.*

In the ESH problem, initially all edges of the graph are dirty; during the progression of an ESH strategy, once a searcher occupies a vertex $v \in V_h$, v remains occupied by at least one searcher for the remainder of the search. Suppose that S' is an ESH strategy satisfying the following condition: during the progression of S', (1) if there is a vertex $u \in V_h$ such that all edges incident with u are cleared and u is occupied by at least two searchers, then one searcher stays still on u and the other searcher(s) are removed from u in the next action(s); (2) if there is a vertex $v \in V \setminus V_h$ such that all edges incident with v are cleared and v is occupied by at least one searcher, then all the searchers on v are removed from v in the next actions. Such an ESH strategy S' is called *standard*. Similarly to Lemma 3, we can prove the following property for the ESH problem.

Lemma 4. *For any instance* (T, V_h) *of the* ESH *problem, there is a monotonic standard* ESH *strategy that clears* T *using* $esh(T, V_h)$ *searchers.*

From Lemmas 3 and 4, we can establish the following relations between an ESS strategy and its reverse.

Lemma 5. *Let* (T, V_s) *be an instance of the* ESS *problem and let* (T, V_h) *be an instance of the* ESH *problem such that* $V_s = V_h$. *Then* S *is a monotonic standard* ESS *strategy for* (T, V_s) *if and only if* S^{-1} *is a monotonic standard* ESH *strategy for* (T, V_h). *Furthermore,* $esh(T, V_h) = ess(T, V_s) + |V_h|$.

From Lemma 5, Theorems 1 and 2, we have the following result.

Theorem 6. *Let* (T, V_h) *be an instance of the* ESH *problem with* $|V(T)| = n$. *The search number* $esh(T, V_h)$ *can be computed in* $O(kn)$ *time by modifying Algorithm 1, where* k *is the number of times when Step 4 is run in Algorithm 1.*

The next result follows from Theorem 3 and Lemma 5.

Corollary 1. *The* ESH *problem is monotonic.*

We now consider the FSH problem. We first modify the FSS strategy by adding removing actions after all edges are cleared. An *enhanced* FSS *strategy* is an FSS strategy satisfying the following condition: all additional searchers are placed before the first sliding action and all searchers are removed from the graph after all edges are cleared. An enhanced FSS strategy has the following property.

Lemma 6. *For any instance* (T, V_s) *of the* FSS *problem, there is an enhanced* FSS *strategy that clears* T *using* $fss(T, V_s)$ *searchers.*

In the remainder of this section, we consider the FSH problem, which can be seen as an inversion of the FSS problem. We define an *enhanced* FSH *strategy* to be an FSH strategy satisfying the following condition: all placing actions happen before the first sliding action, and after all edges are cleared, remove searchers from the graph such that no vertex in $V(T) \setminus V_h$ contains a searcher and each vertex in V_h is occupied by exactly one searcher.

Lemma 7. *For any instance (T, V_h) of the FSH problem, there is an enhanced FSH strategy that clears T using $fsh(T, V_h)$ searchers.*

Just like the ESH problem, we can define the reverse actions and the reverse strategy in the same way.

Note that fast search strategies are always monotonic. Similarly to Lemma 5, we can prove the following lemma.

Lemma 8. *Let (T, V_s) be an instance of the FSS problem and let (T, V_h) be an instance of the FSH problem such that $V_s = V_h$. Then S is an enhanced FSS strategy for (T, V_s) if and only if S^{-1} is an enhanced FSH strategy for (T, V_h). Furthermore, $fsh(T, V_h) = fss(T, V_s) + |V_h|$.*

From Lemma 8, we have the following result.

Theorem 7. *Let (T, V_h) be an instance of the FSH problem with $|V(T)| = n$. We can compute $fsh(T, V_h)$ in $O(n)$ time by modifying Algorithm 2.*

6 Edge Searching with Starts and Halts

In this section we study the following edge searching problem.

Edge Searching with Starts and Halts (ESSH)
Instance: A tree T, a nonempty set of start vertices $V_s \subseteq V(T)$, and a nonempty set of halt vertices $V_h \subseteq V(T)$, such that $V_h \cap V_s = \emptyset$, and each $v \in V_s$ is initially occupied by exactly one starting searcher.
Question: What is the minimum number of additional searchers for clearing T in the edge searching model such that once a searcher occupies a vertex $v \in V_h$, v remains occupied by at least one searcher for the remainder of the search, and each vertex from V_s remains clear during the entire search?

The *ESSH number* of T with respect to V_s and V_h, denoted $essh(T, V_s, V_h)$, is defined as the minimum number of additional searchers required to clear T in the ESSH model. An ESSH search strategy is *optimal* if it uses $essh(T, V_s, V_h)$ searchers to clear T in the ESSH model.

Our algorithm for computing $essh(T, V_s, V_h)$ is described in Algorithm 3.
Similar to Lemma 1, Algorithm 3 has the following property.

Lemma 9. *For an instance (T, V_s, V_h) of the ESSH problem, there is an optimal ESSH strategy for (T, V_s, V_h) that contains the actions performed in Steps 3 and 4 of Algorithm 3.*

Algorithm 3. EDGESEARCHSH(T, V_s, V_h) (Edge Search with Starts and Halts)

Input: A tree T, a nonempty set of start vertices V_s and a nonempty set of halt vertices V_h.

Output: $essh(T, V_s, V_h)$.

1: $e_{sh} \leftarrow 0$; $f \leftarrow 0$; $E_c \leftarrow \emptyset$.

2: If $E_c = E(T)$, then *return* e_{sh}.

3: While there exists a moveable searcher on vertex $u \in V_s$: let uv be the only dirty edge incident with u; slide this searcher from u to v to clear the edge uv; $E_c \leftarrow E_c \cup \{uv\}$; if $v \in V_h$, then $V_s \leftarrow V_s \setminus \{u\}$, otherwise, $V_s \leftarrow (V_s \setminus \{u\}) \cup \{v\}$; if v contains two searchers then remove this searcher and $f \leftarrow f + 1$.

4: While there exists a searcher on a clear vertex $v \in V_s$: remove the searcher, $V_s \leftarrow V_s \setminus \{v\}$ and $f \leftarrow f + 1$.

5: Compute $T \ominus V_s$; let $\mathcal{T}$ be the set of the completely dirty trees in $T \ominus V_s$; let $\mathcal{T}_h$ be a subset of $\mathcal{T}$ such that $T' \in \mathcal{T}_h$ if and only if T' contains a halt vertex, let $\mathcal{T}' = \mathcal{T} \setminus \mathcal{T}_h$.

6: If $\mathcal{T}' \neq \emptyset$, then: find the subtree $T_1 \in \mathcal{T}'$ that has the smallest edge search number in $\mathcal{T}'$; clear T_1 using $es(T_1)$ searchers; $E_c \leftarrow E_c \cup E(T_1)$; $e_{sh} \leftarrow \max\{e_{sh}, es(T_1) - f\}$; go to Step 2.

7: Find the subtree $T_1 \in \mathcal{T}_h$ such that T_1 maximizes the difference $esh(T_i, V_h \cap V(T_i)) - |V_h \cap V(T_i)|$ for all $T_i \in \mathcal{T}_h$; clear T_1 using the method from Sect. 5; $E_c \leftarrow E_c \cup E(T_1)$; $e_{sh} \leftarrow \max\{e_{sh}, esh(T_i, V_h \cap V(T_i)) - f\}$; go to Step 2.

Now we show that the strategy used in Algorithm 3 that clears the subtree T_1 in Step 6 or Step 7 is a part of an optimal ESSH strategy.

Lemma 10. *For an instance* (T, V_s, V_h) *of the ESSH problem, there is an optimal ESSH strategy for* (T, V_s, V_h) *that contains the clearing actions performed in Steps 6 and 7 of Algorithm 3.*

By Lemmas 9 and 10, we can use induction to show the main result of this section.

Theorem 8. *For an instance* (T, V_s, V_h) *of the ESSH problem, the value of* e_{sh} *at the termination of Algorithm 3 is equal to* $essh(T, V_s, V_h)$.

From Theorems 2 and 6, we have the following result.

Theorem 9. *If* (T, V_s, V_h) *is an instance of the ESSH problem with* $|V(T)| = n$, *then* $essh(T, V_s, V_h)$ *can be computed in* $O(n^2)$ *time.*

Similarly to Theorem 3, an analysis of the actions used by Algorithm 3 implies the following:

Theorem 10. *The ESSH problem is monotonic.*

7   Fast Searching with Starts and Halts

In this section we consider the following fast searching problem.

Fast Searching with Starts and Halts (FSSH)
Instance: A tree T, a nonempty set of start vertices $V_s \subseteq V(T)$, and a nonempty set of halt vertices $V_h \subseteq V(T)$, such that $V_h \cap V_s = \emptyset$, and each $v \in V_s$ is initially occupied by exactly one starting searcher.
Question: What is the minimum number of additional searchers for clearing T in the fast searching model such that once a searcher occupies a vertex $v \in V_h$, v remains occupied by at least one searcher for the remainder of the search, and each vertex from V_s remains clear during the entire search?

We define the *FSSH number* of T with respect to V_s and V_h, denoted $fssh(T, V_s, V_h)$, as the minimum number of additional searchers required to clear T in the FSSH model. An FSSH search strategy is *optimal* if it uses $fssh(T, V_s, V_h)$ searchers to clear T in the FSSH model.

Algorithm 4. FASTSEARCHSH(T, V_s, V_h) (Fast Search with Starts and Halts)

Input: A tree T, a nonempty set of start vertices V_s, and a nonempty set of halt vertices V_h.
Output: $fssh(T, V_s, V_h)$.

1: Initially $f_{sh} \leftarrow 0$.
2: If V_h contains a vertex whose degree is even, place a searcher on it and $f_{sh} \leftarrow f_{sh}+1$; repeat this step until the condition is not satisfied.
3: If V_s contains a vertex u that is a leaf on a component in T such that this component contains an odd vertex $v \in V_h$, then: delete all edges of the path between u and v from T; delete isolated vertices from T; update V_s and V_h by deleting the vertices that are not on the current T; repeat this step until the condition is not satisfied.
4: If there is a component T' in T that does not contain a halt vertex with odd degree, then: let V_s' be the set of start vertices in T' (V_s' can be an empty set); call FASTSEARCHS(T', V_s') to compute $fss(T', V_s')$; $f_{sh} \leftarrow f_{sh} + fss(T', V_s')$; repeat this step until the condition is not satisfied.
5: If there is a component T' in T that contains a halt vertex with odd degree but does not contain a start vertex with odd degree, then: let V'' be the set of odd halt vertices in T'; call FASTSEARCHS(T', V'') to compute $fss(T', V'')$ (refer to Lemma 8); $f_{sh} \leftarrow f_{sh} + fss(T', V'') + |V''|$; repeat this step until the condition is not satisfied.
6: If $E(T) \neq \emptyset$, then:
  (6.1) arbitrarily select a leaf u and an odd vertex $v \in V_h$ that are in the same component of T;
  (6.2) update T by deleting all edges on the path between u and v from T and deleting isolated vertices from T; update V_s and V_h by deleting the vertices that are not on the current T; $f_{sh} \leftarrow f_{sh} + 1$; go to Step 3.
7: Return f_{sh}.

Our method for computing $fssh(T, V_s, V_h)$ is described in Algorithm 4.

Theorem 11. *For an instance* (T, V_s, V_h) *of the* FSSH *problem, the output* f_{sh} *at the termination of Algorithm 4 is equal to* $fssh(T, V_s, V_h)$.

Theorem 12. *If* (T, V_s, V_h) *is an instance of the* FSSH *problem with* $|V(T)| = n$, *then* $essh(T, V_s, V_h)$ *can be computed in* $O((|V_s| + |V_h|)n)$ *time.*

Acknowledgments. Lusheng Wang's research was supported by GRF grants for Hong Kong Special Administrative Region, P.R. China (CityU 11210119 and CityU 11206120) and National Science Foundation of China (NSFC: 61972329). Boting Yang's research was supported in part by an NSERC Discovery Research Grant, Application No.: RGPIN-2018-06800.

References

1. Alspach, B.: Sweeping and searching in graphs: a brief survey. Matematiche **59**, 5–37 (2006)
2. Bienstock, D.: Graph searching, path-width, tree-width and related problems (a survey). DIMACS Ser. Discrete Math. Theoret. Comput. Sci. **5**, 33–49 (1991)
3. Bienstock, D., Seymour, P.: Monotonicity in graph searching. J. Algorithms **12**, 239–245 (1991)
4. Bonato, A., Nowakowski, R.: The Game of Cops and Robbers on Graphs. American Mathematical Society, Providence (2011)
5. Bonato, A., Yang, B.: Graph searching and related problems. In: Pardalos, P., Du, D.-Z., Graham, R. (eds.) Handbook of Combinatorial Optimization, 2nd edn., pp. 1511–1558. Springer, Heidelberg (2013). https://doi.org/10.1007/978-1-4419-7997-1_76
6. Dereniowski, D., Diner, Ö., Dyer, D.: Three-fast-searchable graphs. Discret. Appl. Math. **161**(13), 1950–1958 (2013)
7. Dyer, D., Yang, B., Yaşar, Ö.: On the fast searching problem. In: Fleischer, R., Xu, J. (eds.) AAIM 2008. LNCS, vol. 5034, pp. 143–154. Springer, Heidelberg (2008). https://doi.org/10.1007/978-3-540-68880-8_15
8. Ellis, J., Sudborough, I., Turner, J.: The vertex separation and search number of a graph. Inf. Comput. **113**, 50–79 (1994)
9. Fomin, F., Thilikos, D.: An annotated bibliography on guaranteed graph searching. Theor. Comput. Sci. **399**(3), 236–245 (2008)
10. Kinnersley, N.: The vertex separation number of a graph equals its path-width. Inf. Process. Lett. **42**, 345–350 (1992)
11. LaPaugh, A.: Recontamination does not help to search a graph. J. ACM **40**, 224–245 (1993)
12. Makedon, F., Papadimitriou, C., Sudborough, I.: Topological bandwidth. SIAM J. Algebraic Discret. Methods **6**, 418–444 (1985)
13. Megiddo, N., Hakimi, S., Garey, M., Johnson, D., Papadimitriou, C.: The complexity of searching a graph. J. ACM **35**, 18–44 (1998)
14. Parsons, T.D.: Pursuit-evasion in a graph. In: Alavi, Y., Lick, D.R. (eds.) Theory and Applications of Graphs. LNM, vol. 642, pp. 426–441. Springer, Heidelberg (1978). https://doi.org/10.1007/BFb0070400
15. Stanley, D., Yang, B.: Fast searching games on graphs. J. Comb. Optim. **22**, 763–777 (2011)

16. Xue, Y., Yang, B.: The fast search number of a cartesian product of graphs. Discret. Appl. Math. **224**, 106–119 (2017)
17. Xue, Y., Yang, B., Zhong, F., Zilles, S.: The fast search number of a complete k-partite graph. Algorithmica **80**(12), 3959–3981 (2018)
18. Xue, Y., Yang, B., Zilles, S., Wang, L.: Fast searching on cactus graphs. J. Comb. Optim. **45**, 1–22 (2023)
19. Yang, B.: Fast edge searching and fast searching on graphs. Theor. Comput. Sci. **412**(12), 1208–1219 (2011)
20. Yang, B., Zhang, R., Cao, Y., Zhong, F.: Search numbers in networks with special topologies. J. Interconnection Netw. **19**, 1–34 (2019)

EFX Allocations Exist for Binary Valuations

Xiaolin Bu, Jiaxin Song, and Ziqi Yu[(✉)]

Shanghai Jiao Tong University, Shanghai, China
{lin_bu,sjtu_xiaosong,yzq.111}@sjtu.edu.cn

Abstract. We study the fair division problem and the existence of allocations satisfying the fairness criterion *envy-freeness up to any item* (EFX). The existence of EFX allocations is a major open problem in the fair division literature. We consider *binary valuations* where the marginal gain of the value by receiving an extra item is either 0 or 1. Babaioff et al. (2021) proved that EFX allocations always exist for binary and *submodular* valuations. In this paper, by using completely different techniques, we extend this existence result to general binary valuations that are not necessarily submodular, and we present a polynomial time algorithm for computing an EFX allocation.

Keywords: Fair Division · EFX · Binary Valuations

1 Introduction

Fair division studies how to allocate heterogeneous resources fairly among a set of agents. It is a classical resource allocation problem that has been widely studied by mathematicians, economists, and computer scientists. It has a wide applications including school choices (Abdulkadiroğlu et al., 2005), course allocations (Budish and Cantillon, 2012), paper review assignments (Lian et al., 2018), allocating computational resources (Ghodsi et al., 2011), etc. Traditional fair division literature considers resources that are *infinitely divisible*. The fair division problem for infinitely divisible resources is also called *the cake-cutting problem*, which dated back to Steinhaus (1948, 1949) and has been extensively studied thereafter (Aumann and Dombb, 2015; Aumann et al., 2013; Brams et al., 2012; Caragiannis et al., 2012; Cohler et al., 2011; Bei et al., 2012, 2017; Tao, 2022; Bu et al., 2023). Among those fairness notions, *envy-freeness* (EF) is most commonly considered, which states that each agent values her own allocated share at least as much as the allocation of any other agent. In other words, each agent does not envy any other agent in the allocation.

Recent research in fair division has been focusing more on allocating *indivisible items*. It is clear that absolute fairness such as envy-freeness may not be achievable for indivisible items. For example, we may have fewer items than agents. Caragiannis et al. (2019b) proposed a notion that relaxes envy-freeness, called *envy-freeness up to any item* (EFX), which requires that, for any pair of

M. Li et al. (Eds.): IJTCS-FAW 2023, LNCS 13933, pp. 252–262, 2023.
https://doi.org/10.1007/978-3-031-39344-0_19

agents i and j, i does not envy j after the removal of *any* item from j's bundle. Despite a significant amount of effort (e.g., (Plaut and Roughgarden, 2020; Chaudhury et al., 2020, 2021; Caragiannis et al., 2019a; Babaioff et al., 2021; Berger et al., 2022; Akrami et al., 2022; Feldman et al., 2023)), the existence of EFX allocations is still one of the most fundamental open problems in the fair division literature. The existence of an EFX allocation is only known for two agents (Plaut and Roughgarden, 2020) or three agents with more special valuations (Chaudhury et al., 2020; Akrami et al., 2022).

For a general number of agents, the existence of EFX allocations is only known for *binary* and *submodular* valuations, also known as *matroid-rank* valuations (Babaioff et al., 2021). Under this setting, Babaioff et al. (2021) designed a mechanism that outputs an EFX allocation and maximizes the *Nash social welfare* at the same time, where the Nash social welfare of an allocation is the product of all the agents' values. Unfortunately, Babaioff et al.'s techniques cannot be extended to the setting with binary valuations that are not necessarily submodular, as it is possible that all Nash social welfare maximizing allocations can fail to be EFX. For example, consider a fair division instance with n agents and $m > n$ items, and the valuations of the n agents are defined as follows:

- for agent 1, she has value 1 if she receives at least $m - n + 1$ items, and she has value 0 otherwise;
- for each of the remaining $n - 1$ agents, the value equals to the number of the items received.

It is easy to check that the valuations are binary in the instance above. However, the allocation that maximizes the Nash social welfare must allocate $m - n + 1$ items to agent 1 and allocate one item to each of the remaining $n - 1$ agents, as this is the only way to make the Nash social welfare nonzero. It is easy to check that such an allocation cannot be EFX if m is significantly larger than n.

As a result, studying the existence of EFX allocations for general binary valuations requires different techniques from Babaioff et al. (2021).

1.1   Our Results

We prove that EFX allocations always exist for general binary valuations which may not be submodular. In addition, we provide a polynomial-time algorithm to compute such an allocation. Our technique is based on the envy-graph procedure by Chaudhury et al. (2021). Chaudhury et al. (2021) proposed a pseudo-polynomial-time algorithm that computes a *partial* EFX allocation such that the number of the unallocated items is at most $n - 1$ and no one envies the unallocated bundle. We show that, for binary valuations, we can always find a complete EFX allocation, and it can be done in a polynomial time. In particular, binary valuations enable some extra update steps which can further allocate the remaining unallocated items while guaranteeing EFX property.

1.2 Related Work

The existence of EFX allocations in general is a fundamental open problem in fair division. Many partial progresses have been made in the previous literature. Plaut and Roughgarden (2020) showed that EFX allocations exist if agents' valuations are identical. By a simple "I-cut-you-choose" protocol, this result implies EFX allocations always exist for two agents. Chaudhury et al. (2020) showed that EFX allocations exist for three agents if the valuations are additive (meaning that the bundle's value equals to the sum of the values to the individual items). The existence of EFX allocations for three agents is extended to slightly more general valuation functions by Feldman et al. (2023).

For binary valuations, Babaioff et al. (2021) showed that EFX allocations always exist if the valuation functions are submodular. More details about this work have already been discussed in the introduction section, and this is the work most relevant to our paper.

Since the existence of EFX allocations is a challenging open problem, many papers study partial EFX allocations. Chaudhury et al. (2021) proposed a pseudo-polynomial-time algorithm that computes a partial EFX allocation such that the number of the unallocated items is at most $n - 1$ and no one envies the unallocated bundle. For additive valuations, Caragiannis et al. (2019a) showed that it is possible to obtain a partial EFX allocation that achieves 0.5-approximation to the optimal Nash social welfare; Berger et al. (2022) showed that, for four agents, we can have a partial EFX allocation with at most one unallocated item.

Budish (2011) proposed *envy-freeness up to one item* (EF1), which is weaker than EFX. It requires that, for any pair of agents i and j, *there exists* an item from j's allocated bundle whose hypothetical removal would keep agent i from envying agent j. It is well-known that EF1 allocations always exist and can be computed in polynomial time (Lipton et al., 2004; Budish, 2011).

Other than fairness, previous work has also extensively studied several efficiency notions including social welfare, Nash social welfare, Pareto-optimality, and their compatibility with fairness (Barman et al., 2018; Caragiannis et al., 2019b; Murhekar and Garg, 2021; Aziz et al., 2023; Barman et al., 2020; Bu et al., 2022; Bei et al., 2021b; Caragiannis et al., 2019a; Chaudhury et al., 2020). We will not further elaborate on it here.

2   Preliminaries

Let N be the set of n agents and M be the set of m indivisible items. Each agent i has a valuation function $v_i : 2^M \to \mathbb{R}_{\geq 0}$ to specify her utility to a bundle, and we abuse the notation to denote $v_i(\{g\})$ by $v_i(g)$. Each valuation function v_i is normalized and monotone:

- Normalized: $v_i(\emptyset) = 0$;
- Monotone: $v_i(S) \geq v_i(T)$ whenever $T \subseteq S$.

We say a valuation function v_i is *binary* if $v_i(\emptyset) = 0$ and $v_i(S \cup \{g\}) - v_i(S) \in \{0, 1\}$ for any $S \subseteq M$ and $g \in M \setminus S$. In this paper, we will consider exclusively binary valuation functions. Note that we do not require the valuation function to be additive.

An allocation $\mathcal{A} = (A_1, A_2, \ldots, A_n)$ is a partition of M, where A_i is allocated to agent i. A partial allocation $\mathcal{A} = (A_1, A_2, \ldots, A_n, P)$ is also a partition of M, where P is the set of unallocated items. Notice a partial allocation is an allocation if and only if $P = \emptyset$. It is said to be *envy-free* if $v_i(A_i) \geq v_i(A_j)$ for any i and j. However, an envy-free allocation may not exist when allocating indivisible items, for example, when $m < n$. In this paper, we consider a relaxation of envy-freeness, called *envy-freeness up to any item* (EFX) (Caragiannis et al., 2019b).

Definition 1 (EFX). *An allocation $\mathcal{A} = (A_1, \ldots, A_n)$ satisfies envy-freeness up to any item (EFX) if $v_i(A_i) \geq v_i(A_j \setminus \{g\})$ for any i, j and any item $g \in A_j$.*

Definition 2 (Strong Envy). *Given a partial allocation $(A_1, \ldots, A_n, P)$, we say that i **strongly envy** j if $v_i(A_i) < v_i(A_j \setminus \{g\})$ for some $g \in A_j$. Notice that we can extend this definition to a complete allocation by setting $P = \emptyset$.*

It is clear that an allocation is EFX if and only if there is no strong envy between every pair of agents.

Given an allocation $\mathcal{A}$, its *utilitarian social welfare*, or simply *social welfare*, is defined by

$$\mathcal{USW}(\mathcal{A}) = \sum_{i=1}^{n} v_i(A_i).$$

2.1 Envy-Graph

Definition 3 (Envy-Graph). *Given a partial allocation $\mathcal{A} = \{A_1, A_2, \ldots, A_n, P\}$, the envy graph $G = (V, E)$ is defined as follows. Each vertex in V is considered as an agent. For agents $i, j \in N$, $(i, j) \in E$ if and only if i envies j, i.e. $v_i(A_i) < v_i(A_j)$.*

Definition 4 (Cycle-Elimination). *For a cycle $u_0 \rightarrow u_1 \rightarrow \cdots u_{k-1} \rightarrow u_0$, the **cycle-elimination procedure** is performed as follows. Each agent u_i receives the bundle from u_{i+1} for $i \in \{0, 1, \ldots, k-1\}$ (indices are modulo k). This process terminates when there is no cycle in the envy-graph.*

It can be verified that a cycle-elimination step will not break the EFX property, since all bundles remain unchanged, what we have done is just exchanging these bundles. Also, the social welfare will increase since every agent in the cycle gets the bundle which is of higher value than her previous one.

Definition 5 (Source agent). *An agent is called a **source agent** if the in-degree of her corresponding vertex in the envy-graph is 0. In other words, an agent is a source agent if no one envies her bundle.*

2.2   Binary Valuations and Pre-Envy

Recall that we focus on binary valuations in this paper. When valuation functions are binary, there are some special properties compared with general valuation functions. In this case, the marginal gain of a single item is 0 or 1.

Proposition 1. *In an EFX allocation, if i envies j, then $v_i(A_j) - v_i(A_i) = 1$.*

Proof. Since i envies j, $v_i(A_j) - v_i(A_i) > 0$. Also, suppose $v_i(A_j) - v_i(A_i) \geq 2$. Since we focus on the binary valuation profile, for any $g \in A_j$

$$v_i(A_i) + 2 \leq v_i(A_j) \leq v_i(A_j \setminus \{g\}) + 1.$$

However, by the definition of EFX, we have $v_i(A_i) \geq v_i(A_j \setminus \{g\})$ for any $g \in A_j$, which leads to a contradiction. □

Proposition 2. *Suppose j does not envy i's bundle, i.e., $v_j(A_j) \geq v_j(A_i)$. If j envies i's bundle after adding an item g to it, then $v_j(A_j) = v_j(A_i)$*

Proof. If not so, then $v_j(A_j) \geq v_j(A_i) + 1$. Then $v_j(A_i \cup \{g\}) \leq V_j(A_i) + 1 \leq V_j(A_j)$, which contradicts to the fact that j will envy i's bundle after adding an item to it. □

By Proposition 2, suppose a new edge (j, i) appears in the envy graph after adding an item to i's bundle. Then, if we do not add this item, we have $v_j(A_j) = v_j(A_i)$. To better illustrate this observation, we introduce the following notion and consider this new relationship in the envy graph.

Definition 6 (Pre-Envy). *Suppose $i, j \in N$ are two agents. We say j **pre-envies** i, if $v_j(A_j) = v_j(A_i)$. In the envy graph, j pre-envies i will be represented by $j \dashrightarrow i$.*

In the binary valuation profile, we can jointly consider envy and pre-envy. Let $G = (V, E, E')$, where V is the set of all agents, E is the set of all envy relationships and E' are the pre-envy ones. If edges in $E \cup E'$ form a cycle, we can also use the cycle-elimination procedure.

As a remark, for a cycle with pre-envy edges only, it may still exist after applying the cycle-elimination procedure. For the cycle with pre-envy edges, we do not always eliminate it. Our algorithm only eliminates this type of cycles under some particular scenarios. We will provide more details in the next section.

As another remark, our notion of *pre-envy edge* is the same as the *equality edge* in the paper (Bei et al., 2021a). We choose a different word in this paper as we will sometimes use "pre-envy" as a verb.

3   Existence of EFX Allocations

In this section, we prove our main result.

Theorem 1. *For any binary valuation profile $(v_1, \ldots, v_n)$, there exists an EFX allocation, and it can be computed by a polynomial-time algorithm.*

3.1   The Main Algorithm

We describe our algorithm in Algorithm 1.

Algorithm 1: Computing an EFX allocation with binary marginal gain
---
    Output: an EFX allocation $(A_1, A_2, \ldots, A_n, P)$ with $P = \emptyset$

1   Let $A_i = \emptyset, i \in N$, and $P = M$;
2   **while** $P \neq \emptyset$ **do**
3 | find an arbitrary item $g \in P$;
4 | $(A_1, A_2, \ldots, A_n, P) \leftarrow \texttt{Update}(\mathcal{A}, g)$;
5 | Update the envy graph;
6 | Perform the cycle-elimination procedure (Definition 4);
7   **end**

The algorithm starts with the allocation $\mathcal{A} = (A_1, \ldots, A_n, P)$ where $A_1 = \cdots = A_n = \emptyset$. In each iteration, it will consider an unallocated item g and invoke the update function $\texttt{Update}(\mathcal{A}, g)$. In particular, we attempt to allocate g to a source agent by applying one of the two update rules U_0 and U_1. After that, we update the envy graph and perform the cycle-elimination procedure to guarantee the existence of source agents.

Rule U_0. If there exists a source agent $i \in N$, such that no agent will strongly envy $A_i \cup \{g\}$. In this case, U_0 just allocates g to this source agent i.

Rule U_1. If U_0 fails, then for every source agent i, adding g to i's bundle will cause at least one strong envy. Suppose $\{s_1, \ldots, s_k\}$ is the set of source agents. By assumption, after adding g to s_1's bundle, some agents will strongly envy s_1.

    Before describing the update rule U_1, we first define some notions which are used later.

Definition 7 (Safe Bundle and Maximal Envious Agent). *Suppose a partial allocation* $\mathcal{A} = (A_1, A_2, \ldots, A_n, P)$, *an agent i and an item $g \in P$ satisfy that*

(a) $\mathcal{A}$ *is a partial EFX allocation, and*
(b) adding g to A_i causes someone strongly envy i.

*We say that $S \subseteq A_i \cup \{g\}$ is a **safe bundle** with respect to $\mathcal{A}, i, g$, if*

1. *There exists $j \in N$ such that j envies S.*
2. *For each agent j that envies the bundle $A_i \cup \{g\}$ (i.e., $v_j(A_i \cup \{g\}) > v_j(A_j)$), j does not strongly envy S (i.e., for all items $s \in S$, $v_j(S \setminus \{s\}) < v_j(A_j)$).*

*The **maximal envious agent** is one of the agents who envies S.*

    Intuitively, a safe bundle S is a *minimal* subset of $A \cup \{g\}$ such that someone still envies S. The minimal property guarantees that no one will strongly envy S.

    We also remark that, according to our definition, the maximal envious agent of $\mathcal{A}, i, g$ may be agent i herself.

Lemma 1. *For every triple $(\mathcal{A}, i, g)$ satisfying (a) and (b) in Definition 7, a safe bundle $S \subseteq A_i \cup \{g\}$ and the corresponding maximal envious agent a always exists, and they can be found in a polynomial time.*

We defer the proof of Lemma 1 to Sect. 3.2.

By Lemma 1, we can always find $S \subseteq A_{s_1} \cup \{g\}$ and the maximal envious agent $c_1 \in N$, such that

- Agent c_1 envies S, i.e., $v_{c_1}(S) > v_{c_1}(A_{c_1})$.
- No agent strongly envies S.

Since agent c_1 does not envy s_1 before g is added to A_{s_1} (as s_1 is a source agent) and envies s_1 after the addition of g, according to Proposition 2, agent c_1 pre-envies s_1 in the allocation $\mathcal{A}$ (where g has not been added yet). We will use $c_1 \dashrightarrow_g s_1$ to denote this special pre-envy edge where c_1 is a maximal envious agent for $\mathcal{A}, s_1, g$. We say that $\dashrightarrow_g$ is a maximal envy edge.

Since we have assumed adding g to each source agent causes strong envy, by Lemma 1, for each source agent $s_i \in \{s_1, \ldots, s_k\}$, there exists an agent c_i such that $c_i \dashrightarrow_g s_i$. It is possible that $s_i \dashrightarrow_g s_i$. In this case, we add a self-loop to the envy graph.

In the lemma below, we show that, if the pre-condition of U_0 fails, then there must exist a cycle that only consists of edges in the original G and the maximal envy edges.

Lemma 2. *For a partial EFX allocation $\mathcal{A}$ and an unallocated item g, if $A_i \cup \{g\}$ is strongly envied by someone for every source agent i, there must exist a cycle containing at least one source agent that only consists of edges in the original G and the maximal envy edges. Moreover, the cycle can be found in polynomial time.*

Proof. We have shown that, for any s_i, there exists c_i such that $c_i \dashrightarrow_g s_i$. We start from s_1, and we find c_1 such that $c_1 \dashrightarrow_g s_1$. We find the source s_2 from which c_1 is reachable, and we find $c_2 \dashrightarrow_g s_2$. We keep doing this: whenever we are at c_i, we find the source s_{i+1} from which c_i is reachable; and we find c_{i+1} such that $c_{i+1} \dashrightarrow_g s_{i+1}$. Notice that we are traveling backward along a path that consists only of graph edges and maximal envy edges. We can keep traveling until we find a source that has been already visited before, in which case we have found a cycle.

It is also easy to check that the above procedure can be done in polynomial time.   □

Next, we find an arbitrary source agent s in the cycle described in Lemma 2. We apply Lemma 1 and find a safe bundle $S \subseteq A_s \cup \{g\}$. We replace A_s by S, and perform an operation that is similar to cycle-elimination: let each agent in the cycle receives the bundle of the next agent. In particular, the agent preceding s is the one that maximally envy s; she will receive S and get a higher value.

The lemma below justifies the correctness of U_1.

Lemma 3. *After applying the update rule* U_1, *the partial allocation remains EFX.*

Proof. In the update, a bundle A_s is replaced by S. Definition 7 implies that no one strongly envies S and the agent in the cycle preceding s receives a higher value by getting S. Each remaining agent in the cycle receives a weakly higher value. Thus, the resultant allocation is EFX after the update for the same reason that the cycle-elimination procedure does not destroy the EFX property.

Algorithm 2: The update rules for allocation $\mathcal{A}$ and item $g \in P$

1 **Function** Update*(allocation* $\mathcal{A} = (A_1, \ldots, A_n, P)$, *item* $g \in P$*)*
2   if there exists a source agent i such that adding g to A_i does not break the EFX property, perform $U_0(\mathcal{A}, g)$;
3   otherwise, perform $U_1(\mathcal{A}, g)$;
4 **end**
5 **Function** $U_0(\mathcal{A}, g)$
6   Allocate g to i: $A_i \leftarrow A_i \cup \{g\}$;
7   Update pool: $P \leftarrow P \setminus \{g\}$;
8 **end**
9 **Function** $U_1(\mathcal{A}, g)$
10   find a cycle C described in Lemma 2;
11   for a source agent s on C, find the safe bundle S for $(\mathcal{A}, s, g)$ by Lemma 1;
12   let $A_s \leftarrow S$ and $P \leftarrow P \cup A_s \cup \{g\} \setminus S$;
13   let each agent on C receive the bundle of the next agent;
14 **end**

3.2   Proof of Theorem 1

We first prove the fact that a safe bundle always exists and can be calculated in polynomial time. Notice that the updating rule U_1 makes sense if and only if Lemma 1 holds.

Proof of Lemma 1: Suppose $SE := \{a_1, a_2, \ldots, a_k\}$ is the set of agents who strongly envy i after adding g to A_i. In other words, they strongly envy $A_i \cup \{g\}$. Firstly, we claim that there exists $S \subseteq A_i \cup \{g\}$ such that

$$v_{a_1}(S) > v_{a_1}(A_{a_1}) \geq v_{a_1}(S \setminus \{g'\}) \quad \text{for any } g' \in S.$$

In other words, a_1 envies S but does not strongly envy S.

The set S can be found by letting agent a_1 iteratively remove an item from $A_i \cup \{g\}$ until agent a_1 does not strongly envy the bundle. If S does not exist, it must be that 1) agent a_1 strongly envies some $S' \subseteq A_i \cup \{g\}$, and 2) removing any single item g' from S' will cause a_1 no longer envy this bundle, i.e.,

$$v_{a_1}(S' \setminus \{g'\}) \leq v_{a_1}(A_{a_1}).$$

However, this contradicts to the fact that a_1 strongly envies S'. Hence, such S always exists.

Update SE such that it is the set of agents who strongly envy S. By the above procedure, a_1 is no longer in SE. If SE is empty now, then S is the safe bundle and a_1 is the corresponding maximal envious agent. If SE is not empty, then choose $a' \in SE$ and do the same procedures above. After this, S and SE will be further reduced and a' is no longer in SE. Since the size of SE is decreased at least by 1 every round, there exists an agent $a_j \in SE$ who is the last one to remove items from the bundle. Suppose $S \subseteq A_i \cup \{g\}$ to be the final version of the bundle after item removals. Then S is the safe bundle and $a_j \in SE$ is the corresponding maximal envious agent. The following algorithm illustrates these processes. Here we use S, a to store the last version of the bundle and the last agent who remove items, respectively.

Algorithm 3: Computing safe bundle and corresponding the maximal envious agent

---

Input: A partial allocation $\mathcal{A}$, i, g
Output: a safe bundle S and a maximal envious agent a
1 $S \leftarrow A_i \cup \{g\}$;
2 let SE be the set of agents who strongly envies S;
3 **while** SE *is non-empty* **do**
4   Find an index j, such that $a_j \in SE$;
5   While a_j strongly envies S, ask a_j to remove an item from S such that $v_{a_j}(A_{a_j}) < v_{a_j}(S)$;
6   If some items in S have been removed in the process above, update $a \leftarrow a_j$;
7   Remove a_j from SE;
8 **end**
9 **return** S, a

Notice that this algorithm will perform at most n rounds, as at least one agent is removed from SE in each iteration. Each round obviously costs a polynomial time. Then the safe bundle and the maximal envious agent can be found in polynomial time. □

Proof of Theorem 1: Firstly, the social welfare is increased by at least 1 after each application of $U_1(\mathcal{A}, g)$ in update$(\mathcal{A}, g)$: for U_1, Definition 7 and Lemma 1 ensure that the agent receiving the safe bundle S gets a strictly higher value than before. Secondly, U_0 can be applied for at most m times between two applications of U_1, as there can be at most m items in P and U_0 decreases $|P|$ by exactly 1. Since the social welfare can be at most mn, the algorithm terminates in at most mn applications of U_1. Therefore, the algorithm terminates with at most m^2n applications of U_0 or U_1, which runs in a polynomial time.

Also, the EFX property is preserved after each round. Condition U_0 clearly preserves the EFX by its definition. Lemma 3 guarantees that the allocation remains EFX after applying U_1.

Hence, an EFX allocation always exists and it can be found in polynomial time. □

4 Conclusion

In this paper, we studied the existence of an EFX allocation under a binary valuation profile. In particular, we proved that such an EFX allocation always exists and proposed a polynomial-time algorithm to compute it. Compared with the general valuation, "pre-envy" makes sense in binary valuation profiles and gives us extra properties to cope with EFX.

References

Abdulkadiroğlu, A., Pathak, P.A., Roth, A.E.: The New York City high school match. Am. Econ. Rev. **95**(2), 364–367 (2005)

Akrami, H., Chaudhury, B.R., Garg, J., Mehlhorn, K., Mehta, R.: EFX allocations: Simplifications and improvements. arXiv preprint arXiv:2205.07638 (2022)

Aumann, Y., Dombb, Y.: The efficiency of fair division with connected pieces. ACM Trans. Econ. Comput. **3**(4), 1–16 (2015)

Aumann, Y., Dombb, Y., Hassidim, A.: Computing socially-efficient cake divisions. In: Proceedings of the International Conference on Autonomous Agents and Multi-Agent Systems (AAMAS), pp. 343–350 (2013)

Aziz, H., Huang, X., Mattei, N., Segal-Halevi, E.: Computing welfare-maximizing fair allocations of indivisible goods. Eur. J. Oper. Res. **307**(2), 773–784 (2023)

Babaioff, M., Ezra, T., Feige, U.: Fair and truthful mechanisms for dichotomous valuations. In: Proceedings of the AAAI Conference on Artificial Intelligence, vol. 35, pp. 5119–5126 (2021)

Barman, S., Bhaskar, U., Shah, N.: Optimal bounds on the price of fairness for indivisible goods. In: Chen, X., Gravin, N., Hoefer, M., Mehta, R. (eds.) WINE 2020. LNCS, vol. 12495, pp. 356–369. Springer, Cham (2020). https://doi.org/10.1007/978-3-030-64946-3_25

Barman, S., Krishnamurthy, S.K., Vaish, R.: Finding fair and efficient allocations. In: Proceedings of the ACM Conference on Economics and Computation (EC), pp. 557–574 (2018)

Bei, X., Chen, N., Hua, X., Tao, B., Yang, E.: Optimal proportional cake cutting with connected pieces. In: Proceedings of AAAI Conference on Artificial Intelligence (AAAI), pp. 1263–1269 (2012)

Bei, X., Chen, N., Huzhang, G., Tao, B., Wu, J.: Cake cutting: envy and truth. In: IJCAI, pp. 3625–3631 (2017)

Bei, X., Li, Z., Liu, J., Liu, S., Lu, X.: Fair division of mixed divisible and indivisible goods. Artif. Intell. **293**, 103436 (2021a)

Bei, X., Lu, X., Manurangsi, P., Suksompong, W.: The price of fairness for indivisible goods. Theory Comput. Syst. **65**(7), 1069–1093 (2021b)

Berger, B., Cohen, A., Feldman, M., Fiat, A.: Almost full EFX exists for four agents. In: Proceedings of the AAAI Conference on Artificial Intelligence, vol. 36, pp. 4826–4833 (2022)

Brams, S.J., Feldman, M., Lai, J., Morgenstern, J., Procaccia, A.D.: On maxsum fair cake divisions. In: Proceedings of the AAAI Conference on Artificial Intelligence (AAAI), pp. 1285–1291 (2012)

Bu, X., Li, Z., Liu, S., Song, J., Tao, B.: On the complexity of maximizing social welfare within fair allocations of indivisible goods. arXiv preprint arXiv:2205.14296 (2022)

Xiaolin, B., Song, J., Tao, B.: On existence of truthful fair cake cutting mechanisms. Artif. Intell. **319**(2023), 103904 (2023). https://doi.org/10.1016/j.artint.2023.103904

Budish, E.: The combinatorial assignment problem: approximate competitive equilibrium from equal incomes. J. Polit. Econ. **119**(6), 1061–1103 (2011)

Budish, E., Cantillon, E.: The multi-unit assignment problem: theory and evidence from course allocation at harvard. Am. Econ. Rev. **102**(5), 2237–2271 (2012)

Caragiannis, I., Gravin, N., Huang, X.: Envy-freeness up to any item with high nash welfare: the virtue of donating items. In: Proceedings of the ACM Conference on Economics and Computation (EC), pp. 527–545 (2019a)

Caragiannis, I., Kaklamanis, C., Kanellopoulos, P., Kyropoulou, M.: The efficiency of fair division. Theory Comput. Syst. **50**(4), 589–610 (2012)

Caragiannis, I., Kurokawa, D., Moulin, H., Procaccia, A.D., Shah, N., Wang, J.: The unreasonable fairness of maximum nash welfare. ACM Trans. Econ. Comput. **7**(3), 1–32 (2019b)

Chaudhury, B.R., Garg, J., Mehlhorn, K.: EFX exists for three agents. In: Proceedings of the ACM Conference on Economics and Computation (EC), pp. 1–19 (2020)

Chaudhury, B.R., Kavitha, T., Mehlhorn, K., Sgouritsa, A.: A little charity guarantees almost envy-freeness. SIAM J. Comput. **50**(4), 1336–1358 (2021)

Cohler, Y.J., Lai, J.K., Parkes, D.S., Procaccia, A.D.: Optimal envy-free cake cutting. In: Proceedings of AAAI Conference on Artificial Intelligence (AAAI), pp. 626–631 (2011)

Feldman, M., Mauras, S., Ponitka, T.: On optimal tradeoffs between EFX and nash welfare. arXiv preprint arXiv:2302.09633 (2023)

Ghodsi, A., Zaharia, M., Hindman, B., Konwinski, A., Shenker, S., Stoica, I.: Dominant resource fairness: fair allocation of multiple resource types. In: Proceedings of USENIX Symposium on Networked Systems Design and Implementation (NSDI) (2011)

Lian, J.W., Mattei, N., Noble, R., Walsh, T.: The conference paper assignment problem: using order weighted averages to assign indivisible goods. In: Proceedings of the AAAI Conference on Artificial Intelligence (AAAI), pp. 1138–1145 (2018)

Lipton, R., Markakis, E., Mossel, E., Saberi, A.: On approximately fair allocations of indivisible goods. In Proceedings of the ACM Conference on Electronic Commerce (EC), pp. 125–131 (2004)

Murhekar, A., Garg, J.: On fair and efficient allocations of indivisible goods. In: Proceedings of the AAAI Conference on Artificial Intelligence (AAAI), pp. 5595–5602 (2021)

Plaut, B., Roughgarden, T.: Almost envy-freeness with general valuations. SIAM J. Discrete Math. **34**(2), 1039–1068 (2020)

Steinhaus, H.: The problem of fair division. Econometrica **16**(1), 101–104 (1948)

Steinhaus, H.: Sur la division pragmatique. Econometrica **17**(1949), 315–319 (1949)

Tao, B.: On existence of truthful fair cake cutting mechanisms. In: Proceedings of the 23rd ACM Conference on Economics and Computation, pp. 404–434 (2022)

Maximize Egalitarian Welfare for Cake Cutting

Xiaolin Bu[ID] and Jiaxin Song[✉][ID]

Shanghai Jiao Tong University, Shanghai, China
{lin_bu,sjtu_xiaosong}@sjtu.edu.cn

Abstract. A major problem in *cake-cutting* is how to both *fairly* and *efficiently* allocate the cake. *Egalitarian welfare*, which prioritizes agents with the worst utilities, is a compelling notion that provides guarantees for both fairness and efficiency. In this paper, we investigate the complexity of finding a maximized egalitarian welfare (MEW) allocation when all the value density functions are *piecewise-constant*. We design an FPT (fixed-parameter tractable) algorithm (with respect to the number of the agents) for computing an MEW allocation when all the bundles are requested to be contiguous. Furthermore, we show that this problem is NP-hard to approximate to within any constant factor.

Keywords: Cake-cutting · Egalitarian Welfare · Fair Division

1 Introduction

The *cake-cutting* problem is one of the most fundamental research problems explored in social science, economics, computer science, and other related fields (Steinhaus 1948, 1949; Aumann and Dombb 2015; Brams et al. 2012; Cohler et al. 2011; Bei et al. 2012, 2017; Tao 2022). Its primary objective is to *fairly* distribute a *heterogeneous* and *divisible* resource (also called a *cake*) among a group of agents, each with their own preferences for different portions of the resource. Although the setting seems simple, the problem becomes interesting and nontrivial when specific properties (*e.g.*, fairness, efficiency, truthfulness, etc.) must be met by the output allocation.

How to design an allocation that obtains a high *efficiency* when *fairness* is guaranteed is an important research question of cake-cutting and has garnered significant interest due to its wide range of applications (*e.g.*, network routing, public traffic, etc.). Many previous studies have attempted to investigate the correlation between fairness and efficiency. For example, there are some topics like the price of fairness (Bertsimas et al. 2011; Roughgarden 2010; Caragiannis et al. 2012), the complexity of computing the most efficient allocation subject to fairness constraints (Cohler et al. 2011; Bei et al. 2012; Aumann et al. 2013). The term "efficiency" mentioned above mostly refers to *social welfare*, which is defined as the sum of the agents' utilities, and the above-mentioned papers

© Springer Nature Switzerland AG 2023
M. Li et al. (Eds.): IJTCS-FAW 2023, LNCS 13933, pp. 263–280, 2023.
https://doi.org/10.1007/978-3-031-39344-0_20

formulate the problem as a constrained optimization problem where a fairness notion is served as a constraint and the social welfare is the object that we would like to optimize.

Other than considering a combination of two separate notions (one for efficiency and one for fairness), there are other allocation criteria/measurements that embed both fairness and efficiency. *Egalitarian welfare*, defined as the minimum utility among all agents, is a measurement of both efficiency and fairness. It is a natural measurement of allocation efficiency. Moreover, from a fairness perspective, acquiring excess value for an arbitrary individual may not contribute to the improvement of egalitarian welfare, as it primarily focuses on the agent with the lowest value, so we tend to allocate the remaining resources to this specific agent to make the allocation fairer. In addition, an allocation that maximizes the egalitarian welfare also satisfies some common fairness criteria such as *proportionality*, where an allocation is proportional if each agent receives a share that is worth at least $1/n$ fraction of her total value of the entire cake (where n is the number of the agents).[1] In our work, we study the complexity of computing an allocation with *maximum egalitarian welfare* (MEW) in the setting where all the value density functions are piecewise-constant and all the agents are hungry.

Comparison with Other Notions. Other than egalitarian welfare, other notable notions concerning both efficiency and fairness are *Nash welfare* and *leximin*.

Nash welfare is defined as the product of agents' utilities. It is known that an allocation maximizing the Nash welfare satisfies the fairness notion *envy-freeness* (Kelly 1997), which is stronger than the proportionality mentioned earlier. Compared with Nash welfare, egalitarian welfare puts more focus on the least happy agent.

An allocation is *leximin* if it maximizes the utility to the least happy agent, and, subject to this, it maximizes the utility to the second least happy agent, and so on. Clearly, a leximin allocation always maximizes egalitarian welfare, and the notion of leximin places further requirements on the other agents. However, in many settings, maximizing egalitarian welfare is already NP-hard. We can then study the design of *approximation algorithms* for maximizing egalitarian welfare. Unlike egalitarian welfare which has a single objective to maximize, it is unclear how to define the approximation version of leximin.

1.1   Related Work

Fairness Notions. Apart from egalitarian welfare, a well-studied fairness notion is *proportionality*, which states that each agent receives at least an average value from her perspective. When there are two agents, proportionality can be easily achieved through the *"I cut, you choose"* algorithm, where the first agent cuts

---

[1] It is well known that a proportional allocation always exists even if we require each agent to receive a connected piece (see, e.g., (Dubins and Spanier 1961)). Therefore, there exists an allocation with egalitarian welfare of at least $1/n$, and the allocation with optimal egalitarian welfare is proportional.

the cake into two pieces that she thinks to be equal, then the second agent chooses one piece that she thinks of higher value. For any number of agents, two well-known algorithms to guarantee proportional are Dubins-Spanier Dubins and Spanier (1961) devised in 1961 and Even-Paz Even and Paz (1984) in 1984.

Another fairness notion is *envy-freeness*, which means each agent prefers her bundle over any other agent. It is a stronger notion than proportionality, and an envy-free allocation is also proportional. For two agents, the "I cut, you choose" still works. However, Dubins-Spanier and Even-Paz algorithms for general number of agents are not envy-free. For three agents, an elegant algorithm to compute an envy-free allocation is Selfridge-Conway (Brams and Taylor 1995). Furthermore, it has been confirmed that for any number of agents, an envy-free allocation always exists Aziz and Mackenzie (2017), even if only $n - 1$ cuts are allowed (Brams and Taylor 1995).

The third wide-studied notion is equitability. Incomparable to envy-freeness or proportionality, each agent is assigned a piece of the same value. An equitable allocation can be achieved by Austin moving-knife procedure for two agents, and if only one cut is allowed, it can be calculated with full knowledge of the partners' valuations Jones (2002); Brams et al. (2006). Further, if more than two cuts are allowed, we can achieve an equitable allocation that is also envy-free Barbanel and Brams (2011). Austin moving-knife and full revelation procedure can also be extended to any number of agents, while the latter one still works under contiguous constraints. However, equitability and envy-freeness are not compatible under such constraints for three or more agents Brams et al. (2006).

Price of Fairness. The price of fairness (POF) is defined as the worst-case ratio between the optimal welfare obtained and the maximum welfare obtained by a fair allocation. In Caragiannis et al. (2012)'s work, they have shown, the price of envy-freeness and proportionality for two agents is $8 - 4\sqrt{3}$. For n agents, they show the price of proportionality is $\Theta(\sqrt{n})$ and the price of equitability is $\Theta(n)$. The price of proportionality directly implies that the lower bound of the price of envy-freeness is $\Omega(\sqrt{n})$ as envy-freeness implies proportionality. Bertsimas et al. (2011) further shows the upper bound of the price of envy-freeness is $O(\sqrt{n})$, concluding the price of envy-freeness is also $\Theta(\sqrt{n})$.

MEW in Fair Division. In cake-cutting, the problem of computing MEW (Maximum Egalitarian Welfare) has been proven to have a 2-inapproximation ratio by Aumann et al. (2013).[2] With the number of agents being the parameter, they also developed an FPT PTAS (Polynomial Time Approximation Scheme) for egalitarian welfare objectives. For indivisible goods, the best-known polynomial-time algorithm can achieve an approximation factor of $O\left(\sqrt{n}\log^3 n\right)$ (Asadpour and Saberi 2010), and this problem has been proven to have a 2-inapproximation ratio (Aumann et al. 2013).

---

[2] The 2-inapproximation result is in the arXiv version of the paper Aumann et al. (2013).

1.2 Our Results

In our paper, we mainly study the complexity of computing MEW allocation in two settings. In the first setting, each bundle is not required to be *contiguous* (i.e. each agent could receive a lot of scattered intervals). This case is relatively straightforward, and we demonstrate that the MEW allocation can be directly obtained through linear programming. In the second setting, a stricter constraint is imposed where each bundle must be contiguous. Despite the moving-knife procedure being able to output an allocation with $\frac{1}{n}$ egalitarian welfare, it fails to extend for computing an MEW allocation. We design an FPT algorithm with respect to the number of agents for computing an MEW allocation, which improves the previous result of FPT PTAS by Aumann et al. (2013). Finally, we prove that this problem is NP-hard to approximate to within any constant factor by a reduction from 3-SAT, which significantly improves the 2-inapproximability by Aumann et al. (2013).

2 Preliminaries

Let $[n] = \{1, \ldots, n\}$. In the cake-cutting problem, the cake is modeled as an interval $[0, 1]$ and is required to allocate to a set of $N = [n]$ agents. Each agent i has a non-negative value density function f_i over the cake $[0, 1]$. In particular, agent i's utility v_i to a subset X of the cake is defined as $v_i(X) = \int_X f_i(x)dx$. Throughout this paper, we assume her value density function is piecewise-constant, that is, f_i is constant on each contiguous interval separated by a finite number of breakpoints. We also assume the agents are *hungry* and *normalized*, that is, f_i is strictly positive at any point of $[0, 1]$ and for any $i \in [n]$, $v_i([0, 1]) = 1$.

An allocation $\mathcal{A} = (A_1, \ldots, A_n)$ is a partition of the cake, where $A_i \cap A_j = \emptyset$ for any i, j and $\cup_{i=1}^n A_i = [0, 1]$. Among those bundles, bundle A_i is allocated to agent i. We say an allocation $\mathcal{A} = (A_1, \ldots, A_n)$ is *equitable* if all the agents get the exact same utility (by their own valuations). Formally, for each $i, j \in [n]$, $v_i(A_i) = v_j(A_j)$. Bundle A_i is called *contiguous* if it contains a single interval. An allocation with contiguous pieces requires that each bundle in the allocation is contiguous, so such an allocation contains only $n - 1$ cuts.

The egalitarian welfare ($\mathcal{EW}$) of an allocation $\mathcal{A}$ is defined as

$$\mathcal{EW}(\mathcal{A}) \triangleq \min_{i \in [n]} v_i(A_i).$$

Clearly, if an allocation $\mathcal{A}$ is *proportional* (i.e. $v_i(A_i) \geq \frac{1}{n} v_i([0, 1])$), then $\mathcal{EW}(\mathcal{A})$ is at least $\frac{1}{n}$. A common approach for computing a proportional allocation is called *moving-knife procedure*, which is defined as follows,

Definition 1 (Moving-knife Procedure). *We set a threshold $\theta = \frac{1}{n}$, and let a knife moves from the left of the cake to the right. An agent calls 'stop' when her value to the interval left to the knife reaches θ, and we cut the cake and allocate it to this agent.*

In our paper, we mainly study the problem of computing an allocation that maximizes egalitarian welfare (MEW). We respectively discuss two scenarios when the bundles could be non-contiguous and all of them must be contiguous.

Our technique for solving the above problem includes linear programming, and it is known that an optimal vertex solution can be found in polynomial time.

Lemma 1 (Güler et al. (1993)). *For a linear program* $\max\{\mathbf{c}^\top\mathbf{x} : \mathbf{Ax} \leq \mathbf{b}, \mathbf{x} \geq \mathbf{0}\}$, *an optimal solution (if exists) can be found in polynomial time.*

To warm up, let us consider a simple scenario where each bundle may not be contiguous. In Theorem 1, we demonstrate that the MEW allocation can always be efficiently obtained via linear programming in polynomial time.

Theorem 1. *If all the bundles are not required to be contiguous, an MEW allocation can be found in polynomial time.*

Proof. First, assume all the different breakpoints of $f_1, \ldots, f_n$ are $0 = p_0 < p_1 < \ldots < p_m = 1$, where $m \in \mathbb{Z}^+$. Let variable $x_{i,k}$ represent the fractional ratio that agent i receives from the interval $[p_{k-1}, p_k]$. Consider the following linear program:

$$\max \quad \mathcal{EW}(\mathcal{A})$$
$$\textbf{subject to } x_{1,k} + \ldots + x_{n,k} = 1, \qquad\qquad k \in [m] \quad (1)$$
$$\sum_{k=1}^{m} x_{i,k} v_i \left([p_{k-1}, p_k]\right) \geq \mathcal{EW}(\mathcal{A}), \qquad\qquad i \in [n] \quad (2)$$
$$0 \leq x_{i,k} \leq 1, \qquad\qquad i \in [n], k \in [m] \quad (3)$$

Within the above linear program, the constraints (1) ensures each interval is exactly allocated and the constraints (2) ensure each agent's utility is no less than the objective function $\mathcal{EW}(\mathcal{A})$. Clearly, $\{x_{i,k}\}_{i\in[n],k\in[m]}$ could describe an allocation and it is MEW. Since the linear program can be solved in polynomial time, the theorem concludes.   □

3   Maximize Egalitarian Welfare with Contiguous Pieces

In this section, we will discuss our results when only $n - 1$ cuts are allowed (i.e. each agent's bundle should be contiguous). As mentioned before, the moving-knife procedure could find a proportional and contiguous allocation, whose egalitarian welfare is already $\frac{1}{n}$. Inspired by this, a natural idea to compute an MEW allocation is to first guess the value of the egalitarian welfare, then adopt the moving-knife algorithm where θ is set as the guessed value. If the moving-knife procedure works, we can trivially achieve a PTAS algorithm. Unfortunately, when the threshold θ exceeds $\frac{1}{n}$, moving-knife may fail to find an allocation with egalitarian welfare θ when such allocation exists.

$$f_1(x) = \begin{cases} 2 - \epsilon, & x \in [0, \frac{1}{4}] \cup [\frac{3}{4}, 1] \\ \epsilon, & x \in (\frac{1}{4}, \frac{3}{4}). \end{cases} \qquad f_2(x) = f_3(x) = \begin{cases} \frac{4}{3} - \frac{2\epsilon}{3}, & x \in [0, \frac{3}{4}] \\ 2\epsilon, & \text{otherwise.} \end{cases}$$

Counter-Example of Moving-Knife. Consider a counter-example with three agents. Assume ϵ is sufficiently small, and we are given the value density functions as follows,

In the example, the optimal allocation is $A_1 = [\frac{3}{4}, 1]$, $A_2 = [0, \frac{3}{8}]$ and $A_3 = [\frac{3}{8}, \frac{3}{4}]$ with MEW as $\frac{1}{2} - \frac{\epsilon}{4}$. However, it cannot be found through the moving-knife algorithm, as agent 1 first calls stop at $\frac{1}{4}$, and assume we allocate the second interval with value $\frac{4}{3} - \frac{\epsilon}{3}$ to agent 2. When $\epsilon \to 0$, agent 3 can only receive the remaining part with value $\left(\frac{4}{3} - \frac{2\epsilon}{3}\right) \times \left(\frac{3}{4} - \frac{1}{4} - \frac{3}{8}\right) = \frac{1}{6} - \frac{\epsilon}{12} < \frac{1}{2} - \frac{\epsilon}{4}$. Further, in Sect. 3.2, we provide a constant-inapproximability result.

3.1 Agents with Fixed Permutation

We first consider an easy case when the order of agents receiving the cake is fixed, that is, for two agents i and j where $i < j$ in the permutation, i needs to receive a bundle before j. We show that the optimal allocation could be found in polynomial time. An observation is that we could directly extend it to the general case of constant n by enumerating all the permutations and handling each of them.

Next, we begin to present our algorithm for a fixed permutation. As shown in Algorithm 1, we design a linear programming-based algorithm to compute an optimal allocation. In our latter analysis, without loss of generality, we assume the fixed permutation is just $(1, 2, \ldots, n)$. We define a series of knifes $\{\delta_0, \delta_1, \ldots, \delta_n\}$ where $\delta_i \in [0, 1]$, and let $\delta_0 = 0$. For each agent i, we consider allocating the interval $[\delta_{i-1}, \delta_i]$ to her. Denote all the different breakpoints of $f_1, \ldots, f_n$ by $0 < p_0 < p_1 < \cdots < p_m = 1$, where $m \in \mathbb{Z}^+$. Let d_i be the distance between knife δ_i and its right-closest breakpoint $p_{\sigma(i)}$ (i.e. $d_i \triangleq \min_{p_k} (p_k - \delta_i)$) where $p_k > \delta_i$, and $p_{\sigma(i)} \triangleq \arg\min_{p_k} (p_k - \delta_i)$). Specifically, if $\delta_i = 1$, by this definition, $p_{\sigma(i)}$ and d_i would be 1 and 0, respectively.

$$\textbf{max} \quad \hat{y} \qquad (4)$$
$$\textbf{subject to } 0 \le x_i \le d_i, \qquad\qquad i \in [n] \quad (5)$$
$$v_{1,\sigma(1)}x_1 = \hat{y}, \qquad (6)$$
$$v_{i+1,\sigma(i+1)}x_{i+1} - v_{i+1,\sigma(i)}x_i = \hat{y}, \qquad i \in [n-1]. \quad (7)$$

Now, we move each knife δ_i from left to right within distance d_i to achieve a maximum increase in egalitarian welfare using linear program 4. In the linear program, we use $\hat{y}$ to denote the increase of egalitarian welfare and use x_i to denote the distance that knife δ_i moves. For simplicity, we denote agent i's value to an interval $[x_j, x_{j+1}]$ with a constant value density as $v_{i,j+1}$. The linear

---

Algorithm 1: Algorithm for computing maximum egalitarian welfare

---

Input: Each agent i's value density function f_i with all the breakpoints $p_{i \in [m]}$, and a permutation of agents $(i_1, i_2, \ldots i_n)$.

Output: An allocation $\mathcal{A}$ that maximizes egalitarian welfare.

1 Let $y \leftarrow 0$ denote the maximum egalitarian welfare under this permutation;
2 Let $\delta_k \leftarrow 0$ denote the knife position for agent i_k for each $k \in [n]$;
3 Let $\hat{y} \leftarrow \infty$ denote the increment of y within each iteration;
4 **while** $\hat{y} \neq 0$ **do**
5   Let $p_{\sigma(k)}$ denote the right-closest breakpoint of δ_k and $d_k \leftarrow p_{\sigma(k)} - \delta_k$;
6   Run `linear program` 4 to solve the optimal $\hat{y}$ and x_k;
7   Update $\delta_k \leftarrow \delta_k + x_k$ and $y \leftarrow y + \hat{y}$;

8 Let $A_i \leftarrow [\delta_{i-1}, \delta_i]$ for each $i \in [n]$;
9 **return** $\mathcal{A} = (A_1, \ldots, A_n)$

---

program is subject to, first, each agent's utility to her bundle has the same increase $\hat{y}$, hence, the output allocation is equitable. Second, to compute such an increase for each agent, we need to focus on the interval with constant value density, so the maximum distance that δ_i moves cannot exceed d_i. If δ_i moves and y increases, we update d_i and repeat the process. If $\hat{y} = 0$, we further prove y is the MEW value.

Theorem 2. *Algorithm 1 will output an MEW allocation for a fixed permutation of agents in polynomial time.*

Proof. Let $\text{OPT} = ([0, o_1], \ldots, [o_{n-1}, o_n])$ represent the optimal allocation. We first claim that OPT is both equitable and unique via the following two lemmas.

Lemma 2. *OPT is equitable.*

Proof. For the sake of contradiction, we assume there exist $i \neq j \in [n]$ such that $v_i([o_{i-1}, o_i]) \neq v_j([o_{j-1}, o_j])$. Hence, there exist two adjacent agents such that one has the smallest utility and the other has a higher utility. Otherwise, the utilities of all the agents would be equal, which violates our assumption. Without loss of generality, we assume agent i and agent $i + 1$ are such two agents. Due to the intermediate value theorem, we could find o'_i such that $v_i([o_{i-1}, o'_i]) = v_{i+1}([o'_i, o_{i+1}])$. This operation clearly improves agent i's utility and does not decrease agent $i + 1$'s utility to the original minimum. By repeating this operation, we can improve egalitarian welfare, which contradicts the assumption of the optimal solution. Note that the case where agent n has the smallest utility can be handled in a similar way. $\square$

Lemma 3. *An equitable allocation is also unique.*

Proof. By contradiction, we assume there are two equitable allocations, and denote their cut points as $\{x_0 = 0, x_1, \ldots, x_n = 1\}$ and $\{y_0 = 0, y_1, \ldots, y_n = 1\}$ respectively. Since the allocations are different, we can find a minimal interval

$[x_i, x_j]$ and $[y_i, y_j]$ such that $x_i = y_i, x_j = y_j$, and for any $k \in [i+1, j-1]$, $x_k \neq y_k$. Without loss of generality, we assume $x_{i+1} < y_{i+1}$. We consider the following two cases.

1. For all $k \in [i+2, j-1]$, $x_k < y_k$. Since $[x_i, x_{i+1}]$ is a subset of $[y_i, y_{i+1}]$ and each agent is hungry, we have $v_{i+1}([x_i, x_{i+1}]) < v_{i+1}([y_i, y_{i+1}])$. Similarly, $v_j([x_{j-1}, x_j]) > v_j([y_{j-1}, y_j])$. However, since the two allocations are equitable, we have $v_{i+1}([x_i, x_{i+1}]) = v_j([x_{j-1}, x_j])$ and $v_{i+1}([y_i, y_{i+1}]) = v_j([y_{j-1}, y_j])$, which leads to a contradiction.
2. Otherwise, there exists at least one index $k \in [i+2, j-1]$ such that $x_k > y_k$, We further assume k is the smallest index, i.e. for all $\ell \in [i+2, k-1]$, $x_\ell < y_\ell$. We already know $v_{i+1}([x_i, x_{i+1}]) < v_{i+1}([y_i, y_{i+1}])$. As $[x_{k-1}, x_k]$ is a superset of $[y_{k-1}, y_k]$, we have $v_j([x_{k-1}, x_k]) > v_j([y_{k-1}, y_k])$. Same to the analysis in the above case, this leads to a contradiction.

Combining the two above cases, the lemma is concluded.   □

Back to the original proof. Due to our previous description, the utilities of these agents keep the same during running Algorithm 1. If $\delta_n = 1$ when the algorithm terminates, according to Lemma 2 and Lemma 3, the output allocation must be the unique optimal solution. Otherwise, there will be an interval of cake left, and $\hat{y}$ will be zero. We claim this case cannot happen and δ_n will reach 1 when Algorithm 1 terminates.

Lemma 4. *If $\hat{y} = 0$, then $\delta_n = 1$.*

Proof. According to the definition of linear program 4, if $\hat{y} = 0$, that implies $x_1 = \ldots = x_n = 0$. If the proposition is false (i.e. $\delta_n < 1$), our main idea is to add a sufficiently small constant to each of x_i and $\hat{y}$ so that the solution remains feasible but has a larger objective value. Let x_i' (for $i = 1, \ldots, n$) and $\hat{y}'$ be the new values of x_i and $\hat{y}$. Consider the following three constants λ, μ, ϵ and the new objective value $\hat{y}'$,

$$\lambda = \max_{i \in [n-1]} \left\{ \frac{v_{i+1, \sigma(i)}}{v_{i+1, \sigma(i+1)}} \right\}, \mu = \max_{i \in [n-1]} \left\{ \frac{1}{v_{i+1, \sigma(i+1)}} \right\},$$

$$\epsilon = \min_{i \in [n]} \left\{ \frac{d_i}{2\lambda^{i-1}}, \frac{1}{v_{1, \sigma(1)}} \cdot \frac{d_i(1-\lambda)}{2\mu(1-\lambda^{n-1})} \right\}, \hat{y}' = v_{1, \sigma(1)}\epsilon.$$

Next, we set $x_1' = \varepsilon$ and $x_{i+1}' = \frac{v_{i+1, \sigma(i)}}{v_{i+1, \sigma(i+1)}} x_i' + \frac{\hat{y}'}{v_{i+1, \sigma(i+1)}}$ for $i = 1, \ldots, n-1$. Clearly, $\hat{y}'$ is strictly positive. After that, we claim that $(x_1', \ldots, x_n', \hat{y}')$ is also an feasible solution. It is not hard to verify the constraints (6) and (7) are satisfied by the definition of x_i' and $\hat{y}'$. Additionally, since $x_1' = \epsilon \leq \frac{1}{2} \min_{i \in [n]} \left(\frac{d_i}{\lambda^{i-1}} \right) \leq \frac{1}{2} d_1 \leq d_1$. For $i = 1, \ldots, n-1$, constraints (5) can be verified by the following inequality:

$$x'_{i+1} = \frac{v_{i+1,\sigma(i)}}{v_{i+1,\sigma(i+1)}}x'_i + \frac{\hat{y}'}{v_{i+1,\sigma(i+1)}} \le \lambda x'_i + \mu \hat{y}' \le \dots$$

$$\le \lambda^i x'_1 + \mu\left(1 + \dots + a_0^{i-1}\right)\hat{y}' \qquad \text{(By the definition of } a_0 \text{ and } b_0\text{)}$$

$$\le \frac{1}{2}d_{i+1} + \mu\left(1 + \dots + \lambda^{i-1}\right)\hat{y}' \qquad \text{(By the first part of the definition of } \epsilon\text{)}$$

$$\le \frac{1}{2}d_{i+1} + \frac{1}{2}d_{i+1} = d_{i+1}. \qquad \text{(By the second part of the definition of } \epsilon\text{)}$$

Thus, $(x'_1, \dots, x'_n, \hat{y}')$ is also feasible and has a larger objective value, leading to a contradiction. □

Due to Lemma 4 and our prior statements, the output allocation is equitable. Since the optimal allocation is unique and also equitable, we conclude that our algorithm achieves the optimal allocation.

Finally, we show this algorithm also runs in polynomial time. Within each iteration, we will find an optimal solution at a vertex of the feasible region. Since there are $3n$ constraints of the linear program and the vertex has $n+1$ dimensions, then at least one of the constraints (5) is tight. Suppose x_i satisfies $x_i = 0$ or d_i. If $x_i = 0$, then $x_i, \dots, x_1$ would be all zero, and $\hat{y}$ would also be zero, which contradicts to the assumption of $\hat{y} \neq 0$. Hence, $x_i = d_i$ and δ_i will be updated to $p_{\sigma(i)}$ at the end of this iteration.

We observe that all the knives would keep moving rightward during the algorithm and at least one would encounter a breakpoint within each iteration. Since there are only m breakpoints, the number of total iterations is at most $n \cdot m$. Within each iteration, the time complexity of solving the linear program is also polynomial. Thus, the overall time complexity is polynomial. □

Note that for general number of agents, we can still enumerate all the permutations and adopt Algorithm 1. This directly leads to the following result.

Corollary 1. *For general cases without any constraint on the permutation of the agents, the problem of computing an MEW allocation can be solved within a complexity of fixed-parameter tractable with respect to the number of agents n.*

3.2 Agents Without Permutation Constraints

In this section, we present the inapproximability results for a general number of agents and no constraints on the permutation of the agents. To illustrate our reduction, we first provide proof of the inapproximability of 2 in Theorem 3. Although it has been demonstrated by Aumann et al. (2013), we also give our proof here. Following that, we expand upon this theorem and adapt it to demonstrate a c-inapproximation ratio in Theorem 4, where c can be any constant.

Theorem 3 (Aumann et al. (2013)). *The problem of computing an MEW allocation with contiguous pieces is NP-hard to approximate to within factor 2 for a general number of agents with no constraint on permutation.*

We present a reduction from the 3-SAT problem. Given a 3-SAT instance Φ with m clauses and $3m$ literals, we construct an instance of maximizing egalitarian welfare with contiguous pieces with four types of agents: *literal agents, clause agents, logic agents,* and *blocking agents.* Let $\Phi_{i\in[m]}$ denote the i-th clause and $\Phi^{j\in[3m]}$ denote the j-th literal in Φ. Before presenting the formal construction of our reduction. We first show some high-level ideas as follows.

Ideas of Construction. For each clause, we introduce a clause agent c_i and three literal agents $\ell_{3i-2}, \ell_{3i-1}, \ell_{3i}$ for the literals within it. Then, for each literal agent $j \in \{3i+1, 3i+2, 3i+3\}$, let her have three disjoint valued pieces of cake $I_j^{(1)}, I_j^{(2)}, I_j^{(3)}$ such that there is a spacing between every two pieces. We denote those spacing by $s_{3i-2}^{(1)}, s_{3i-2}^{(2)}, s_{3i-1}^{(1)}, s_{3i-1}^{(2)}, s_{3i}^{(1)}, s_{3i}^{(2)}$. Within those spacing, we design to put other agents' valued intervals.

First, as shown in Fig. 1, we let the clause agent c_i have three valued pieces within $s_{3i-2}^{(2)}, s_{3i-1}^{(2)}, s_{3i}^{(2)}$. For each of the three literal agents, we refer to the first two of her valued pieces $I_j^{(1)}, I_j^{(2)}$ including the spacing $s_j^{(1)}$ as her *true interval* and $I_j^{(2)}, I_j^{(3)}$ including the spacing $s_j^{(2)}$ as her *false interval.* If Φ is satisfiable, we aim to let each literal agent receive her true interval $I_j^{(1)}, s_j^{(1)}, I_j^{(2)}$ if the corresponding literal is true in the assignment and false interval $I_j^{(2)}, s_j^{(2)}, I_j^{(3)}$ if the corresponding literal is false. If Φ is unsatisfiable, there always exists a clause such that all three literals are assigned false under any assignment. If we insist that the allocation represents a valid boolean assignment for Φ, there will exist a clause such that the literal agents of it will all receive their false intervals at the same time. That will cause the value received by the corresponding clause agent to be zero, which will lead to egalitarian welfare being zero. Hence, we could no longer make the allocation represent a valid boolean assignment.

To guarantee the consistency and validity of the assignment, we further add some constraints by introducing logic agents. There are two types of logic agents. The first type of logic agent ensures any literal agent ℓ_j cannot get a complete interval including $I_j^{(1)}, I_j^{(2)}$ and $I_j^{(3)}$ (so that we cannot assign both "false" and "true" to a single literal). Otherwise, there will exist a logic agent receiving zero utility. The second type of logic agent guarantees the consistency of the literals. For example, if a variable x occurs twice as the form of x in two literals Φ^{j_1} and Φ^{j_2}, then we introduce two logic agents t_x^1 and t_x^2. The logic agent t_x^1 has two valued pieces within $s_{j_1}^{(1)}$ and $s_{j_2}^{(2)}$, and t_x^2 has two valued pieces within $s_{j_1}^{(2)}$ and $s_{j_2}^{(1)}$. In this case, the literal agent ℓ_{j_2} cannot receive her false interval if ℓ_{j_1} receives her true interval, and the literal agent ℓ_{j_2} cannot receive her true interval if ℓ_{j_1} receives her false interval. Otherwise, t_x^1 or t_x^2 will receive a utility of zero. The case is similar when the literal agent ℓ_{j_1} receives her false interval.

Construction. The detailed construction is defined as follows:

- For each clause Φ_i, we construct a clause agent c_i.
- For each literal Φ^j, we construct a literal agent ℓ_j.

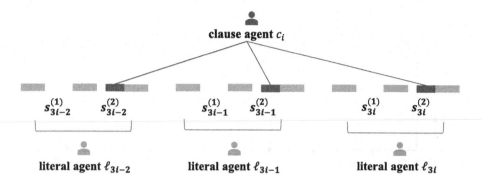

Fig. 1. Construction of clause agents and literal agents.

- Logic agents further contain two types: (1) For each literal Φ^j, we construct a logic agent t_j. (2) If a variable x as well as its negation $\bar{x}$ appear totally k times where $k > 1$, we construct $2(k-1)$ logic agents $t_x^1, \ldots, t_x^{2k-1}$.
- Further, we construct $9m$ blocking agents $d_1, \ldots, d_{9m}$.

For simplification, we extend the cake from interval $[0, 1]$ to $[0, 36]$ while each agent's value to the cake remains normalized. We refer to the interval $[0, 18m]$ as the *actual cake*, and $[18m, 36m]$ as the *dummy cake*. Each agent's value density function to the cake is constructed as follows.

Let $\epsilon > 0$ be a sufficiently small real number. In order to simplify the description of the value density function, we will only define the parts where the density is not equal to ϵ. In fact, in our later argument, we will regard ϵ as 0. It will not affect our proof of inapproximability ratio.

Figure 2 illustrates our construction of the value density functions for both literal agents and clause agents on the actual cake. For clarity, for each literal agent $\ell_{j \in [3m]}$, she has three disjoint valued intervals on the actual cake.

$$f_{\ell_j}(x) = \frac{1}{6}, \text{ for } x \in [6j-6, 6j-5] \cup [6j-4, 6j-3] \cup [6j-2, 6j-1].$$

Let $\hat{k}$ be the maximum number of occurrences of any variable x in Φ, formally,

$$\hat{k} \triangleq \max_x \sum_{j \in [3m]} \mathbf{1}\left(\Phi^j = x \vee \Phi^j = \bar{x}\right).$$

Then, we evenly divide the spacing interval $[6j-5, 6j-4]$ into $\hat{k}$ disjoint sub-intervals and $[6j-3, 6j-2]$ into $\hat{k}+1$ disjoint sub-intervals. For each clause agent $c_{i \in [m]}$, she only has value to three disconnected intervals. When $x \in [18i - 14 - \frac{1}{k+1}, 18i-14] \cup [18i-8-\frac{1}{k+1}, 18i-8] \cup [18i-2-\frac{1}{k+1}, 18i-2]$, $f_{c_i}(x) = \frac{k+1}{3}$.

As shown in Fig. 3, for each logic agent t_j of the first type constructed from Φ^j, we define her value density function as follows,

$$f_{t_j}(x) = \begin{cases} \frac{\hat{k}}{2}, & x \in [6j-5, 6j-5+\frac{1}{k}] \\ \frac{k+1}{2}, & x \in [6j-3, 6j-3+\frac{1}{k+1}]. \end{cases}$$

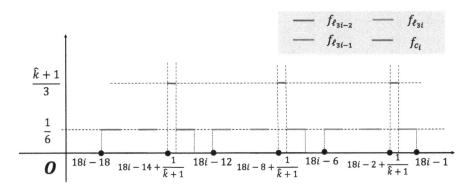

Fig. 2. Value density function of literal agents and clause agents.

For each logic agent of the second type, we consider her corresponding variable y. Assume y and $\bar{y}$ appears k times totally in Φ. We only focus on the case where $k > 1$. Assume y first appears in literal Φ^a, and without loss of generality, assume its (including $\bar{y}$) s-th appearance $(2 \leq s \leq k)$ is in literal Φ^b. We construct two literal agents of the second type t_y^{2s-3} and t_y^{2s-2}.

If $\Phi^b = y$, we define their value density functions as

$$f_{t_y^{2s-3}}(x) = \begin{cases} \frac{\hat{k}}{2}, & x \in [6a - 5 + \frac{s-1}{k}, 6a - 5 + \frac{s}{k}] \\ \frac{\hat{k}+1}{2}, & x \in [6b - 3 + \frac{s-1}{k+1}, 6b - 3 + \frac{s}{k+1}], \end{cases}$$

$$f_{t_y^{2s-2}}(x) = \begin{cases} \frac{\hat{k}+1}{2}, & x \in [6a - 3 + \frac{s-1}{k+1}, 6a - 3 + \frac{s}{k+1}] \\ \frac{\hat{k}}{2}, & x \in [6b - 5 + \frac{s-1}{k}, 6b - 5 + \frac{s}{k}]. \end{cases}$$

If $\Phi^b = \bar{y}$, we swap the second interval of the two agents.

$$f_{t_y^{2s-3}}(x) = \begin{cases} \frac{\hat{k}}{2}, & x \in [6a - 5 + \frac{s-1}{k}, 6a - 5 + \frac{s}{k}] \\ \frac{\hat{k}}{2}, & x \in [6b - 5 + \frac{s-1}{k}, 6b - 5 + \frac{s}{k}], \end{cases}$$

$$f_{t_y^{2s-2}}(x) = \begin{cases} \frac{\hat{k}+1}{2}, & x \in [6a - 3 + \frac{s-1}{k+1}, 6a - 3 + \frac{s}{k+1}] \\ \frac{\hat{k}+1}{2}, & x \in [6b - 3 + \frac{s-1}{k+1}, 6b - 3 + \frac{s}{k+1}]. \end{cases}$$

In addition to the above intervals for literal agents, each literal agent also has three disconnected valued intervals on the dummy cake. When $x \in [18m + 6j - 6, 18m + 6j - 5] \cup [18m + 6j - 4, 18m + 6j - 3] \cup [18m + 6j - 2, 18m + 6j - 1]$, $f_{\ell_j}(x) = \frac{1}{6}$ (Fig. 4).

The blocking agents are designed to block the valuable intervals of the above agents. Each blocking agent has value to one piece between them:

$$f_{d_j}(x) = 1, \text{ for } x \in [18m + 2j - 1, 18m + 2j] \text{ and } j \in [1, 9m].$$

Under this construction, the integral of the value density function for each agent over the entire cake is equal to 1. Now we are ready to provide the formal proof.

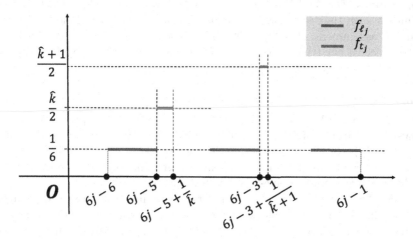

Fig. 3. The value density function of t_j. In this case, literal agent ℓ_j (whose value density function is colored red) cannot receive the three intervals $[6j-6, 6j-5]$, $[6j-4, 6j-3]$ and $[6j-2, 6j-1]$ at the same time. Otherwise, the utility of logic agent t_j (whose value density function is colored green) will be zero. (Color figure online)

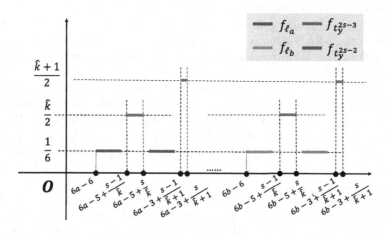

Fig. 4. The value density functions of t_y^{2s-3} and t_y^{2s-2} when literals Φ^a, Φ^b are both y. In this case, when literal agent ℓ_a (whose value density function is colored red in the above figure) receives her true intervals, literal agent ℓ_b (whose value density function is colored yellow) is forbidden to receive her false intervals. Otherwise, logic agent t_y^{2s-3} (whose value density function is colored green) will have zero utility. For a similar reason, literal agents ℓ_a and ℓ_b cannot respectively receive their true intervals and false intervals at the same time. Otherwise, logic agent t_y^{2s-2} (whose value density function is colored blue) will have zero utility. The construction of the case when $\Phi^a = y$ and $\Phi^b = \bar{y}$ is similar. (Color figure online)

Proof. Suppose the 3-SAT instance is a yes instance (i.e., there exists a valid assignment such that $\Phi = true$). We construct an allocation as follows.

In the assignment, if a literal Φ^j is assigned *true*, then we let the literal agent ℓ_j receive her true interval $[6j - 6, 6j - 3]$. If the clause agent $c_{\lceil j/3 \rceil}$ has not received any interval, we let her take the interval $\left[6j - 2 - \frac{1}{k+1}, 6j - 2\right]$. Further, we let all the logic agents that have value on $\left[6j - 3, 6j - 2 - \frac{1}{k+1}\right]$ receive their valued intervals. Otherwise, if Φ^j is assigned *false*, we let ℓ_j receive her false interval $[6j - 4, 6j - 1]$, and let all the logic agents receive their valued intervals on $[6j - 5, 6j - 4]$. Further, we let each blocking agent receive her valued interval. For each contiguous interval that remains unallocated, we allocate it to an arbitrary agent who receives the interval adjacent to it.

Under this allocation, it is straightforward to find that each literal agent receives value $\frac{1}{3}$, while each blocking agent receives a value of 1. For each clause Φ_i, there exists at least one literal $\Phi^{j \in [3i-2,3i]}$ assigned *true*. Therefore, the interval received by each clause agent c_i will be at least $\frac{1}{3}$. For each logic agent t_j of the first type, since ℓ_j only receives one of $[6j - 5, 6j - 4]$ and $[6j - 3, 6j - 2]$, she can always receive an interval with value $\frac{1}{2}$. For each logic agent of the second type t_y^j, denote the two intervals that she has value to by $J_{y_1}^j$ and $J_{y_2}^j$. Assume y appears k times, then $k - 1$ logic agents can receive $J_{y_1}^j$ and obtain value $\frac{1}{2}$ on the first appearance of y. Since the assignment is consistent, by our construction, the other $k - 1$ agents can always receive $J_{y_2}^j$ with value $\frac{1}{2}$.

Hence, the maximum egalitarian welfare under a yes instance is at least $\frac{1}{3}$.

Suppose the 3-SAT instance is a no instance, that is, there is no assignment such that $\Phi = true$. We prove the maximum egalitarian welfare will be at most $\frac{1}{6}$ by contradiction.

Assuming the maximum egalitarian welfare is more than $\frac{1}{6}$, then, each literal agent ℓ_j needs to receive a length of more than 2 on the cake (since the maximum density of f_{ℓ_j} is $\frac{1}{6}$). She cannot receive anything from the dummy cake, otherwise, at least one blocking agent will receive a utility of zero. Therefore, we only consider the case that each literal agent receives only part of the actual cake, hence she will surely receive an interval containing either $[6j - 5, 6j - 4]$ or $[6j - 3, 6j - 2]$.

Denote the corresponding variable of Φ^j as y. To ensure each logic agent of the second type receives more than $\frac{1}{6}$, all the literal agents ℓ_a where $\Phi^a = y$ cannot receive $[6j - 3, 6j - 2]$ and all the literal agents ℓ_b where $\Phi^b = \bar{y}$ cannot receive $[6j - 5, 6j - 4]$. Moreover, to ensure each clause agent receives more than $\frac{1}{6}$, at least one literal agent in each clause cannot receive $[6i - 3, 6i - 2]$. If such allocation exists, we can construct an assignment that if the literal agent ℓ_j receives $[6j - 5, 6j - 4]$, we assign $\Phi^j = true$. If the literal agent ℓ_j receives $[6j - 3, 6j - 2]$, we assign $\Phi^j = false$. The assignment is consistent according to the above analysis, and a literal is assigned *true* in each clause.

Hence, we conclude Φ is satisfiable, which leads to the contradiction. Therefore, the inapproximability is at least $\frac{\frac{1}{3}}{\frac{1}{6}} = 2$ and the theorem holds.   □

Theorem 4. *The problem of computing MEW with contiguous pieces is NP-hard to approximate within any constant factor for a general number of agents with no constraint on permutation.*

Construction. We follow the same idea as the construction in the above example on the actual cake while the dummy cake is not needed in the following construction. Let $r \in \mathbb{Z}^+$ be a constant integer.

- For literal Φ^j, we still construct a literal agent ℓ_j. Instead of constructing three disconnect intervals that she has value to, we construct $2r + 1$ such disconnect intervals. Denote these intervals by $\mathcal{I}_j = \{I_j^{(1)}, \ldots, I_j^{(2r+1)}\}$, and the interval that splits $I_j^{(k)}$ and $I_j^{(k+1)}$ by $s_j^{(k)}$. All these intervals are disjoint with each other. In our new construction, we call the interval from $I_j^{(1)}$ to $I_j^{(r+1)}$ as the agent ℓ_j's true interval T_j (which begins at the leftmost point of $I_j^{(1)}$ and ends at the rightmost point of $I_j^{(r+1)}$), and the interval from $I_j^{(r+1)}$ to $I_j^{(2r+1)}$ as false interval F_j (which begins at the leftmost point of $I_j^{(r+1)}$ and ends at the rightmost point of $I_j^{(2r+1)}$).
  Our goal is to make a literal agent receive the entire true interval when the 3-SAT instance is a yes instance. In the case of a no instance, we want to limit her bundle to at most one interval of $\mathcal{I}_j$.
- For clause Φ_i, we construct r^3 clause agents instead of only one clause agent in the previous construction. In particular, for each of the r^3 combinations $\{s_{3i-2}^{(j_1)} \subset F_{3i-2}, s_{3i-1}^{(j_2)} \subset F_{3i-1}, s_{3i}^{(j_3)} \subset F_{3i}\}$, we construct a clause agent that has value to a sub-intervals of each of the three disconnect intervals. The design of the clause agent in this context is adapted from the previous proof. If the three literal agents of the same clause all receive their false interval, there will always exist a clause agent receiving zero utility.
- For each literal, we construct r^2 logic agents of the first type. In particular, for each of the r^2 combinations $\{s_i^{(j_1)} \subset T_i, s_i^{(j_2)} \subset F_i\}$, we construct a logic agent of the first type that has value to a sub-intervals of each of the two disconnected intervals. These agents are also used to prevent any literal agent from receiving her true and false intervals at the same time.
- For each variable z, assume it first appears in Φ^i. If it appears more than once, for each of its appearances in Φ, we construct $2r^2$ logic agents of the second type. In particular, if $\Phi^j = z$, the first r^2 agents have value to the sub-intervals of each combination $\{s_i^{(i')} \subset T_i, s_j^{(j')} \subset F_j\}$, and the latter r^2 agents have value to the sub-intervals of each combination $\{s_i^{(i')} \subset F_i, s_j^{(j')} \subset T_j\}$. Otherwise, if $\Phi^j = \bar{z}$, the first r^2 agents have value to the sub-intervals of each combination $\{s_i^{(i')} \subset T_i, s_j^{(j')} \subset T_j\}$, and the latter r^2 agents have value to the sub-intervals of each combination $\{s_i^{(i')} \subset F_i, s_j^{(j')} \subset F_j\}$.

After constructing these three types of agents, if there are n' agents that have value to a sub-interval of $s_i^{(j)}$, we divide $s_i^{(j)}$ into n' disjoint intervals, and let each agent has value only on one sub-interval. Moreover, for each agent, if there are multiple intervals that she has value to, we let her value to each of these intervals equal and normalized to 1 in total.

Formal Proof. Assume Φ is a yes instance, then there exists a valid assignment such that $\Phi = true$. We construct an allocation $\mathcal{A}$ from a satisfying assignment as follows.

We traverse j from 1 to $3m$. If the literal Φ^j is assigned *true*, then we let literal agent ℓ_j receive her true interval T_j (i.e., $\mathcal{A}_{\ell_j} = T_j$). If a corresponding clause agent or a logic agent has not received any interval, we let her receive a sub-interval of $s_j^{(k)} \in F_j$ that she has value to. If Φ^j is assigned *false*, we let ℓ_j receive her false interval F_j (i.e., $\mathcal{A}_{\ell_j} = F_j$). We also let each of the corresponding logic agents that has value on T_j and has not received any bundle receive an interval that she has value to. For each contiguous interval that remains unallocated, we allocate it to an arbitrary agent who receives the interval adjacent to it.

Then we show that $\mathcal{EW}(\mathcal{A})$ is at least $\frac{1}{3}$. Since each literal agent takes either her true interval or false interval, she will receive $\frac{1}{2}$ in $\mathcal{A}$. Since $\Phi = true$ under this assignment, in each clause, there is at least one literal that is assigned *true*, so all the r^3 clause agents can receive an interval from this literal agent's false interval and obtain value $\frac{1}{3}$. Each logic agent of the first type can receive value $\frac{1}{2}$ as the corresponding literal agent can only receive either her true interval or her false interval. Since the assignment is consistent, each logic agent of the second type can also receive $\frac{1}{2}$ for the same reason as we have shown in the 2-inapproximation result. Therefore, the maximum egalitarian welfare under a yes instance is $\frac{1}{3}$.

Now we consider the case that Φ is a no instance, so no assignment can satisfy $\Phi = true$. We prove $\mathcal{EW}(\mathcal{A})$ will be no more than $\frac{1}{2r+1}$ for any allocation $\mathcal{A}$ by contradiction.

Suppose the maximum egalitarian welfare is more than $\frac{1}{2r+1}$, we show that there will exist a satisfying assignment. Assume for each literal agent ℓ_i, $|I_i^{(j)}| = |s_i^{(j)}| = 1$. Then, each literal agent needs to receive an interval J_i with $|J_i| > 2$, $|J_i \cap \mathcal{I}_i| > 1$ and $|J_i \setminus \mathcal{I}_i| \geq 1$. J_i cannot intersect with both T_i and F_i, otherwise, there will be at least one logic agent of the first type that receives 0. We assume $\Phi^i = z$ and, without loss of generality, $J_i \subseteq T_i$. Then, to avoid any logic agent of the second type receiving 0, for any j such that $\Phi^j = z$, we need to allocate an interval $J_j \subseteq T_j$ to ℓ_j using the similar analysis in the 2-inapproximation result. If $\Phi^j = \bar{z}$, ℓ_j needs to receive an interval $J_j \subseteq F_j$. Further, to avoid any clause agent receiving 0, at least one literal agent corresponding to this clause cannot receive any sub-interval from $s_j^{(r+1)}$ to $s_j^{(2r)}$. Hence, she needs to receive a bundle on the true interval. Then, we consider the following assignment. For a literal agent ℓ_i, if $J_i \subseteq T_i$, we assign Φ^i as *true*. If $J_i \subseteq F_i$, we assign Φ^i as *false*. From the above analysis, the assignment is consistent, and there is a *true* literal in each clause, indicating $\Phi = true$ under this assignment and leading to a contradiction.

We have shown that the inapproximability factor is $\frac{1/3}{1/(2r+1)} = \frac{2r+1}{3}$. Since r can be any constant, Theorem 4 holds. $\square$

4   Conclusion and Future Work

In this paper, we consider the problem of maximizing egalitarian welfare for hungry agents with and without contiguous constraints respectively. In the absence of contiguous constraints, the problem can be readily solved using linear programming. With contiguous constraints, there exists a non-trivial algorithm to output an optimal allocation for agents with fixed permutation in polynomial time, which indicates a fixed parameter tractable algorithm with respect to the number of agents n. Moreover, if a fixed permutation is not required, we provide a constant-inapproximability result.

As we consider hungry agents in our work, an interesting future direction is to generalize the setting to agents without hungry assumption. An intuitive idea is to first apply our algorithm, however, as an agent may have a value of 0 to the next interval, the egalitarian welfare we calculate may not increase and the knives may not move. In this case, we may deploy another linear program to maximize the summation of the distance of each knife's position while maintaining that each agent's utility to her contiguous piece does not decrease. Further, we may consider a more general case where the constraints for contiguous are subjective to the agents, which means some agents can receive a non-contiguous bundle while others need to receive a contiguous one.

References

Asadpour, A., Saberi, A.: An approximation algorithm for max-min fair allocation of indivisible goods. SIAM J. Comput. **39**(7), 2970–2989 (2010). https://doi.org/10.1137/080723491

Aumann, Y., Dombb, Y.: The Efficiency of Fair Division with Connected Pieces. ACM Trans. Econ. Comput. **3**(4), 1–16 (2015)

Aumann, Y., Dombb, Y., Hassidim, A.: Computing socially-efficient cake divisions. In: Proceedings of the International Conference on Autonomous Agents and Multi-Agent Systems (AAMAS), pp. 343–350 (2013)

Aziz, H., Mackenzie, S.: A discrete and bounded envy-free cake cutting protocol for any number of agents. arXiv:1604.03655 [cs.DS] (2017)

Barbanel, J.B., Brams, S.J.: Two-person cake-cutting: the optimal number of cuts. Available at SSRN 1946895 (2011)

Bei, X., Chen, N., Hua, X., Tao, B., Yang, E.: Optimal proportional cake cutting with connected pieces. In: Proceedings of AAAI Conference on Artificial Intelligence (AAAI), pp. 1263–1269 (2012)

Bei, X., Chen, N., Huzhang, G., Tao, B., Wu, J.: Cake cutting: envy and truth. In: IJCAI, pp. 3625–3631 (2017)

Bertsimas, D., Farias, V.F., Trichakis, N.: The price of fairness. Oper. Res. **59**(1), 17–31 (2011)

Brams, S.J., Feldman, M., Lai, J., Morgenstern, J., Procaccia, A.D.: On maxsum fair cake divisions. In: Proceedings of the AAAI Conference on Artificial Intelligence (AAAI), pp. 1285–1291 (2012)

Brams, S.J., Jones, M.A., Klamler, C., et al.: Better ways to cut a cake. Not. AMS **53**(11), 1314–1321 (2006)

Brams, S.J., Taylor, A.D.: An envy-free cake division protocol. Am. Math. Monthly **102**(1), 9–18 (1995). http://www.jstor.org/stable/2974850

Caragiannis, I., Kaklamanis, C., Kanellopoulos, P., Kyropoulou, M.: The efficiency of fair division. Theory Comput. Syst. **50**(4), 589–610 (2012)

Cohler, Y.J., Lai, J.K., Parkes, D.C., Procaccia, A.D.: Optimal envy-free cake cutting. In: Proceedings of AAAI Conference on Artificial Intelligence (AAAI), pp. 626–631 (2011)

Dubins, L.E., Spanier, E.H.: How to cut a cake fairly. Am. Math. Monthly **68**(1P1), 1–17 (1961)

Even, S., Paz, A.: A note on cake cutting. Discret. Appl. Math. **7**(3), 285–296 (1984)

Güler, O., den Hertog, D., Roos, C., Terlaky, T., Tsuchiya, T.: Degeneracy in interior point methods for linear programming: a survey. Ann. Oper. Res. **46**(1), 107–138 (1993)

Jones, M.A.: Equitable, envy-free, and efficient cake cutting for two people and its application to divisible goods. Math. Mag. **75**(4), 275–283 (2002)

Kelly, F.: Charging and rate control for elastic traffic. Eur. Trans. Telecommun. **8**(1), 33–37 (1997)

Roughgarden, T.: Algorithmic game theory. Commun. ACM **53**(7), 78–86 (2010)

Steinhaus, H.: The problem of fair division. Econometrica **16**(1), 101–104 (1948)

Steinhaus, H.: Sur la division pragmatique. Econometrica **17**(1949), 315–319 (1949)

Tao, B.: On existence of truthful fair cake cutting mechanisms. In: Proceedings of the 23rd ACM Conference on Economics and Computation, pp. 404–434 (2022)

Stackelberg Strategies on Epidemic Containment Games

Tingwei Hu[ID], Lili Mei[✉][ID], and Zhen Wang[ID]

School of Cyberspace, Hangzhou Dianzi University, Hangzhou, China
{retrieve,wangzhen}@hdu.edu.cn, meilili@zju.edu.cn

Abstract. In this paper, we discuss epidemic containment games, where agents are vertices of a graph G. Each agent has two strategies: being vaccinated or not. The cost of each agent for being vaccinated is 1. Consider an induced subgraph G_s consisting of all non-vaccinated agents in G. If an agent is isolated in G_s then her cost is 0; otherwise her cost is infinity. Each agent wants to minimize her cost. It is a pure Nash Equilibrium (NE) if every agent does not want to change her strategy unilaterally. Naturally, the NE with the minimum (maximum) total cost is defined as the best (worst) NE. In this paper, we focus on Stackelberg games, where a leader injects some agents compulsorily. Due the budget constraint, we require that the total cost of the leader is strictly less than the total cost of the best NE. The rest agents which is a group called followers perform a worst NE on the rest of graph.

We first show that it is NP-hard to find the minimum total cost including the leader and the followers' strategies. Then we turn to consider the effectiveness of Stackelberg strategies, which means that the total cost of the leader and the followers is within the worst NE without the leader. We mainly show that there exist effective Stackelberg strategies for graphs including one-degree vertices or a pair of vertices having inclusive social relationships. Finally, a complete characterization of effectiveness on bipartite graphs is established.

Keywords: Epidemic containment games · Stackelberg strategies · Effectiveness

1 Introduction

Epidemics or computer viruses have caused enormous economic loss, such as COVID-19 spreading in the past years and the famous WannaCry Ransomware 2017. For individuals, it might spend a lot of money or time to recover from an infection. Fortunately, by taking simple precautions like injecting vaccines or installing firewalls, the great expenditure can be avoided. However, not everyone

---

This work was supported by Shanghai Key Laboratory of Pure Mathematics and Mathematical Practice [Project NO. 18dz2271000], and National Natural Science Foundation of China [Project NO. 62176080, 12201594].

M. Li et al. (Eds.): IJTCS-FAW 2023, LNCS 13933, pp. 281–294, 2023.
https://doi.org/10.1007/978-3-031-39344-0_21

would voluntarily adopt these measures. Once all their neighbors have taken the measures, they can be safe without taking any measure. In this paper, we discuss an epidemic containment game on networks.

The virus spreads in a social network or a computer network. Each edge represents a possible transmission path for the virus. Each vertex can either inject vaccines/install firewalls or do nothing. If a vertex takes the former way then her cost is set to be 1. If a vertex does nothing and her neighbours all injecting vaccines/installing firewalls, which means she is safe, then the cost is 0; otherwise, she is infected, and the cost is infinity. Everyone aims at minimizing her cost. A scenario, where no vertex can make her cost less by changing her action unilaterally, is named as a Nash Equilibrium (NE). It's easy to see that finding a NE is equivalent to finding a maximal independent set. Hence, finding a minimum/maximum total cost NE is NP-hard, since finding a maximum/minimum maximal independent set is NP-hard [6]. To simplify the statement, in the context, let $MinNE$ and $MaxNE$ be the injected vertices in the NEs with the minimum and maximum total cost, respectively.

In this paper, we mainly focus on Stackelberg games. The game consists of two stages: A leader injects some vertices in advance, then the rest vertices perform a Nash Equilibrium in the rest network. It is easy to see that the total cost is minimized if the leader injects all vertices in $MinNE$. Therefore, the budget of the leader is restricted to strictly less than $|MinNE|$. This paper considers the worst case where the rest vertices always perform the worst NE in the network removing the vertices injected by the leader. We call the action of the leader as the Stackelberg strategy, who aims at minimizing the total cost (the total cost of the leader and the followers).

Our Contributions. We first show that finding an optimal Stackelberg strategy is NP-hard. Then we propose the concept of effectiveness, where the total cost involving the leader and the followers should be no more than the cost of $MaxNE$ on the original network. We find that some networks do not exist effective Stackelberg strategies. Hence, we mainly focus on what kinds of networks have effective Stackelberg strategies. Our results are as follow.

- Injecting a vertex who has one-degree neighbors is an effective Stackelberg strategy in networks including one-degree vertices;
- In graphs including a couple of vertices with inclusive social relationships, injecting all the neighbors of the vertex who has fewer neighbors in the couple is an effective Stackelberg strategy;
- Finally, we demonstrate an exhaustive characterization of effective Stackelberg strategies in bipartite graphs $G = (V_1 \cup V_2, E)$. If $|MinNE| = 1$ or the graph is a complete bipartite graph with $|V_1| = |V_2|$, there do not exist effective Stackelberg strategies; if the graph is a complete graph with different number of vertices in two sides, without loss of generality, assume that $|V_1| < |V_2|$, then injecting any subsets of vertices in V_2 with the cardinality no more than $\min\{|V_1| - 1, |V_2| - |V_1|\}$ are effective Stackelberg strategies;

Otherwise, i.e., the bipartite graph is not complete and $|MinNE| > 1$, there exist an effective Stackelberg strategy. The effective Stackelberg strategies are demonstrated in Algorithm 1

Related Work. In our game, finding a Max/Min NE is equivalent to finding a minimum maximal/maximum independent set. Some useful bounds and approximation algorithms are given in [2,3,8,9]. Meanwhile, our game is a special case of epidemic containment games based on spectral radius of the components [5,17]. In our game, the spectral radius of non-vaccinated vertices is no more than 1. Another highly-related problem is the Nodal Spectral Radius Minimization (NSRM) problem whose goal is to reduce the spectral radius under a threshold by removing the fewest vertices proposed by Mieghem et al. [19]. They introduce a heuristic algorithms based on the components of the first eigenvector. Later, Saha et al. [18] give an $O(log^2 n)$-approximation algorithm for the NSRM problem.

There are other different ways to model the spread of epidemics and the behaviors of vertices in the networks. Games with interdependent actions are often modeled as Interdependent Security games [10,12]. One way to model the propagation of an epidemic is that the virus starts from a vertex randomly and spreads along paths consisting of non-vaccinated vertices [1]. It is shown that a centralized solution can give a much better total cost than an equilibrium solution. Kumar et al. [15] study the setting in [1] with appending a hop-limit on the spread of infection. In some models, individuals either reduce the communication range to suppress the spread of epidemic or maintain the range but take more risks [13]. Based on economic incentives, insurance is also considered in [7,16].

Organization. In the reminder of the paper, the Stackelberg epidemic containment games are well defined in Sect. 2. In Sect. 3, the hardness of the game is discussed and the concept of effectiveness is proposed. In the next section, we give a full characterization of the effectiveness for graphs including one-degree vertices, graphs including couples of inclusive social relationships and bipartite graphs. The conclusion and future work are in Sect. 5.

2   Preliminaries

2.1   Epidemic Containment (EC) Games

The virus spreads in an undirected graph $G(V, E)$, where V represents the vertex (agent) set and E represents the edge set. Let $|V| = n$ and $|E| = m$. A vertex v_i is a neighbor of v_j if there is an edge between them. Let $\mathcal{A} = \{a_1, a_2, ..., a_n\} \in \{0, 1\}^n$ be a strategy profile of agents. Each agent i has two strategies: $a_i = 1$ if the agent is vaccinated, otherwise $a_i = 0$. Let $c_i(\cdot)$ be the cost function of agent i. If $a_i = 1$, then the cost $c_i(\cdot) = 1$. Otherwise, there are two kinds of costs with related to the strategies of her neighbors. If all neighbors of v_i is vaccinated, then v_i is isolated and its cost is zero. In the other case, the cost of agent i is set as positive infinity. To sum up, the cost of agent i is

$$c_i(\mathcal{A}) = \begin{cases} 1, & \text{if } a_i = 1 \\ 0, & \text{if } a_i = 0 \text{ and } a_j = 1 \ \forall j \in N(i), \\ +\infty, & \text{otherwise} \end{cases} \qquad (1)$$

where $N(i)$ is the set of all the neighbors of v_i.

Each agent wants to minimize her cost. Thus, an agent will inject herself if one neighbor is non-vaccinated. A vaccinated agent will regret and switch to be non-vaccinated, when she finds that all neighbors has been vaccinated. That is, if one of the neighbors is non-vaccinated, the vertex will choose to be vaccinated; if all neighbors are vaccinated, the vertex will choose to be non-vaccinated. Eventually, all vertices reach a stable state, i.e. a Nash equilibrium (NE), where no one can benefit from switching her strategy.

Definition 1. *A strategy profile $\mathcal{A} = \{a_1, a_2, ..., a_n\}$ is a Nash Equilibrium (NE) if for any agent i, and any strategy a_i' for agent i, we have $c_i(a_i, \boldsymbol{a}_{-i}) \leq c_i(a_i', \boldsymbol{a}_{-i})$, where $\boldsymbol{a}_{-i}$ is the strategy profile of all agents without i.*

We define the social cost as $sc(\mathcal{A}) = \sum_{i=1}^{n} c_i(\mathcal{A})$. Note that in any NE, agents actually only have two kinds of cost, 0 and 1. Thus the social cost of a NE is exactly the number of vertices vaccinated. Due to the fact that there are probably more than one NE in a graph, we denote the NE with the largest number of vaccinated vertices as $MaxNE$ and the NE with the fewest number of vaccinated vertices as $MinNE$. With a slight abuse of notation, we also use the symbol NE ($MaxNE/MinNE$) to represent the set of vaccinated vertices in the Nash Equilibrium. Let $|NE|$ ($|MaxNE|$, $|MinNE|$) denote the number of vaccinated vertices and also the social cost.

Lemma 1.   *In a NE, the non-vaccinated set I is a maximal independent set.*

Proof. A single edge with two ends of non-vaccinated vertices would make the cost infinity. Thus, the set of non-vaccinated vertices I should be an independent set of G. Suppose that I is not maximal in a NE. Then there exists an agent i such that all her neighbors and herself have been vaccinated. Then agent i will regret being vaccinated, which contradicts that this is a NE.   □

Then we have, finding a Max/Min NE is equivalent to obtain a minimum maximal/maximum independent set. Two NEs on an instance are shown in Fig. 1.

2.2   Stackelberg Epidemic Containment Games

In this subsection, we focus on Stackelberg epidemic containment games (Stackelberg EC games), where a leader compulsorily injects part of the agents in advance, then followers perform $MaxNE$ on the rest of the graph.

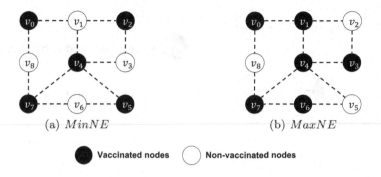

(a) $MinNE$ (b) $MaxNE$

● Vaccinated nodes   ○ Non-vaccinated nodes

Fig. 1. Two NEs on the above network. (a) $|MinNE| = 5$. (b) $|MaxNE| = 6$.

Note that after the followers take action, for each non-vaccinated agent, there does not exist non-vaccinated agent. Hence, the costs of a vaccinated agent and a non-vaccinated agent are 1 and 0, respectively. The only difference is that the vertices vaccinated by the leader can not change their strategies to reduce the personal costs. Denote the vertices vaccinated by the leader as L. Let G' denote the rest graph by removing L and the edges adjacent to L from G. We define the vertex set of followers as $F = V - L$. Let the vertices vaccinated in G' be S' and the vertices non-vaccinated be $I' = F - S'$. To simplify the statement, in the context, let $MinNE_D$ and $MaxNE_D$ denote the best and worst NE with respect to graph D. If we use the notation without subscript, it denotes the corresponding NE in original graph G.

If the budget is sufficient, the leader can always inject $MinNE$ to achieve the optimal solution. Thus we only focus on the case that leader has a limited budget strictly less than $|MinNE|$.

According to the above statement, the **Stackelberg EC game** is well defined. The game aims to minimize the social cost, which is the number of vaccinated vertices by the leader and followers, i.e. $|L| + |S'|$. Firstly, the leader compulsorily injects a set L. The cost of injecting L should be strictly smaller than $|MinNE|$ on G. Then, the followers always perform a $MaxNE$ on the graph G', i.e. $S' = MaxNE_{G'}$. An instance of Stackelberg games is demonstrated in Fig. 2.

3   Hardness

Finding a maximum independent set ($MinNE$) and a minimum maximal independent set ($MaxNE$) are showed to be NP-hard [6]. Next we show that finding an optimal strategy profile in our Stackelberg games is also NP-hard. Before that, we give a lower bound of the social cost.

Lemma 2.   *The minimum social cost in the Stackelberg EC game is no less than* $|MinNE|$.

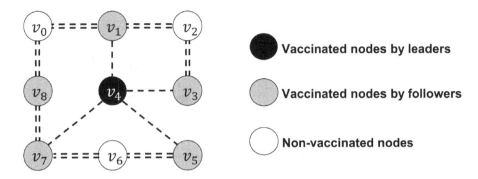

Fig. 2. An instance of Stackelberg EC games. Single dashed lines are removed by the leader. Double dashed lines are removed by followers. The leader injects $\{v_4\}$. Then the followers may inject $\{v_1, v_3, v_5, v_7, v_8\}$. The social cost is 6.

Proof. In Stackelberg EC games, the strategy of the leader leads to two cases. One is that some vertices vaccinated by the leader are regretted but can't change their actions. The other is that no one regrets thus it's also a NE in EC games. If there are regretted vertices, then in each time we switch the action of a regretted vertex until we can't find one. Then the strategy profile must be a NE for no one regrets. The cost of this NE is certainly not less than the $|MinNE|$. Thus the cost before switching is not less than the $|MinNE|$.   □

Lemma 2 implies that if we find a feasible solution of the leader which makes a social cost $|MinNE|$, then the solution is optimal. Furthermore, given any graph to find $MinNE$, if we are able to construct another graph of Stackelberg games, where $MinNE$ is the only optimal solution, then finding an optimal solution of our games is NP-hard.

Theorem 1. *Finding a strategies set $\mathcal{A}$ of the minimum total cost is NP-hard.*

Proof. We prove by reduction from the $MinNE$ problem in EC games. Given an instance of EC games on $G(V, E)$, we construct a three-layer graph $G^*(V^*, E^*)$ of Stackelberg games. $Layer_1$ is exactly the same as G. For every vertex in G, we add a pendant vertex (one-degree vertex) to it and these pendant vertices constitute $Layer_2$. Similarly, we add a pendant vertex to every vertex in $Layer_2$ and get $Layer_3$. The new graph G^* is showed in Fig. 3(a). Note that each vertex in $Layer_2$ or $Layer_3$ has a matching vertex in $Layer_1$. The second and third layers can be seen as copies of the vertices of $Layer_1$ without edges.

Denote the $MinNE$ (vaccinated vertices) in $Layer_j$ as S_j, and the set of non-vaccinated vertices in $Layer_j$ as I_j. According to Lemma 2, the minimum social cost is no less than the $MinNE$ on G^*. Two steps are followed in the next paragraph. Firstly, we give a feasible leader's strategy including S_1, with a social cost of exactly $|MinNE_{G^*}|$. It means that an optimum strategy has been found. Secondly, we demonstrate that any strategies at the cost of $|MinNE_{G^*}|$ must include S_1. In other words, if the optimum strategy is found, the $MinNE$ problem on G is solved.

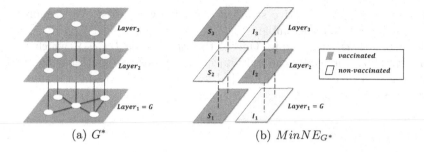

(a) G^* (b) $MinNE_{G^*}$

Fig. 3. (a) The constructed graph $G*$. (b) $MinNE$ on $G*$.

One of the feasible strategy is that the leader injects S_1 and I_2, i.e. $L = S_1 \cup I_2$. After L is removed, the followers' graph G' is divided into three parts, I_1, I_3 and $S_2 \cup S_3$. The sets I_1 and I_3 are both independent sets, for they only connect to L on G^*. The set $S_2 \cup S_3$ is in a graph with $|S_2|$ edges. For each edge, one end is in $Layer_2$, the other end is in $Layer_3$. The only NE that followers could perform on G' is to inject one end for every edge in $S_2 \cup S_3$, such as $MaxNE_{G'} = S_3$. The social cost is $|L| + |S_3| = |V| + |S_1|$. Next, it will be shown that the set $S_1 \cup I_2 \cup S_3$ is a $MinNE$ on G^*, as Fig. 3(b) showed. For every edge between $Layer_2$ and $Layer_3$, one of its end must be vaccinated, that is $|V|$ in total. And on $Layer_1$, the cost $|S_1|$ is optimum because it's $MinNE$ on G. On the other hand, if we obtain an optimum strategies set $\mathcal{A}$ and the cost $|V| + |S_1|$, knowing that the cost on $Layer_2$ and $Layer_3$ is at least $|V|$, then the cost on $Layer_1$ should be at most $|S_1|$. And we also know that the cost on $Layer_1$ is no less than $|MinNE| = |S_1|$. Thus, if we find a strategies set $\mathcal{A}$ at the minimum cost, we must find a $MinNE$ on G, which is NP-hard. □

Finding the optimum solution can be very hard. Then, the leader would like to know whether her compulsory strategy increases the social cost of the original case. So we next introduce the concept of effectiveness.

Definition 2. *A strategy profile of the leader is effective if the total cost of the Stackelberg EC game is no more than the cost of $MaxNE$ on G.*

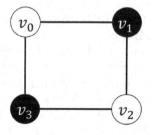

Fig. 4. A counter-example of no effective strategies.

Unfortunately, an effective strategy doesn't always exist. We demonstrate this by giving a counter-example.

Take Fig. 4 as an example. One of the $MinNE$ is $\{v_0, v_2\}$, which is also $MaxNE$. Thus, the leader can only inject one vertex. No matter which vertex is vaccinated by the leader, either $\{v_0, v_2\}$ or $\{v_1, v_3\}$ will be vaccinated by followers. $sc > |MaxNE|$.

4   Effectiveness

In this section, we mainly focus on the existence of effective Stackelberg strategies. Three kinds of graphs with effective strategies are given, which are graphs with one-degree vertices, graphs with couples of inclusion social relationships, and bipartite graphs. Note that isolated vertices can be ignored because they can be removed from the original graph.

4.1   Graphs Including One-Degree Vertices

One-degree vertices are usually referred to as pendant vertices. We will show that any graph including a pendant vertex has an effective strategy if $|MinNE| > 1$. In the next lemma, we show if the neighbour of a pendant vertex is in a NE, then we have an efficient strategy. Let $NE(S, I)$ denote a NE which injects vertices in S and vertices in I is non-vaccinated.

Lemma 3. *If $|MinNE| > 1$ and there exists a $NE(S, I)$ with a pendant vertex $y \in I$ and its neighbor $x \in S$, then $L = \{x\}$ is an effective strategy.*

Proof. We denote δ as the difference between $|MaxNE|$ and $|NE(S, I)|$, i.e. $\delta = |MaxNE| - |NE(S, I)|$. After compulsorily vaccinating x, the rest should perform their $MaxNE$ on G'. There is a set $T_1 \in S$ that turns to be non-vaccinated and there is a set $T_2 \in I$ that turns to be vaccinated. In order to group vaccinated vertices together, we say that T_1 and T_2 is swapped when they come to the other set. Then the $MaxNE(S', I')$ on G' is performed, where $|S'| = |S| - 1 + |T_2| - |T_1|$.

If $|T_2| - |T_1| \leq \delta$, then the social cost is no more than $|MaxNE|$, for $sc = |L| + |F| = 1 + |S'| \leq |S| + \delta = |MaxNE|$. The strategy $L = \{x\}$ is effective.

If $|T_2| - |T_1| > \delta$, we prove it by contradiction. Assume that $L = \{x\}$ is not an effective strategy. We try to find a $|NE|$ on G that is larger than $|MaxNE|$, so that the contradiction holds. But before that, we first show $y \in I'$. Suppose that $y \in S'$, y must regret being vaccinated for its only neighbor x is vaccinated. Then (S', I') can not be a NE. Thus $y \in I'$. Next it is easy to show that $(S' \cup x, I')$ is a NE. We have already known that (S', I') performs a $MaxNE$ on G', which means I' is a maximal independent set. Now adding x back to the S', I' is still maximal because x has at least one neighbor $y \in I'$. The number of vaccinated vertices in this new NE is $|S'| + 1 = |S| + |T_2| - |T_1| > |S| + \delta = |MaxNE|$, which is contradict to the fact that $MaxNE(S, I)$ is maximum.   □

Through Lemma 3, we show that if there is a case that x appears in the vaccinated set S of some NE and y appears in the non-vaccinated set I, then $L = \{x\}$ is effective. Next, it only remains to show that we can always find a NE where $y \in I$ and $x \in S$.

Theorem 2. *If $|MinNE| > 1$ and there exists a pendent vertex y and its neighbor x, then $L = \{x\}$ is an effective strategy.*

Proof. It's easy to see that we can always find a $NE(S, I)$ where $y \in I$ and $x \in S$. A simple algorithm is given to find the maximal independent set I including y. Initially, $I = \{y\}$. Each round we select a vertex that has no neighbor in I, until we cannot find one. Then I is a maximal independent set and $y \in I$, $x \notin I$. $\quad\square$

It is easy to see that whether $MinNE$ is greater than one. Special graphs ($|V| > 2$), such as trees, paths, absolutely have vertices of one degree, so that effective strategies always exist.

4.2   Graphs Including a Couple of Inclusive Social Relationships

In this subsection, we consider graphs with a couple of inclusive social relationships. In this graph $G = (V, E)$, there exist two vertices x and y satisfying that $(x, y) \in E$ and $N(y) - \{x\} \subseteq N(x)$. This type of graphs is frequent, like as one spouse knows everyone the other knows.

Theorem 3. *If any couple of vertices x, y satisfy the following three conditions: (1) there is an edge between x and y, (2) all neighbors of y except x are a subset of the neighbors of x, i.e. $N(y) - \{x\} \subseteq N(x)$, and (3) $N(y) < |MinNE|$, then there exists an effective strategy $L = N(y)$.*

Proof. Let $k = |N(y)|$. If $k = 1$, then y is a pendant vertex and x is its only neighbor. According to Theorem 2, we already have that $L = \{x\}$ is an effective strategy. Let the graph with $k = 1$ be G_1. A new graph G_k is obtained from G_{k-1} by adding a new vertex l_k which at least connects to x and y. It will be shown that $L_k = \{x, l_2, ..., l_k\} = N(y)$ is an effective strategy on G_k ($k \geq 2$). In other words, given any graphs G_k satisfying the three conditions, we can conversely obtain $G_{k-1}, ..., G_1$ by vaccinating $l_k, ..., l_2$, and easily know $L_k = N(y)$ is an effective strategy.

Let G'_k be the graph after removing L_k on G_k. In G_1, $L_1 = \{x\}$ is an effective strategy, that is $sc = 1 + |MaxNE_{G'_1}| \leq |MaxNE_{G_1}|$. For G_k ($k \geq 2$), by adding $k - 1$ vaccinated vertices, the social cost has increased by $k - 1$. We get $sc = k + |MaxNE_{G'_1}| \leq |MaxNE_{G_1}| + k - 1$. The rest is to show $|MaxNE_{G_1}| + k - 1 \leq |MaxNE_{G_k}|$. It's impossible that both x and y are vaccinated in $MaxNE_{G_1}$. In that case, y will regret. Due to one of them must belong to the non-vaccinated set, their neighbor l_j ($2 \leq j \leq k$) can not be in the non-vaccinated set. Then the maximal independent set on G_1 is still maximal on G_k. The set $MaxNE_{G_1} \cup \{l_2, ..., l_k\}$ is a NE on G_k and its cost $|MaxNE_{G_1}| + k - 1$ is no more than $|MaxNE_{G_k}|$. $\quad\square$

Remark: In Theorem 3, note that the graph must satisfy $N(y) < |MinNE|$, which is easy to satisfy in an enormous network.

4.3   Bipartite Graphs

In this subsection, we research on a kind of special graphs, bipartite graphs. Let V_1, V_2 be the two parts of the graph. Note that finding a $MinNE$ on bipartite graphs can be in polynomial time [11,14], while finding a $MaxNE$ is NP-hard [4].

We will give a complete characterization of effective strategies in bipartite graphs, which is showed by Lemmas 4, 5, 6. We first show the existence of effective strategies in complete bipartite graphs.

Lemma 4. *For a complete bipartite graph, if $|V_1| \neq |V_2|$ and $|MinNE| > 1$, there must exist an effective strategy. If $|V_1| = |V_2|$, any strategies are ineffective.*

Proof. It's easy to see that $MaxNE$ on a complete graph is always the larger part. If $|V_1| \neq |V_2|$, we simply let $|V_2|$ be the larger one, i.e. $MaxNE = V_2$. To be effective, the leader only need to inject a number of vertices ranging from one to $\min\{|MinNE| - 1, |V_2| - |V_1|\}$ in V_2. Then the followers would inject the rest of vertices in V_2 as the $MaxNE_{G'}$. The social cost is $|V_2| = |MaxNE|$.

If $|V_1| = |V_2|$, there are two possible cases for L. One is that L belongs to only one of the sets ($L \subseteq V_1$ or $L \subseteq V_2$). Then $MaxNE_{G'}$ is the larger part on the rest graph, i.e. $MaxNE_{G'} = V_1$ or $MaxNE_{G'} = V_2$. The social cost is $|L| + |V_1|$, which must be greater than $|V_1|$.

The other case is that L belongs to both sets ($L \subseteq V_1$ and $L \subseteq V_2$). Then $MaxNE_{G'}$ following is still the larger part on the rest graph. No matter $MaxNE_{G'}$ is on which side, the part including $MaxNE_{G'}$ must all be vaccinated. In addition, there are some vertices vaccinated in the other part. Thus the social cost must be greater than $|V_1|$.   □

Next, for general bipartite graphs, we consider two cases based on one side or two sides of the $MinNE$. We first discuss the case that $MinNE$ is on two sides.

Lemma 5. *For a bipartite graph, if $MinNE \cap V_1 \neq \emptyset$, $MinNE \cap V_2 \neq \emptyset$, then there exists an effective strategy.*

Proof. Let S_1, S_2 be the vaccinated sets of $MinNE$ and I_1, I_2 be the non-vaccinated sets of $MinNE$, i.e. $S_1 = MinNE \cap V_1$, $S_2 = MinNE \cap V_2$, $I_1 = V_1 - L_1$, $I_2 = V_2 - L_2$. We will prove that $L = S_1$ or $L = S_2$ is an effective strategy. Since the proofs of two cases are the same, let $L = S_1$ for simplicity. Now we have divided the bipartite graph into four sets by $MinNE$ and L is compulsorily vaccinated. Recall that the vaccinated set of $MaxNE'_G$ is denoted as S'. It is easy to see that S' cannot include any vertices in I_2. As parts of non-vaccinated vertices in $MinNE$, I_1 and I_2 are independent of each other. On G', the set L has been removed by the leader and S_2 is at the same side with I_2 in a bipartite graph. Thus I_2 turns out to be an independent set on G' which cannot be vaccinated. Clearly, only three cases are left for S': (1) $S' \subseteq S_2$, (2) $S' \subseteq I_1$ and (3) some vertices are in S_2, the others are in I_1.

In case 1, even if the set S' is all the vertices in S_2, the social cost is $|L_1| + |L_2| = |MinNE| \leq |MaxNE|$. In case 2, even if the set S' is all the vertices in I_1, the social cost is $|L_1| + |I_1| = |V_1| \leq |MaxNE|$. In case 3, when the set S' includes some vertices in I_1 and some vertices in S_2, we show that the set $L \cup S'$ is a NE on G which certainly satisfies $|L| + |S'| \leq |MaxNE|$. It is known that $I' = V - L - F$ is a maximal independent set on G'. The rest is to show it's still maximal on G. Any vertices in L must have at least one neighbor in I_2. If they don't, the only set they connect is S_2 and would regret in $MinNE = L_1 \cup L_2$. Thus no one in L is independent of I' so that I' is still maximal. Above all, there always exist an effective strategy. □

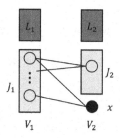

Fig. 5. Schematic diagrams of Lemma 6.

Finally, we discuss that $MinNE$ is on one side.

Lemma 6.   *For a bipartite graph which is not a complete bipartite graph, if $MinNE \subseteq V_1$ or $MinNE \subseteq V_2$, there must exist an effective strategy.*

Proof. Since the way of naming V_1 and V_2 is arbitrary, we're just going to let $MinNE \subseteq V_2$. For a bipartite graph, it's easy to see that $MinNE \subseteq V_2$ is actually equivalent to $MinNE = V_2$. If there is a vertex in V_2 but not in the vaccinated set of $MinNE$, then it must be zero-degree and should be ignored. Because the graph G is not a complete graph, we are always able to find a vertex $x \in V_2$ that does not connect to all vertices in V_1 but just connect to a proper set of V_1.

Denote the neighbors of x as J_1. The vertices in $V_2 \backslash x$ which only have neighbors in J_1 are denoted as J_2. Let the remaining vertices be $L_1 = V_1 \backslash J_1$, $L_2 = V_2 \backslash (J_2 \cup x)$. None of them is empty for G is not complete. Fig. 5 shows a diagram of this case.

Next we show that $L = L_2$ is an effective strategy. Recall that S' is the $MaxNE$ on G', I' is the non-vaccinated set on G'. Our goal is to prove that the set $L \cup S'$ is a NE on G so that immediately the social cost $|L| + |S'| \leq |MaxNE|$. We have known that I' is a maximal independent set on G'. It only remains to prove I' is still maximal on G. Firstly, L_1 must be in I', for any vertices in L_1 is independent of J_1 and $J_2 \cup x$. Secondly, any vertices in L_2 have at least one neighbor in L_1, otherwise, they only connect to J_1 and would be grouped to J_2 by denotation. Thus, no one in L_2 is independent of I', so that I' is still a maximal independent set on G.   □

Algorithm 1: Effective strategies on bipartite graphs

Input: A bipartite graph G(V,E) with two parts V_1 and V_2.
Output: An effective strategy L.

1 Find a $MinNE$ on G;
2 **if** $|MinNE| = 1$ **return** False;
3 **if** G *is complete* **then**
4 | **if** $|V_1| = |V_2|$ **return** False;
5 | **else**
6 | **return** L is any one node in the larger part;
7 **else**
8 | **if** $MinNE \cap V_1 \neq \emptyset$ *and* $MinNE \cap V_2 \neq \emptyset$ **then**
9 | | **return** $L = MinNE \cap V_1$.
10 | **else**
11 | | Find a vertex $x \in MinNE$ satisfying $N(x) \neq V_j$ ($j = 1$ *or* 2);
12 | | Find a vertex set J_2, where $\forall y \in J_2$, $N(y) \subseteq N(x)$;
13 | | **return** $L = MinNE - J_2 - x$.

Above all, we have discussed all cases in bipartite graphs demonstrated by Algorithm 1. It is worth to note that in bipartite graph the maximum independent set can be found in polynomial time [14] which implies that Algorithm 1 is polynomial.

Theorem 4. *For a bipartite graph which is $|MinNE| = 1$ or a complete bipartite graph with $|V_1| = |V_2|$, any strategies are ineffective. For other bipartite graphs, there must exist an effective strategy.*

5   Conclusion and Future Work

We study a Stackelberg EC game on networks. It is shown that finding an optimal Stackelberg strategy is NP-hard. Further, the effectiveness of the leader's strategy is studied on three kinds of graphs. For graphs with one-degree vertices, an effective strategy always exists by compulsorily vaccinating any neighbor of the one-degree vertex. For graphs with couples of inclusive social relationships, an effective strategy exists by vaccinating all neighbors of the one with fewer neighbours. For bipartite graphs, we give a exhaustive of the existence of the effective Stackelberg strategies.

It is an interesting direction to consider the effective Stackelberg strategies in general graphs. Besides it is also another direction to study approximation Stackelberg strategies on networks aiming at minimizing the total cost.

References

1. Aspnes, J., Chang, K., Yampolskiy, A.: Inoculation strategies for victims of viruses and the sum-of-squares partition problem. J. Comput. Syst. Sci. **72**(6), 1077–1093 (2006)
2. Balliu, A., Brandt, S., Hirvonen, J., Olivetti, D., Rabie, M., Suomela, J.: Lower bounds for maximal matchings and maximal independent sets. J. ACM (JACM) **68**(5), 1–30 (2021)
3. Berman, P., Fürer, M.: Approximating maximum independent set in bounded degree graphs. In: Proceedings of the Fifth Annual ACM-SIAM Symposium on Discrete Algorithms (SODA), pp. 365–371 (1994)
4. Corneil, D.G., Perl, Y.: Clustering and domination in perfect graphs. Discret. Appl. Math. **9**(1), 27–39 (1984)
5. Ganesh, A.J., Massoulié, L., Towsley, D.F.: The effect of network topology on the spread of epidemics. In: Proceedings of the 24th IEEE Annual Joint Conference of the Computer and Communications Societies (INFOCOM), pp. 1455–1466 (2005)
6. Garey, M.R., Johnson, D.S.: Computers and Intractability. Freeman San Francisco Press (1979)
7. Grossklags, J., Christin, N., Chuang, J.: Secure or insure? A game-theoretic analysis of information security games. In: Proceedings of the 17th International Conference on World Wide Web (WWW), pp. 209–218 (2008)
8. Halldórsson, M., Radhakrishnan, J.: Greed is good: approximating independent sets in sparse and bounded-degree graphs. In: Proceedings of the Twenty-Sixth Annual ACM Symposium on Theory of Computing (STOC), pp. 439–448 (1994)
9. Haviland, J.: On minimum maximal independent sets of a graph. Discret. Math. **94**(2), 95–101 (1991)
10. Heal, G., Kunreuther, H.: IDS models of airline security. J. Conflict Resolut. **49**(2), 201–217 (2005)
11. Kashiwabara, T., Masuda, S., Nakajima, K., Fujisawa, T.: Generation of maximum independent sets of a bipartite graph and maximum cliques of a circular-arc graph. J. Algorithms **13**(1), 161–174 (1992)
12. Kearns, M., Ortiz, L.E.: Algorithms for interdependent security games. In: Proceedings of the 16th International Conference on Neural Information Processing Systems (NeurPIS), pp. 561–568 (2003)
13. Khouzani, M., Altman, E., Sarkar, S.: Optimal quarantining of wireless malware through reception gain control. IEEE Trans. Autom. Control **57**(1), 49–61 (2011)
14. Korte, B.H., Vygen, J., Korte, B., Vygen, J.: Combinatorial Optimization. Springer, Heidelberg (2011)
15. Kumar, V.A., Rajaraman, R., Sun, Z., Sundaram, R.: Existence theorems and approximation algorithms for generalized network security games. In: Proceedings of the 30th IEEE International Conference on Distributed Computing Systems (ICDCS), pp. 348–357 (2010)
16. Lelarge, M., Bolot, J.: Economic incentives to increase security in the internet: the case for insurance. In: Proceedings of the 28th IEEE Annual Joint Conference of the Computer and Communications Societies (INFOCOM), pp. 1494–1502 (2009)

17. Saha, S., Adiga, A., Kumar, V.S.A.: Equilibria in epidemic containment games. In: Proceedings of the 28th AAAI Conference on Artificial Intelligence (AAAI), pp. 777–783 (2014)
18. Saha, S., Adiga, A., Prakash, B.A., Kumar, V.S.A.: Approximation algorithms for reducing the spectral radius to control epidemic spread. In: Proceedings of the 2015 SIAM International Conference on Data Mining (SDM), pp. 568–576 (2015)
19. Van Mieghem, P., et al.: Decreasing the spectral radius of a graph by link removals. Phys. Rev. E **84**(1), 016101 (2011)

Author Index

© Springer Nature Switzerland AG 2023
M. Li et al. (Eds.): IJTCS-FAW 2023, LNCS 13933, pp. 295–296, 2023.
https://doi.org/10.1007/978-3-031-39344-0

Printed in the United States
by Baker & Taylor Publisher Services